Energy, Environment, and Sustainability

Series Editor

Avinash Kumar Agarwal, Department of Mechanical Engineering, Indian Institute of Technology Kanpur, Kanpur, Uttar Pradesh, India

AIMS AND SCOPE

This books series publishes cutting edge monographs and professional books focused on all aspects of energy and environmental sustainability, especially as it relates to energy concerns. The Series is published in partnership with the International Society for Energy, Environment, and Sustainability. The books in these series are edited or authored by top researchers and professional across the globe. The series aims at publishing state-of-the-art research and development in areas including, but not limited to:

- Renewable Energy
- Alternative Fuels
- Engines and Locomotives
- Combustion and Propulsion
- Fossil Fuels
- Carbon Capture
- Control and Automation for Energy
- Environmental Pollution
- Waste Management
- Transportation Sustainability

Review Process

The proposal for each volume is reviewed by the main editor and/or the advisory board. The chapters in each volume are individually reviewed single blind by expert reviewers (at least four reviews per chapter) and the main editor.

Ethics Statement for this series can be found in the Springer standard guidelines here https://www.springer.com/us/authors-editors/journal-author/journal-author-helpdesk/before-you-start/before-you-start/1330#c14214

More information about this series at https://link.springer.com/bookseries/15901

Avinash Kumar Agarwal · Hardikk Valera
Editors

Potential and Challenges of Low Carbon Fuels for Sustainable Transport

Springer

Editors
Avinash Kumar Agarwal
Department of Mechanical Engineering
Indian Institute of Technology Kanpur
Kanpur, India

Hardikk Valera
Department of Mechanical Engineering
Indian Institute of Technology Kanpur
Kanpur, India

ISSN 2522-8366 ISSN 2522-8374 (electronic)
Energy, Environment, and Sustainability
ISBN 978-981-16-8413-5 ISBN 978-981-16-8414-2 (eBook)
https://doi.org/10.1007/978-981-16-8414-2

This Springer imprint is published by the registered company Springer Nature Singapore Pte Ltd.
The registered company address is: 152 Beach Road, #21-01/04 Gateway East, Singapore 189721, Singapore

Preface

Effort for reduction of global emission level is currently one of the prime areas of concern for the research community across the globe. Particularly due to more stringent standards of emission control, the prevailing diesel engines are on the verge of losing their permission to operate.

The International Society for Energy, Environment and Sustainability (ISEES) was founded at the Indian Institute of Technology Kanpur (IIT Kanpur), India, in January 2014 to spread knowledge/awareness and catalyze research activities in the fields of energy, environment, sustainability, and combustion. Society's goal is to contribute to the development of clean, affordable, and secure energy resources and a sustainable environment for society and spread knowledge in the areas mentioned above and create awareness about the environmental challenges the world is facing today. The unique way adopted by ISEES was to break the conventional silos of specializations (engineering, science, environment, agriculture, biotechnology, materials, fuels, etc.) to tackle the problems related to energy, environment, and sustainability in a holistic manner. This is quite evident by the participation of experts from all fields to resolve these issues. ISEES is involved in various activities such as conducting workshops, seminars, and conferences in the domains of its interests. Society also recognizes the outstanding works of young scientists, professionals, and engineers for their contributions in these fields by conferring them awards under various categories.

The Fifth International Conference on 'Sustainable Energy and Environmental Challenges' (V-SEEC) was organized under the auspices of ISEES from December 19–21, 2020, in virtual mode due to restrictions on travel because of the ongoing COVID-19 pandemic situation. This conference provided a platform for discussions between eminent scientists and engineers from various countries, including India, Spain, Austria, Bangladesh, Mexico, USA, Malaysia, China, UK, Netherlands, Germany, Israel, and Saudi Arabia. At this conference, eminent international speakers presented their views on energy, combustion, emissions, and alternative energy resources for sustainable development and a cleaner environment. The conference presented two high-voltage plenary talks by Dr. V. K. Saraswat, Honorable Member, NITI Ayog, on 'Technologies for Energy Security and Sustainability' and Prof. Sandeep Verma, Secretary, SERB, on 'New and Equitable R&D Funding Opportunities at SERB.'

The conference included nine technical sessions on topics related to energy and environmental sustainability. Each session had 6–7 eminent scientists from all over the world, who shared their opinion and discussed the trends for the future. The technical sessions in the conference included emerging contaminants: monitoring and degradation challenges; advanced engine technologies and alternative transportation fuels; future fuels for sustainable transport; sustainable bioprocessing for biofuel/non-biofuel production by carbon emission reduction; future of solar energy; desalination and wastewater treatment by membrane technology; biotechnology in sustainable development; emerging solutions for environmental applications and challenges and opportunities for electric vehicle adoption. Five hundred plus participants and speakers from all over the world attended this three days conference.

The conference concluded with a high-voltage panel discussion on 'Challenges and Opportunities for Electric Vehicle Adoption,' where the panelists were Prof. Gautam Kalghatgi (University of Oxford), Prof. Ashok Jhunjhunwala (IIT Madras), Dr. Kelly Senecal (Convergent Science), Dr. Amir Abdul Manan (Saudi Aramco), and Dr. Sayan Biswas (University of Minnesota, USA). Prof. Avinash Kumar Agarwal, ISEES, moderated the panel discussion. This conference laid out the roadmap for technology development, opportunities, and challenges in energy, environment, and sustainability domain. All these topics are very relevant for the country and the world in the present context. We acknowledge the support received from various agencies and organizations for the successful conduct of the fifth ISEES conference V-SEEC, where these books germinated. We want to acknowledge SERB (special thanks to Dr. Sandeep Verma, Secretary) and our publishing partner Springer (special thanks to Ms. Swati Meherishi).

The editors would like to express their sincere gratitude to a large number of authors from all over the world for submitting their high-quality work on time and revising it appropriately at short notice. We would like to express our special gratitude to our prolific set of reviewers, Dr. Chetan Kumar Patel, Dr. Kanit Wattanavichien, Dr. Atul Dhar, Dr. Gyorgy Sazabdos, Dr. Felix Leach, Dr. Luigi De Simio, Dr. Suhan Park, Dr. Cedik Jakub, Dr. Anirudh Gautam, Dr. Josef Bradac, Mr. Puneet Bansal, Dr. Harsh Goyal, Dr. Veeresh Babu A., Dr. R. Thirumaleswara Naik, Dr. Bhaskar Tamma, Dr. Yuanxian Zhu, Dr. Bireswar Paul, Dr. Luca Marchitto, Dr. Antonio Ficarella, Dr. Jyoti Prasad Chakraborty, Dr. Bhawna Verma, Mr. Ramnarayan Meena, and Mr. Deepak Agarwal who reviewed various chapters of this monograph and provided their valuable suggestions to improve the manuscripts.

This book is focused on low-carbon fuels, which seem to be preferable class of fuels for internal combustion engines (ICEs). It is simple understanding that if input fuel has a low carbon in its chemical constitute, then tailpipe emissions will have a low-carbon emission. Also, low-carbon fuels can be produced using renewable resources and they are economical compared to conventional fuels. i.e., diesel and gasoline. Low-carbon family has members like biodiesel, Di-ethyl ether, renewable compressed natural gas, ethanol, butanol, etc. Automotive industries and policy makers are clearly understanding that problem is not the engine but fuels used to power the engine. Therefore, researchers are working on the adaptation of low-carbon fuels. This book covers all the possibilities of using low-carbon fuels. Also, it covers the effect of low-carbon fuels on tailpipe emissions. Main take-home message of the chapter will be given pointwise at end of the chapter in tabular format. Readers will learn how to utilize low-carbon fuels in the IC engine as a fuel. This book aims to strengthen the knowledge base dealing with low-carbon fuels as a sustainable transport fuel. Chapters include recent results and are focused on current trends of automotive sector. We hope that the book would greatly interest the professionals and postgraduate students involved in fuels, IC engines, engine instrumentation, and environmental research.

Kanpur, India

Avinash Kumar Agarwal
Hardikk Valera

Contents

Part IV Carbon Neutrality

Editors and Contributors

About the Editors

Prof. Avinash Kumar Agarwal joined IIT Kanpur in 2001. He worked at the Engine Research Center, UW@Madison, the USA as a Post-Doctoral Fellow (1999–2001). His interests are IC engines, combustion, alternate and conventional fuels, lubricating oil tribology, optical diagnostics, laser ignition, HCCI, emissions, and particulate control, 1D and 3D Simulations of engine processes, and large-bore engines. Prof. Agarwal has published 435+ peer-reviewed international journal and conference papers, 70 edited books, 92 books chapters, and 12200+ Scopus and 19000+ Google Scholar citations. He is the associate principal editor of FUEL. He has edited "Handbook of Combustion" (5 Volumes; 3168 pages), published by Wiley VCH, Germany. Prof. Agarwal is a Fellow of SAE (2012), Fellow of ASME (2013), Fellow of ISEES (2015), Fellow of INAE (2015), Fellow of NASI (2018), Fellow of Royal Society of Chemistry (2018), and a Fellow of American Association of Advancement in Science (2020). He is the recipient of several prestigious awards such as Clarivate Analytics India Citation Award-2017 in Engineering and Technology, NASI-Reliance Industries Platinum Jubilee Award-2012; INAE Silver Jubilee Young Engineer Award-2012; Dr. C. V. Raman Young Teachers Award: 2011; SAE Ralph R. Teetor Educational Award-2008; INSA Young Scientist Award-2007; UICT Young Scientist Award-2007; INAE Young Engineer Award-2005.

Prof. Agarwal received Prestigious CSIR Shanti Swarup Bhatnagar Award-2016 in Engineering Sciences. Prof. Agarwal is conferred upon Sir J C Bose National Fellowship (2019) by SERB for his outstanding contributions. Prof. Agarwal was a highly cited researcher (2018) and was in the top ten HCR from India among 4000 HCR researchers globally in 22 fields of inquiry.

Mr. Hardikk Valera is pursuing his Ph.D. from Engine Research Laboratory (ERL), Department of Mechanical Engineering, Indian Institute of Technology (IIT) Kanpur. He has completed his M.Tech. and B.Tech. from National Institute of Technology (NIT) Jalandhar, India, and Ganpat University, respectively. His research interests include methanol-fueled SI engines, methanol-fueled CI engines, optical diagnostics, fuel spray characterization, and emission control from engines.

Contributors

Avinash Kumar Agarwal Engine Research Laboratory, Department of Mechanical Engineering, Indian Institute of Technology Kanpur, Kanpur, India

Sanjukta Banerjee Department of Biotechnology, Indian Institute of Technology Kharagpur, Kharagpur, India

Sawan Bharti Invertis University, Bareilly, India

Balendra V. S. Chauhan Automotive Fuels and Lubricants Application Division, CSIR-Indian Institute of Petroleum, Dehradun, Uttarakhand, India

Byunghchul Choi School of Mechanical Engineering, Chonnam National University, Gwangju, Republic of Korea

Debabrata Das Department of Biotechnology, Indian Institute of Technology Kharagpur, Kharagpur, India

Samar Das Fuel and Combustion Lab, Centre for Energy, Indian Institute of Technology Guwahati, Guwahati, Assam, India

Subhanker Dev The Automotive Research Association of India (ARAI), Pune, India

Liyun Fan Chengdu WIT Electronic Fuel System Co., Ltd., Chengdu, China

Akshay Garg Department of Mechanical Engineering, College of Engineering Roorkee, Roorkee, India

Ananta K. Ghosh Department of Biotechnology, Indian Institute of Technology Kharagpur, Kharagpur, India

Siddharth Jain Department of Mechanical Engineering, College of Engineering Roorkee, Roorkee, India

Pankaj Kalita Fuel and Combustion Lab, Centre for Energy, Indian Institute of Technology Guwahati, Guwahati, Assam, India

Debangsu Kashyap Fuel and Combustion Lab, Centre for Energy, Indian Institute of Technology Guwahati, Guwahati, Assam, India

Babalola Aisosa Oni Department of Chemical Engineering, Covenant University, Ota, Ogun State, Nigeria;
Department of Chemical Engineering, China University of Petroleum, Changping, Beijing, People's Republic of China

Samuel Eshorame Sanni Department of Chemical Engineering, Covenant University, Ota, Ogun State, Nigeria

Selçuk Sarıkoç Department of Motor Vehicles and Transportation Technologies, Tasova Yuksel Akin Vocational School, Amasya University, Amasya, Turkey

Cagatayhan Sevim Naval Architecture and Maritime Faculty, Yildiz Technical University, Istanbul, Turkey

Hardikk Valera Engine Research Laboratory, Department of Mechanical Engineering, Indian Institute of Technology Kanpur, Kanpur, India

Ajitanshu Vedrantam Universidad de Alcalá, Alcalá de Henares, Spain

Yuanxian Zhu Chengdu WIT Electronic Fuel System Co., Ltd., Chengdu, China

Burak Zincir Maritime Faculty, Istanbul Technical University, Istanbul, Turkey

Part I
General

Chapter 1
Introduction of Potential and Challenges of Low Carbon Fuels for Sustainable Transport

Avinash Kumar Agarwal and Hardikk Valera

Abstract Low carbon fuels seem to be a preferable class of fuels for Internal Combustion Engines (ICEs). It is the simple understanding that if input fuel has low carbon content in its constituents, tailpipe emissions will have lower carbonaceous emissions. Also, low carbon fuels can be produced using renewable resources, and they are economical than conventional fuels. The low carbon fuel family has members like biodiesel, Di-methyl ether, Di-ethyl ether, renewable compressed natural gas, Ethanol, Butanol etc. Automotive industries and policymakers clearly understand that the problem is not the engines but the fuels used to power the engines. Therefore, researchers are working on the adaptation of low carbon fuels. This book covers all possibilities of using low carbon fuels. Also, it covers the effect of low carbon fuels on tailpipe emissions. This book aims to strengthen the knowledge base dealing with low carbon fuels as a sustainable transport fuel.

Keywords Low carbon fuels · Ethanol · Butanol · Carbon neutrality

Owing to the increasing global demand for energy, consumption of fossil fuels has increased; however, the utilisation of such fuels negatively affects the environment (Agarwal 2007). This has also led to significant global warming, climate change, rising sea levels, and depletion of carbon-based fuels. Petroleum-based energy supplements are responsible for these problems mainly. More specifically, environmental concerns have peaked after the critical CO_2 level in the atmosphere has been breached. Thus, now more than ever, it is imperative to increase the percentage of clean, renewable energy sources to address the energy demands. Low carbon fuels for internal combustion engines (ICEs) are essential to reduce emissions and improve engine performance and combustion characteristics. The possibilities of low carbon fuel utilisation in ICEs are discussed in this book.

The first section of this book has one chapter, which is the introduction chapter.

A. K. Agarwal (✉) · H. Valera
Engine Research Laboratory, Department of Mechanical Engineering, Indian Institute of Technology Kanpur, Kanpur 208016, India
e-mail: akag@iitk.ac.in

A. V. Agarwal and H. Valera (eds.), *Potential and Challenges of Low Carbon Fuels for Sustainable Transport*, Energy, Environment, and Sustainability,
https://doi.org/10.1007/978-981-16-8414-2_1

The second section of the book focuses on the low carbon fuels' production and fuel injection aspects. This section has three chapters. The first chapter of this section reviews alternative biofuels for ICE, namely alcohols and biodiesel. The advantages and disadvantages of alcohols (e.g., methanol, ethanol, butanol) have been assessed. The properties of biodiesel, its production potential in Turkey and the World, biodiesel production processes such as dilution method, pyrolysis method, microemulsion method, supercritical method, transesterification method, and their standards are evaluated. The second chapter of this section focuses on the production of bioethanol from the microalgal feedstock. Compared to conventional feedstock (corn/sugar beet), microalgae have a superior potential in ethanol yield: 40–140 ML ha^{-1} versus 3–6 ML ha^{-1}, and it can be a promising alternative for the future. Various approaches have been presented to convert algal biomass to fuels, feed, and chemicals, but none is feasible economically. Therefore, it is prudent to develop a biorefinery concept where multiple products can be recovered from the same algal biomass. It is pertinent to focus on developing multiple products from the same biomass to address present bio-economy challenges. In this chapter, the recent developments on the algal refinery concept have been studied, focussing on integrated bioethanol production under a biorefinery paradigm. The review also covers the technical challenge that inhibits the realisation of algal biorefinery and provides insight into overcoming the technical hurdles. The last chapter of this section discusses the fuel delivery system for alternative fuel-powered engines. High-pressure direct injection of methanol is preferred for heavy-duty engines. In this case, the conventional fuel delivery system needs modifications. For example, a higher flow capacity is needed for an increased flow rate due to lower energy density. Additional cooling elements are needed to prevent vaporisation. A special coating should be applied to the surface of the pump plunger and control valve to improve the lubricity because those components are made of anticorrosive material. For natural gas, the dual-fuel mode is an effective combustion mode. Natural gas introduced from the intake port and mixed with air homogenously is ignited by injecting a small amount of diesel when the piston approaches the end of the compression stroke. Recently, a more advanced fuel delivery system has been developed, which uses one injector for diesel and the other for the natural gas (or a co-axial injector with an inner nozzle hole for diesel and outer nozzle holes for natural gas). It has been reported that diesel injected for ignition can be reduced to 5% of the total fuel by using this new system. This dual-fuel mode has been used for methanol and ammonia and could be used in the mainstream alternative fuel engines discussed at length in this chapter.

The third section focuses on alcohol as fuel/additives. This section has three chapters. The first chapter of this section explores the performance, combustion, and emission characteristics of various transportation modes using primary alcohols, such as methanol, ethanol, propanol, and butanol, as alternative fuels. The primary transportation modes include passenger cars, commercial vehicles, ships, locomotives, non-road vehicles, and alternative power systems employing fuel cells powered by alcohol. The effects of using alcohols as alternative fuels with carbon–neutral characteristics, combustion characteristics, and emissions such as NO_x, CO, HC, PM, and CO_2 are also discussed. The second chapter focuses on advances in using ethers and alcohols as additives for improving biofuel properties for SI engines. The commercialisation of biofuels as alternative fuels to conventional diesel for application as transport-fuels for diesel engines is fast becoming attainable owing to the merits offered by the inclusion of significant alcohol (ethanol) quantities and a member of the "ether" group (dimethyl ether) as additives and property-improvers for biofuels obtained from biomass. These additives are fuels that have lower viscosities, flash-points, flammability etc. Hence they affect spray atomisation and moderation in the densities and viscosities of biofuels towards ensuring their suitability for use in IC engines. Biofuels need to be improved for deriving improved engine performance, emission, and combustion characteristics to meet market specifications. This requires suitable fuel-modifiers, which need to be tested for their material compatibily before being used as fuels in ICEs. The mixing ratio of the added components with biofuels is also to be given utmost attention as alcohols such as ethanol and ether (dimethyl ether) are known for their higher volatilities, which in turn regulates the BTE and combustion potential of these fuels, all aimed at improving the cetane number of indices of the blended fuels. The third chapter focuses on the influence of oxygenated fuels/additives in biofuel-powered CI engines. The improvements in physiochemical properties of biodiesel/diesel-alcohol blended fuels and their influence on the engine emission characteristics are covered in this chapter.

The fourth and the last section of the book focuses on carbon neutrality in the last three chapters. The first chapter of the section focuses on future sustainable transport fuels for Indian heavy-duty vehicles. Heavy-duty vehicles fitted with ICE have been instrumental in helping modern civilisation in meeting economic and social goals. In this pursuit, an enormous burden has been put on the environment. Now humans are facing the challenges to limit and reduce the environmental burden over time. Many technology concepts are discussed worldwide and projected as sustainable solutions for the future heavy-duty transport sector. However, several considerations for a nation like India need careful analysis before adopting a particular transport solution on a large scale. Several strategies needed to make the transport sector efficient and environment friendly in India are discussed. The second chapter focuses on the potential and challenges of using biodiesel in a CI engine. This chapter reviews various thermophysical and chemical properties of biodiesels and their significant effects on the combustion characteristics such as cylinder pressure, rate of pressure rise, mean gas temperature, and engine performance parameters, including brake specific fuel consumption (BSFC), brake thermal efficiency (BTE), and brake power (BP). Moreover, engine emissions and various emission control techniques, i.e., blending of fuels, hardware modifications in engines, are thoroughly reviewed. The literature reports different refined oils and feedstocks to produce biodiesel utilising different methods, such as pyrolysis, wet washing, blending or direct use, transesterification, and microemulsions. Still, considerable efforts are needed to commercialise biodiesels due to their properties of higher viscosity, corrosivity, and less stable nature. So, the work concludes with an optimised approach to consider the potentiate biodiesel as a fuel in CI engines. The third chapter of this section focuses on the possibilities of biodiesel and renewable diesel as a drop-in fuel for decarbonised maritime transport. The purpose of this chapter is to understand the steps that IMO has taken for decarbonisation. It also explains the fuel properties, raw materials, and production methods of biodiesel and renewable diesel, an alternative marine fuel, and determine its usability and suitability by comparing it with other promising alternative fuels. This chapter also summarises the pros and cons of the decarbonising maritime industry.

This book is therefore divided into four different sections: (I) General, (II) Production and Fuel Injection Aspects, (III) Alcohol as a Fuel/Additive and (IV) Carbon Neutrality. Specific chapters covered in the manuscript include:

- Introduction of Potential and Challenges of Low Carbon Fuels for Sustainable Transport
- Some of the Biofuels for Internal Combustion Engines: Alcohols and Biodiesel
- Production of Bioethanol from Microalgal Feedstock: A Circular Biorefinery Approach
- Fuel Delivery System for Alternative Fuel Engines—A Review
- Alcohols As Alternative Fuels for Transport
- Advances In the Use of Ethers and Alcohols as Additives for Improving Biofuel Properties for SI Engines
- Influence of Oxygenated Fuel and Additives in Biofuel Run Compression Ignition Engine
- Future Sustainable Transport Fuels for Indian Heavy Duty Vehicles
- Potential and Challenges of Using Biodiesel in A Compression Ignition Engine
- Biodiesel and Renewable Diesel as A Drop-in Fuel for Decarbonized Maritime Transportation.

Reference

Agarwal AK (2007) Biofuels (alcohols and biodiesel) applications as fuels for internal combustion engines. Prog Energy Combust Sci 33(3):233–271. https://doi.org/10.1016/j.pecs.2006.08.003

Part II
Production and Fuel Injection Aspects

Chapter 2
Some of the Bio-fuels for Internal Combustion Engines: Alcohols and Biodiesel

Selçuk Sarıkoç

Abstract Alternative biofuels for the internal combustion engines is one of the important options to decrease exhaust emissions, improving engine performance and combustion characteristics. The purpose of this book chapter is to review alternative biofuels such as alcohols and biodiesel. In this regard, the advantages and disadvantages of alcohols (e.g. methanol, ethanol, butanol) have been assessed. Furthermore, the properties of biodiesel, its production potential in Turkey and the World, biodiesel production process and production methods such as dilution method, pyrolysis method, microemulsion method, supercritical method, transesterification method, and their standards were evaluated. This book chapter gives information about alternative biofuels for internal combustion engines that substantially affect the engine performance, combustion process, and exhaust emission characteristics.

Keywords Alternative fuels · Internal combustion engines · Alcohols · Biodiesel

Abbreviations

ASTM	American Society for Testing Material
CO	Carbon monoxide
CNS	Czech Standards Institute
DI	Direct injection
DIN	German Institute for Standardization
EN	European norm
HC	Hydrocarbon
H_2SO_4	Sulfuric acid
ID	Ignition delay
KOH	Potassium hydroxide
NaOH	Sodium hydroxide

S. Sarıkoç (✉)
Department of Motor Vehicles and Transportation Technologies, Tasova Yuksel Akin Vocational School, Amasya University, Amasya, Turkey

A. V. Agarwal and H. Valera (eds.), *Potential and Challenges of Low Carbon Fuels for Sustainable Transport*, Energy, Environment, and Sustainability,
https://doi.org/10.1007/978-981-16-8414-2_2

NO_x Oxides of nitrogen
UNI Italian National Unification

2.1 Introduction

Fuels can be generally divided into two classes, alternative fuels and conventional fuels. Conventional fuels are fuels derived from fossil fuels such as petroleum and coal. Fuels such as natural gas, liquefied petroleum gas (LPG), gasoline, jet fuel, kerosene, diesel and fuel oil are fossil fuels derived from petroleum. The oil crisis in the 1970s, the health problems and environmental problems in the following years lead to exploring alternative fuels. Today, natural gas, liquefied petroleum gas (LPG), synthetic fuels, hydrogen, biofuels, electricity and coal-based liquid fuels are considered alternative fuels (Beşergil 2009).

In this book chapter, alcohols and biodiesel, which are biofuels, will be discussed as alternative engine fuels. The properties of alcohols and biodiesel will be addressed along with their advantages and disadvantages, production potential in Turkey and the World, production process and production methods, and their standards.

2.2 Alcohols

2.2.1 Methanol, Ethanol, Butanol

Alcohols are produced in large quantities and economically from agricultural products and wastes. It is an alternative fuel widely used in internal combustion engines in countries where it is produced, especially in Brazil. Alcohols are usually obtained artificially from the fermentation of vegetable wastes (ethanol, C_2H_5OH) or the chemical process of coal (methanol, CH_3OH) which can be produced from non-petroleum fuels.

The high octane number of alcohols increases knock resistance, hence it can be used in spark-ignited engines at various rates. Alcohols and its blends with gasoline difficulty of mixing with gasoline at high rates; It creates problems such as phase separation, watering of the fuel and corrosion on engine parts. Therefore, up to 10–15 of alcohol can be blended with gasoline without any change in the engine, with minimum problems thereby improving the characteristics of the gasoline. Another advantage of alcohol mixtures is that they have higher latent heat of evaporation than gasoline, thus increasing the volumetric efficiency by reducing the filling temperature as a result of the removal of heat from the intake air (Safgönül et al. 1995). Table 2.1 summarized the properties of methanol, ethanol, butanol, and gasoline.

Methanol is the simplest component of the alcohol group, also known as "wood alcohol". Approximately 186 gallons (704 m^3) of methanol can be produced from

Table 2.1 Properties of methanol, ethanol, butanol and gasoline (Beşergil 2009; McAllister et al. 2011; Rajesh Kumar and Saravanan 2016)

Properties	Methanol (CH_3OH)	Ethanol (C_2H_5OH)	n-Butanol (C_4H_9OH)	Gasoline (C_8H_{15})
Density, 15 °C (kg/m^3)	791.3	789.4	809.1	750
Molecular weight (kg/kmol)	32.04	46.07	74.12	114.23
Vapor pressure (mmHg)	127	55	7	562.5
Boiling point (°C)	65	78	117.5	30–190
Research octane number (RON)	110	119	–	97
Motor octane number (MON)	92	92	–	86
Cetane number	5	11	17	8
Stoichiometric air/fuel; (kg air/kg fuel)	6.5	9	–	14.7
Lower heating value (MJ/kg, 15 °C)	19.8	26.4	33.09	41.3
Flash point at closed cup (°C)	12	13	29	−45

1 ton of biomass, while approximately 2.83 m^3 of natural gas is used for 1 gallon (3785 L) of methanol. Methanol is seen as the most promising alternative fuel as an alternative to gasoline and is carried out in the research and development as an alternative fuel. It is widely used by mixing with gasoline at rates of M85 (85% methanol, 15% gasoline) and M10 (10% methanol, 90% gasoline). There are some drawbacks to the use of methanol in internal combustion engines. Methanol is an alcohol that has a very corrosive effect on metal parts. However, methanol wastes react with CO_2 and H_2O in the exhaust to form formic acid (CH_2O_2) and carbonic acid (H_2CO_3). When M85 fuel is used, exhaust HC and CO decrease, while NO_x emissions and formaldehyde formations increase approximately 5 times. In addition, methanol in the gasoline-methanol mixture has a high tendency to react with water and separates from gasoline leading to heterogeneous mixture. As this heterogeneous mixture creates different air–fuel ratios, it causes the engine to run irregularly (Beşergil 2009; Pulkrabek 2016). Boiling point, theoretical air/fuel ratio, enthalpy of evaporation, enthalpy of combustion properties of methanol, ethanol, gasoline are given in Table 2.2.

Ethanol is produced by fermentation from maize, sugar beet, sugarcane, cellulose biomass or from ethylene. Fuel mixtures of ethanol, E85 (85% ethanol, 15% gasoline)

Table 2.2 Combustion properties and evaporation of methanol, ethanol, gasoline (Stone 1992)

Fuel type	Boiling point (°C)	Theretical air/fuel ratio	Enthalpy of evaporation (kJ/kg)	Enthalpy of combustion	
				(MJ/kg fuel)	(MJ/kg theoretical blend)
Methanol	65	6.5	1170	22.2	3.03
Ethanol	78.5	9	850	29.7	2.97
Gasoline	25–175	14.5	310	42	2.71

and E10 (10% ethanol, 90% gasoline) are widely used. E10 fuel mixture is also known as gasohol (Pulkrabek 2016).

Butanol is a four-carbon alcohol (Fayyazbakhsh and Pirouzfar 2017). n-butanol is produced from regular chain biomass (biobutanol), often referred to as 'butanol', as well as from fossil fuels (petro-butanol). Although they are produced from different sources, they have the same chemical properties and show similar effects when used in engines (Rajesh Kumar and Saravanan 2016).

As the carbon chain lengths of alcohols increase, the ignition quality of the alcohol generally increases. Therefore, high carbon alcohols $\geq C_4$ attract more attention as biofuels than low carbon alcohols (C_1–C_3) (Zhu et al. 2016). Higher alcohols have greater energy density, higher viscosity, and better mixing ability than short-chain alcohols. In addition, n-butanol is less hygroscopic and less corrosive, making it more compatible with existing fuel delivery systems. n-butanol is a second generation biofuel that can be produced from inedible biomass. In the life cycle analysis of grain-produced n-butanol, it was concluded that while it saves 39–56% more energy than fossil fuels, it reduces greenhouse gas emissions by more than 48% (Li et al. 2016). A comparison of butanol properties with diesel is presented in Table 2.3.

Advantages of butanol: Butanol is not only has a higher cetane number than other alcohols (methanol and ethanol), but also less corrosive. The ignition temperature of butanol is approximately 385 °C, lower than the auto-ignition temperature of methanol (479 °C) and ethanol (434 °C). While the vaporization enthalpy of butanol is 585 kJ/kg, the values of methanol and ethanol are 1100 kJ/kg and 840 kJ/kg, respectively. Butanol can be produced by fermentation of biomass, especially cellulose-rich wood residues, and added to diesel fuel (Fayyazbakhsh and Pirouzfar 2017; Choi et al. 2015). However, butanol can also be produced from biomass sources from which ethanol is produced, such as wheat, grain, maize, sugar beet and sugarcane (Rajesh Kumar and Saravanan 2016). Butanol can be produced from biomass by fermentation as well as by petrochemical processes. Although the majority of butanol is produced from petroleum by petrochemical processes, fermentation butanol production studies have accelerated recently with the merger of biotechnology and fuel industry. In recent years, the progress of biochemical processes in industry and the increase in butanol production from biomass by fermentation have brought biobutanol to the one step forward as a renewable energy source (Smerkowska 2011).

Table 2.3 Properties of diesel and butanol isomers (Rajesh Kumar and Saravanan 2016)

Properties	Diesel	n-Butanol	Sec-Butanol	Iso-Butanol	Tert-Butanol
Molecule structure	–		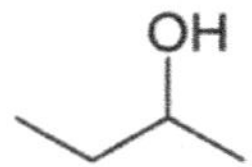	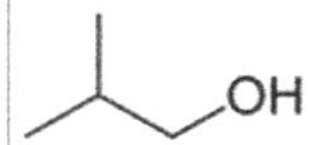	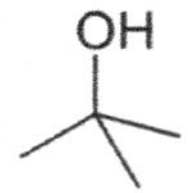
Molecular weight (kg/kmol)	190–211.7	74.12	74.12	74.12	74.12
C (%wt.)	86.13	64.82	64.82	64.82	64.82
H (%wt.)	13.87	13.60	13.60	13.60	13.60
O (%wt.)	0	21.59	21.59	21.59	21.59
Cetane number	52	17	–	–	–
Self-ignition temperature (°C)	254–300	345	406.1	415.6	477.8
Density at 15 °C (kg/m^3)	835	809.7	806	802	789
Lower calorific value (MJ/kg)	42.49	33.09	32.74	33.11	29.79
Evaporation heat (kJ/kg)	270–375	581.4	671	684	511
Vapor pressure (mm Hg)	–	7.2	12.5	8	30.1
Boiling point (°C)	180–360	117.5	99.5	108	82.4

Butanol is less hygroscopic and miscible well with diesel fuel without phase separation and forms a stable mixture. The cetane number of butanol is about 17, and since it is higher than methanol and ethanol, it has a better self-ignition tendency. It has a higher calorific value as it has approximately 25% more energy content compared to ethanol. The stoichiometric air/fuel ratio is closer to diesel fuel. Its lower volatility feature reduces cavitation and vapor fouling. It is less corrosive and can be stored longer in normal tanks. Its high flash point provides safer handling and transportation (Rajesh Kumar and Saravanan 2016).

Butanol has different isomers depending on where in the carbon chain the hydroxyl group (−OH) is attached. The properties of butanol isomers are given in Table 2.3. Butanol alcohols produced from biomass generally tend to form straight chain molecular structures. Synthetic-based butanols can be mixed with diesel at a rate of 5–20% by volume. The homogeneous butanol-diesel fuel mixture can stand stably at temperatures of 5–25 °C without any phase change (Hajba et al. 2011). Studies have shown that a mixture of butanol and diesel fuel is stable and well mixed at room temperature of 40% and above butanol mixtures (Smerkowska 2011).

Although the increase in the number of carbon atoms in the alcohol structure causes a decrease in the oxygen ratio in mass, the fuel values increase such as the cetane number, density, and calorific value. For this reason, n-butanol, one of the higher alcohols, has a higher cetane number, calorific value, viscosity and flame speed than lower alcohols. However, butanol has a lower heat of vaporization, ignition temperature and corrosion risk. In addition, since these higher alcohols have better solvent capacity, they form an easier mixture in diesel and biodiesel. Since butanol contains 21.59% oxygen by weight of its atomic structure, it increases the oxygen ratio of diesel and biodiesel fuel mixtures. Therefore, higher alcohols, especially butanol, are seen as the next generation biofuel potential as they improve the fuel properties of biodiesel (Atmanli 2016).

2.2.2 *Advantages and Disadvantages of Alcohols*

2.2.2.1 Advantages of Alcohols

Alcohols can be obtained from many different sources such as natural and artificial manufactured. Higher octane numbers of alcohols allow working at higher compression ratios, so higher thermal efficiencies can be achieved. The rapid combustion of alcohols and the higher mole ratio of their products than the reactants both increase the efficiency of the cycle. Alcohols contain extra oxygen in their structure so that they require less air for theoretical combustion. Thus, lower emission values can be obtained as alcohols can be burned with lean mixtures. In addition, alcohols have a lower sulfur content and have lower exhaust gas emissions compared to gasoline. The cooling property of the high heat of evaporation of alcohols increases the volumetric efficiency of the engine (Pulkrabek 2016; Stone 1992).

When alcohols are used as fuel additives, they reduce the surface tension and viscosity of the fuel mixture and provide better atomization of the fuel mixture. C_1–C_3 alcohols (methanol, ethanol, and propanol) can be used in engines as a diesel fuel mixture. However, these low carbon number alcohols have disadvantages such as low calorific value in forming mixtures in diesel fuels due to poor and unstable mixtures. Therefore, research has focused on four-carbon C_4 butanol with a longer carbon chain, higher cetane number, and better blend stability with diesel fuel than C_1–C_3 alcohols (Zhu et al. 2016).

2.2.2.2 Disadvantages of Alcohols

The energy content of alcohols is lower than gasoline and diesel. About twice as much alcohol must be used to achieve the same power output with alcohols compared to gasoline. The use of alcohols in engines causes more aldehyde and formaldehyde formation in emissions and increases the formation of these products. The high corrosive effects of alcohols can cause corrosion on copper, brass, aluminum, rubber, and plastic materials in the engine parts and fuel supply system. As a result of long-term use of alcohols, deterioration can be seen in the fuel line, fuel tank, gasket and metallic engine parts. Due to the low vapor pressure of alcohols, there is a high risk of invisible flame formation and ignition during storage and transportation. Its rapid evaporation at low temperatures and pressures can cause vapor plugs to form, especially in the fuel pipeline (Pulkrabek 2016).

The water dissolving ability of alcohols is very high, such as 6000–7000 ppm at 21 °C. Since the solubility decreases in cold weather, water in the mixture attracts some of the alcohol, causing phase separation in the gasoline-alcohol mixture. As a result of this phase separation, an alcohol-poor upper phase and an alcohol-rich water phase are formed. Therefore, the octane number of gasoline decreases and consequently causing knocking. In addition, corrosion occurs in the alcohol-rich water phase regions when phase separation occurs. Vapor pressure is a quantity that quantifies the volatility of the fuel. The vapor pressure of alcohols is much lower than gasoline, which creates a negative effect during the first start of the engine in cold weather. However, it stands out as an advantage as it reduces the risk of fuel explosion and provides lean mixture combustion (Beşergil 2009). Low volatility and energy density of alcohols, being easily miscible with water, and a tendency towards pre-ignition are stepped forward as some of the most important disadvantages of alcohols (Stone 1992).

Consequently, alcohols are promising alternative fuels as an internal combustion engine fuels such as methanol (methyl alcohol) or ethanol (ethyl alcohol), which can be produced naturally from many different sources (Pulkrabek 2016). Among the primary alcohols, biobutanol is seen as the most promising fuel to be an alternative fuel among the methanol and ethanol alcohols (Li et al. 2016).

2.3 Biodiesel

Biodiesel is an environmentally friendly and renewable energy liquid biofuel produced by chemical processes of vegetable, animal or used waste cooking oils (Öğüt and Oğuz 2006). Production potential of biodiesel in Turkey and in the World, biodiesel production and biodiesel production methods, biodiesel fuel properties and fuel quality standards, advantages and disadvantages of biodiesel will be discussed in this section.

2.3.1 Biodiesel Production Potential in Turkey and the World

Biodiesel production in the world is expected to increase by 27% in 2024 compared to 2014, and is forecasted to reach 39 billion liters in a decade at end of the 2024. In Fig. 2.1, the expected values of the change in world biodiesel production and world biodiesel trade amounts between 2008 and 2024 and expected in the future are given. Figure 2.2 shows the estimated percentage distribution of biodiesel production and consumption rates by region in 2024 (OECDiLibrary 2021).

The European Union is expected to be the largest biodiesel producer. Other important countries are Indonesia, USA, Brazil and Argentina. In the European Union, biodiesel production is projected to reach its maximum level in 2020 with 13.6 billion L. Two noteworthy countries here are Indonesia and Argentina. It is estimated that they will produce more than the amount of biodiesel they consume. This means that these two countries will become important biodiesel exporting countries worldwide in the future. In 2024, biodiesel production is expected to decline by 13.1 billion L as the demand for both biodiesel and diesel production is expected to decline. The USA is expected to lose its position as the second largest biodiesel producer in the next ten years. Turkey's biodiesel production, which was approximately 13 million liters between 2012 and 2014, is estimated to be 14 million liters in 2024. All of the

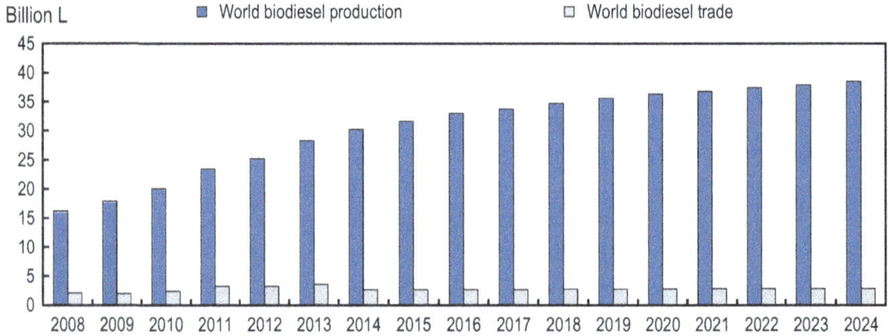

Fig. 2.1 Change of world biodiesel production and trade by years (OECDiLibrary 2021)

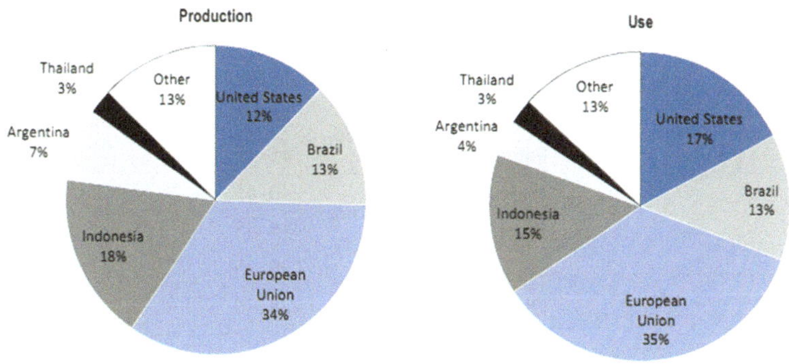

Fig. 2.2 Distribution of biodiesel production and consumption in 2024 by regions (OECDiLibrary 2021)

biodiesel produced in Turkey is used in the domestic market without being exported abroad (OECDiLibrary 2021). The amount of biodiesel produced and consumed in Turkey in 2019 and 2020 is 13.08 and 13.21 million liters, respectively. While the amount of biodiesel production will be estimated at 13.34 million liters by end of the 2021. This amount will be constituted from vegetable oils will be 12.93 million liters, and the amount of biodiesel produced from waste is estimated to be 0.41 million liters (OECD-FOA Agricultural Outlook 2017). However, COVID 19 pandemic has caused limited movements of people around the World. Therefore, a reduction has occurred which was 8.5% in global transport fuel use in 2020 compared to 2019. Furthermore, this effect led to a drop of 8.7% in biofuel use in 2020 compared to 2019. Nevertheless, it is expected the recover the global biofuel demand in 2021 and 2022 (OECDiLibrary 2021).

2.3.2 *Vegetable Oils and Use of Vegetable Oils as the Diesel Engine Fuel*

Vegetable oils are divided into two main classes: oils obtained from oilseeds such as sunflower, sesame, safflower rapeseed, soybean and oil obtained from oily fruits such as palm fruit, olive, almond, hazelnut, walnut, coconut. In Table 2.4, the most produced oil crops in the world and their average oil content are given (Yücel 2008).

The use of vegetable oils as diesel fuel directly in engines has advantages such as portability, energy density being close to diesel fuel (approximately 80% of diesel), easy and fast availability, and being in the renewable energy source class. However, vegetable oils also have disadvantages such as high viscosity, high ignition temperature and tendency of unsaturated fatty acids to oxidize (Koncuk 2008).

The high density, viscosity, boiling point and low volatility properties of vegetable oils can cause a narrowing of the spray angle in diesel engines, thus reducing the fuel atomization. As a result, the combustion efficiency of the fuel is adversely affected.

Table 2.4 The most produced oil crops and their average oil content (Yücel 2008)

Oil plant	Average oil percentage (%)
Coconut	65–68
Sesame	50–55
Palm fruit	45–50
Rapeseed	40–45
Sunflower seeds	35–45
Safflower seed	30–35
Olive (in fruit)	25–30

The use of vegetable oils in direct injection engines for a long time can lead to carbon accumulation and coking as a result of precipitation. This can cause clogged injectors, jamming of oil rings, thinning of the lubricating oil and gelling of the oil as a result of contamination. Especially the deposits formed in the injectors prevent the fuel atomization and adversely affect the combustion. However, low atomization due to high viscosity adversely affects the combustion process and causes incomplete combustion. Although there is a deterioration in performance both as a result of bad atomization and because the energy content of vegetable oil is lower than diesel fuel, an increase in CO and HC emission values and a slight decrease in NO_x value has been observed. As a result of using vegetable oils as fuel in diesel engines, some problems can occur in the short term; there are difficulties in starting up in cold weather, clogging in the fuel filter pipeline and injectors, gum formation, and engine knock problems. Long-term use will lead to problems such as coking on the engine cover, piston, and injectors, engine wear is a result of carbon accumulation. To overcome the problems caused by the high viscosity of vegetable oils in diesel engines, some methods are used for decreasing the viscosity which are discussed below (Koncuk 2008; Özsezen 2007).

2.3.3 Biodiesel Production Process and Production Methods

It is aimed to reduce the flaming, clouding, boiling point as well as viscosity value of the fuel by separating the glycerin in vegetable, animal, and waste cooking oils. When vegetable oils are used as raw material, methods such as dilution, micro-emulsion, thermal cracking (pyrolysis), supercritical method, and transesterification reaction are used to reduce the high viscosity of vegetable oil approximately ten times and to reduce its density to a value close to the density of diesel fuel. Of these methods, the most preferred method in the market and in applications is usually the transesterification method (Koncuk 2008; Özsezen 2007).

The highest cost in biodiesel production is the raw material source with 70.60%. Therefore, the most important factor in reducing the cost of biodiesel is the cheap

Table 2.5 Cost distribution of biodiesel production (Math et al. 2010)

Items	Cost ratio (%)
Raw material	70.60
Maintenance and repair	4
Chemicals	12.60
Energy	2.70
Workmanship	2.50
Depreciation	7.60

and sustainable raw material source. In Table 2.5, the percentage rates of biodiesel production items according to the cost distribution are given (Math et al. 2010).

2.3.3.1 Dilution Method

In the dilution method, vegetable oil is mixed with a certain amount of diesel fuel and its viscosity is reduced so that it is close to diesel engine fuel. This method is mostly applied to vegetable oils such as sunflower, safflower, soybean, rapeseed and waste frying oils (Öğüt and Oğuz 2006; Koncuk 2008).

2.3.3.2 Pyrolysis Method

Pyrolysis, also known as thermal cracking, is the process of converting a substance into another substance in the presence of heat or a catalyst with heat (Koncuk 2008). A chemical change is applied by giving thermal energy to the crude oil under atmospheric conditions. In this chemical process, many components such as alkanes, alkenes, aromatics, and carboxyl acids are formed as the oil is thermally decomposed. Although fuel with a high cetane number and low viscosity is obtained in this method, oxygen is lost in the fuel since it is a process similar to the distillation of crude oil. In this method, not only the oxygen content of the produced biodiesel is lost, but also the cloud point, carbon residue and ash amount formed as a result of combustion is not at the desired standards (Özsezen 2007).

2.3.3.3 Microemulsion Method

Microemulsion is the balance of heterogeneous mixtures with homogeneous appearance, formed by liquid microstructures in the size range of 1–150 mm by one or more immiscible liquids (Koncuk 2008). In this method, crude oil with a very high viscosity can be used directly in diesel engines by reducing its viscosity with various solvents such as ethanol, methanol and butanol. However, it has been observed that the fuel obtained by this method increases the viscosity of the lubricating oil, causes a large amount of carbon accumulation in the injector nozzle and combustion chamber, and the fuel does not fully burn (Özsezen 2007). Besides, the low cetane number of the alcohols used causes the emulsion to have a low cetane number. In addition, phase separation is observed in the mixture in cold conditions, since the stability of the fuel mixture decreases in this method at low temperatures (Öğüt and Oğuz 2006).

2.3.3.4 Supercritical Method

Unlike the transesterification method in the supercritical method, it is a method of producing biodiesel by applying a process at high temperatures such as 350 °C and for a very short time such as 240 s without adding catalyst to the crude oil (Öğüt and Oğuz 2006; Koncuk 2008).

2.3.3.5 Transesterification Method

The most widely used method in the market for the conversion of vegetable oils to biodiesel is transesterification, in other words alcoholysis. The most important feature that distinguishes this method from other methods is that the viscosity of the oil is greatly reduced, while there is no change in the cetane number and oxygen content of the fuel. The raw material source and alcohol used have a significant effect on the fuel properties of the produced biodiesel. Therefore, different raw material sources lead to variation in the combustion calorific value, auto-ignition temperature and emission values of the biodiesel fuel. The cetane number is one of the most important fuel properties that quantifies the ignition ability of the fuel in diesel cycles. Generally, the cetane number of biodiesel fuels is higher than the cetane number of euro diesel fuels, which stands out as the most important feature of biodiesel (Özsezen 2007).

The binding of the ester hydrocarbon chain to another molecule is called the transesterification reaction, and the conversion of this ester form to another form is called. The purpose of biodiesel production is to separate the ester and glycerine from the oil by combining the glycerine in the oil with the catalyst and the liberated esters with the alcohol. Thus, as a result of this reaction, the thick and sticky glycerin in the oil is removed and a thinner and less viscosity biodiesel is obtained (Öğüt and Oğuz 2006). In Fig. 2.3, the transesterification reaction in biodiesel production is given. Here, R_1, R_2 and R_3 represent carbon chains of different fatty acids attached to glycerin. 1 mol of triglyceride turns into 3 mol of fatty acid methyl esters and 1 mol of glycerol by reacting with 3 mol of short-chain alcohol methanol (CH_3OH) and a catalyst. The stoichiometric oil:alcohol molar ratio required for the complete completion of this reaction should be at least 3:1 molar ratio. In the transesterification reaction, it is sufficient to choose 6:1 mol ratio of oil-alcohol to produce biodiesel at the highest rate from vegetable oils with less than 1% free acid content. Reactions with high oil-alcohol molar ratio provide shorter and more efficient biodiesel conversion. Generally, in commercial applications, biodiesel can be produced at an oil-alcohol molar ratio of 6:1 with an efficiency of approximately 95–98% (Özsezen

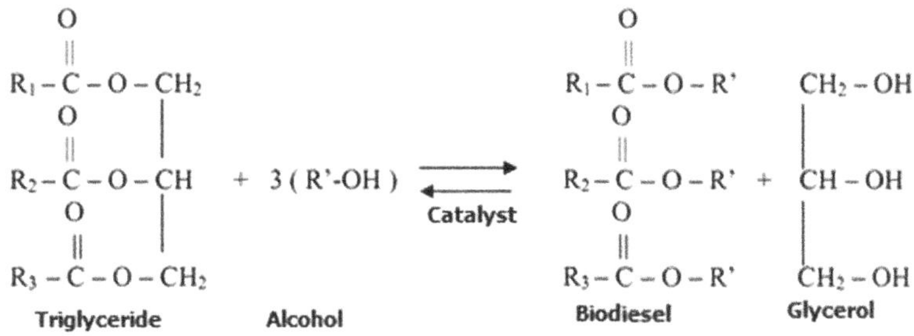

Fig. 2.3 Transesterification reaction in biodiesel production (Şanlı 2014)

2007). Generally, methanol and ethanol alcohols are used in this method. In particular, methanol is preferred as an alcohol in applications due to its rapid reaction with glycerides, easier solution formation with sodium hydroxide, better physical and chemical properties and lower cost (Koncuk 2008). In addition, methyl alcohol provides a more balanced reaction during biodiesel production and is less affected by the water in the oil than ethyl alcohol (Öğüt and Oğuz 2006).

Catalysts are used to increase the transesterification reaction rate and product conversion. Acid-based catalysts such as sulfuric acid, phosphoric acid or organic sulfonic acids or alkaline catalysts with basic character such as NaOH, KOH, carbonates, sodium ethoxide, sodium methoxide are used in the reaction. Alkali catalysts are generally preferred in transesterification reaction applications because they are much faster and more efficient than acid catalyst. Futhermore, in industrial applications, acidic catalysts are not preferred because they have much more corrosive effects on production equipment than alkaline catalysts (Koncuk 2008; Özsezen 2007).

When a basic NaOH catalyst is used at a rate of 1% by mass of the oil at 60 °C, the oil:methyl alcohol ratio is 6:1 mol and the oil is converted to biodiesel at a rate of approximately 95% in one hour. When 1% by mass of oil is sulfuric acid catalyst and 30:1 mol ratio oil:methyl alcohol is used, approximately 90% biodiesel is obtained after 60 h of reaction at 65 °C (Öğüt and Oğuz 2006).

The amount of water in the oil and the free fatty acid ratio, the chemical structure of the alcohol and the type of catalyst, the oil-alcohol molar ratio, the reaction time and the reaction temperature are the most important parameters affecting the biodiesel conversion reaction and its efficiency (Öğüt and Oğuz 2006; Özsezen 2007).

It has been concluded that the oil-alcohol molar ratio is more effective on the reaction process than the type of catalyst that reacts. The types of catalysts used vary according to the free acidity of crude oils. Since the free fatty acid value of refined vegetable oils is low, they can react directly with alkaline catalysts of basic character. However, when the free fatty acid content of waste frying oil and animal fats is high, first a pretreatment process (esterification reaction) is performed with an acidic catalyst. After the free fatty acid value of the crude oil drops below 1% by mass, a transesterification reaction is carried out with an alkaline catalyst. The water in the oil affects the reaction, partially changes the reaction and causes soap formation. The reacting catalyst is wasted for soap formation, reducing the catalytic efficiency of the reaction. Moreover, soap formation also prevents biodiesel from separating from glycerol by increasing the viscosity and sediment formation of biodiesel. Therefore, water is an undesirable substance in the biodiesel raw material source and must be removed from the oil before the reaction. The presence of water in the crude oil causes the catalyst to react in the direction of saponification, as well as increasing the use of catalyst during the biodiesel conversion and reducing the efficiency of the reacting catalyst. In addition, saponification causes gelation of biodiesel, increased viscosity, and more difficult separation of glycerin from biodiesel (Koncuk 2008).

Vegetable oils are an acidic character, while alcohol and catalyst are basic characteristics. Vegetable oils are called fatty acids because they are an acid. The amount of free fatty acid in the raw material is one of the most important parameters affecting the reaction process and its result (Öğüt and Oğuz 2006). In order to complete the transesterification reaction, the free acid value of the crude oil must be at least 3%. It is generally preferred that the free fatty acid value of the glycerin be less than 1% for neutralization with less catalyst. If the free acid ratio value of the crude oil is high, esterification reaction is applied with an acid catalyst such as crude oil sulfuric acid. This process, which is applied to reduce the acid value of the crude oil, is called the pre-treatment process (Koncuk 2008). Acid, peroxide, iodine and saponification number values of biodiesel vary depending on the raw material source (Özsezen 2007).

The temperature change significantly affects the ester conversion rate. In addition, reaction time is one of the important parameters determining ester conversion. At the same time, the type of catalyst used in biodiesel production and the catalyst ratio significantly affect the biodiesel conversion. In the methanolysis process, basic catalysts such as NaOH and KOH can be used between 0.4 and 2% by weight of the crude oil. In applications, the use of alkaline catalysts such as NaOH and KOH at a rate of 1% by weight of oil has yielded successful results. In the study, it was observed that the biodiesel efficiency decreased from 85 to 65% when the amount of NaOH catalyst was increased from 1 to 1.6% by mass. This is due to the fact that too much catalyst reduces the production efficiency as it increases saponification during the reaction (Koncuk 2008; Şanlı 2014). The biodiesel production process occurs at six different steps in the transesterification reaction. These steps are titration, transesterification, separation, washing, drying, and filtration processes are presented in Fig. 2.4 (Sarıkoç 2019).

2.3.4 Biodiesel Fuel Specifications and Standards

At the end of biodiesel production, some catalyst may remain in the ester. If the catalyst used in the reaction is acidic, the catalyst causes abrasive effects on the engine parts, while the alkaline catalysts cause solid ash particles after combustion. Therefore, the produced biodiesel must be purified from water, soap, alcohol, catalyst and glycerin. For this, biodiesel should be checked for compliance with biodiesel

Fig. 2.4 Biodiesel production process with each step (Sarıkoç 2019)

standards after it passes through washing, drying and filtering processes, respectively (Özsezen 2007).

These standards, according to the countries given in Table 2.6, biodiesel quality standards vary from country to country. However, the most widely used and accepted standards in scientific circles are ASTM D 6751 in the USA (American Society for Testing Material) or EN 14214 in Europe (European Norm). The biodiesel quality standards applied in Turkey are TS EN 14213 and TS EN 14214 standards, which are compatible with European standards. In this standard, analyzes such as density, kinematic viscosity, flash point, carbon residue and water content are evaluated in

Table 2.6 Biodiesel quality standards by country (Öğüt and Oğuz 2006; Koncuk 2008)

Parameter	Autria	Czech (CSN)	France (JO)	Gemany (DIN)	Italy (UNI)	USA (ASTM)
Density at 15 °C (g/cm^3)	0.85–0.89	0.87–0.89	0.87–0.89	0.87–0.89	0.86–0.90	–
Kinematic viscosity at 40 °C (mm^2/s)	–	–	–	–	–	1.9–6
Ignition point (°C)	100	110	100	110	100	130
Cold filter plugging point (°C)	0/−5	−5	–	0/−10/−20	–	–
Pour point (°C)	–	–	−10	–	0/−5	–
Cetane number	≥49	≥48	≥49	≥49	–	≥47
Neutralization number (mgKOH/g)	≤0.8	≤0.5	≤0.5	≤0.5	≤0.5	≤0.8
Carbon residue (%)	0.05	0.05	–	0.05	–	0.05
Ester content in massive (%)	–	–	≥96.5	–	≤0.98	–
Total glycerol mass (%)	≤0.24	≤0.24	0.25	≤0.25	–	≤0.24
Iodine number in massive (%)	≤120	–	≤115	≤115	–	–

the physical analysis group. Ester content, acid value, iodine value, glyceride amount and sulfur analyzes are evaluated in the chemical analysis group. Parameters such as free fatty acid ratio, iodine value and water content that directly affect the quality of biodiesel determine the quality of biodiesel raw material. Parameters such as density, flash point, ester content and cetane number are the factors on which physical and chemical properties are effective. The amount of free glycerol, acid, iodine, viscosity values, alcohol and catalyst residues, completion of the esterification (bonding) reaction are the most important factors that determine the production quality. The water content and oxidation stability of biodiesel are the factors that directly affect the post-production biodiesel quality. Water droplets dissolved or suspended in biodiesel cause corrosion in the fuel supply line and injection system, but also lead to microbiological developments, causing acidification and sludge formation in biodiesel (Öğüt and Oğuz 2006; Yücel 2008; Koncuk 2008; Alptekin 2013).

The iodine content of vegetable oils such as cotton, soybean, corn and sunflower grown in Turkey is quite high. For this reason, it is very difficult to meet European EN 14214 standards with biodiesel produced from these oils (Özsezen 2007). In addition, since the acid value and viscosity of the produced biodiesel increases over time, it is recommended to be used within 4–6 months. Transesterification method and the amount of polyunsaturated fatty acids in crude oil are the most important factors that affect and accelerate the change of biodiesel over time (Yücel 2008). Animal oils have longer fatty acid carbon chains and higher cetane numbers than vegetable oils because they contain more saturated molecules (İleri 2012).

Test methods, limits and technical specifications of ASTM D 6751-08 biodiesel standards applied in the USA and EN 14214 biodiesel standards applied in Europe are given in Table 2.7 comparatively.

Table 2.8 shows the variation of important properties of European diesel fuels such as cetane number, cetane index, sulfur content, viscosity and density according to Euro II–V standards. The most striking feature here is that from 1992 (Euro II) to

Table 2.7 Technical specifications of biodiesel in USA and Europe (Wakil et al. 2015)

Technical specifications	USA (ASTM D 6751-08)		Europe (EN 14214)	
	Test method	Limit	Test method	Limit
Kinematic viscosity at 40 °C (mm^2/s)	D 445	1.9–6	EN ISO 3104	3.5–5
Density at 15 °C (kg/m^3)	D 1298	880	EN ISO 3675/12185	860–900
Calorific value (MJ/kg)	–	–	EN 14214	35
Flash point (°C)	D 93	93	EN ISO 3679	Min. 101
Pour point (°C)	D 97	−15 to 16	–	–
Cloud point (°C)	D 2500	−3 to 12	–	–
Cold filter plugging point (°C)	ASTM	Max. +5	EN 14214	–
Cetane number	D 613	Min. 47	EN ISO 5165	Min. 51
Oxidation stability at 110 °C (h)	D 675	Min. 3	EN 14112	Min. 6
Acid value (mg KOH/g)	D 664	Max. 0.5	EN 14104	Max. 0.5
Free glycerin (% in mass, maximum)	D 6584	0.02	EN 14105	0.02
Total glycerin (% in mass, maximum)	D 6584	0.24	EN 14105	0.25
Carbon residue (% in mass, maximum)	D 4530	0.05	EN 10370	0.30
Copper strip corrosion (3 h at 50 °C)	D 130	Max. No. 3	EN 2160	No. 1
Iodine value (g I_2/100 g) max	–	–	EN 14111	120
Water and sediment (% in vol., max.)	D 2709	0.05	EN 12937	0.05
Total sulfur (ppm), max	D 5453	15	EN 20846	10
Phosphorus content (ppm), max	D 4951	10	EN 14107	4

Table 2.8 European diesel fuel specifications (Öğüt and Oğuz 2006)

Specifications	Euro-II 1992	Euro-III 2000	Euro-IV 2005	Euro-V 2010
Cetane number, at least	49	51	–	–
Cetane index, at least	46	46	–	–
Sulfur (ppm)	500	350	50	10 and least
Kinematic viscosity at 40 °C (mm^2/s)	2–4.5	2–4.5	–	–
Density, max. (kg/m^3)	860	845	–	–

2010 (Euro V) the sulfur content of diesel fuel decreased significantly by 1 in 50 or more, according to Euro standards (Öğüt and Oğuz 2006).

Biodiesel is an alternative renewable fuel that is widely used in diesel engines. The most important feature is that it contains less sulfur and aromatics than diesel (Tüccar et al. 2014). The sulfur content of biodiesel is limited to the highest 10 ppm according to EN 14214 biodiesel standards. However, this value is limited to two different classifications, S15 (maximum 15 ppm) and S500 (maximum 500 ppm) in ASTM D 6751 biodiesel standards (Alptekin 2013). Considering the above information, it can be concluded that one of the best ways to reduce the sulfur content of the fuel without losing its lubricating property and to reach Euro standards is to mix biodiesel with diesel fuel. European emission standards and emission values for the diesel vehicle are given in Table 2.9. In addition, European emission standards and emission values for spark-ignited vehicle are presented in Table 2.10.

Table 2.9 European diesel vehicle emission standards and emission (Automobile Technology 2021)

Norm	Euro-I	Euro-II	Euro-III	Euro-IV	Euro-V	Euro-VI
Expiry date	01.07.1992	01.01.1996	01.01.2000	01.01.2005	01.09.2009	01.09.2014
CO (g/km)	3.16	1	0.64	0.5	0.5	0.5
$HC + NO_x$ (g/km)	1.13	0.7/0.9 (DI)	0.56	0.3	0.23	0.17
NO_x (g/km)	0.5	0.25	0.18	0.25	0.18	0.08
PM (particulate matter) (g/km)	–	0.08	0.05	0.025	0.005	0.0045

Table 2.10 European spark-ignited vehicle emission standards and emission (Automobile Technology 2021)

Norm	Euro-I	Euro-II	Euro-III	Euro-IV	Euro-V	Euro-VI
Expiry date	01.07.1992	01.01.1996	01.01.2000	01.01.2005	01.09.2009	01.09.2014
CO (g/km)	3.16	2.2	2.3	1	1	1
$HC + NO_x$ (g/km)	1.13	0.5	–	–	–	–
NO_x (g/km)	0.15	–	0.15	0.08	–	–
HC (g/km)	0.2	–	0.2	0.1	0.1	0.1
PM (particulate matter) (g/km)	–	–	–	–	0.005	0.0045

2.4 The Effect of Biofuels on Economy, Engine Performance, Combustion, and Exhaust Emission Characteristics

Biofuels production cost is highly dependent on the biofuels raw material feed-stock, production process. Biofuels vary geographically so that they are not capable compete with the affordable cost of fossil fuels in some regions for example in Europe, except Brazil or the USA (Randelli 2008). A comparison of biofuels' cost with regard to conventional fossil fuels is presented in Table 2.11.

The cost analysis of biofuels shows differences in some regions that they are higher than fossil fuels. However, biofuels are promising internal combustion engines fuels due to their complementary of fossil fuel, has lower taxes, sustainability, lower exhaust emissions, and lower carbon content fuels.

The effect of biofuels on engine performance, combustion, and exhaust emission characteristics was studied. In this respect, Misra and Murthy (2011), investigated the

Table 2.11 Biofuels and fossil fuels cost comparison (Randelli 2008)

Biofuels	Biofuels costs (€/toe)	Fossil fuel pre-tax cost (€/toe)	Cost difference (€/toe)
Biodiesel (EU)	691	548	143
Biodiesel (EU)	660	548	112
Bioethanol (corn in USA)	338	566	−228
Bioethanol (sugar beet in EU)	671	566	105
Bioethanol (sugarcane in EU)	283	566	−283

effects of fuel additives on the improvement of the cold flow properties of biodiesel. It was concluded that the addition of short-chain alcohols such as ethanol, methanol and butanol to the fuel mixture improves the cold flow properties of biodiesel as well as engine performance and emission values. Ibrahim (2016), investigated the effects of diesel–biodiesel-butanol fuel blends on engine performance, exhaust emission and combustion analysis parameters. It has been revealed that butanol reduces the thermal efficiency of the fuel mixture, increases the specific fuel consumption, significantly reduces the NO emission, reduces the peak cylinder pressure and increases the highest heat release rate. Zheng et al. (2016), investigated the combustion characteristics and exhaust emission parameters of butanol-biodiesel mixtures. As a result, it has been shown that the highest pressure and highest heat release rate values of butanol-biodiesel fuel mixture are higher than both diesel and biodiesel. It was also revealed that butanol significantly reduced CO, HC, NO_x and smoke emissions.

It is clear that butanol usage positively affects biodiesel cold fuel properties. In addition, butanol is considerably affected by the engine performance, combustion process, and exhaust emission values.

Orsi et al. (2016) investigated passenger vehicles in terms of CO_2 emission, energy consumption, and economic cost in different regions using a multi-dimensional well-to-wheels analysis method. It was reported that electric vehicles emit almost zero CO_2 emission and lower operating costs, while electric vehicles have dramatically higher capital costs. In addition, the use of biofuels caused the lowest CO_2 emission. Thus, biofuels can effectively reduce CO_2 emissions so that take the place of fossil fuels. However, biofuels have exhibited more cost with regard to petroleum.

2.5 Conclusions and Recommendations

Alcohols and biodiesel appear as the best fuel option for the alternative fuels that improve the fuel properties of the internal combustion engines and ensure the exhaust emission values in accordance with standards. These alternative fuels cannot be used directly as fuel due to reasons such as physical and chemical properties, low calorific value, corrosion effects on the fuel supply system and engine parts. Alternative fuels cannot be used directly in the engine due to the difficulty of mixing with fossil fuels, phase separation, watering of the fuel, and its corrosive effects. Therefore, they can be used without any problems in the engine by mixing it with fossil fuels at low rates. Thus, alcohols and biodiesel fuels have emerged as an alternative fuel type that complements fossil fuels and improves their fuel properties in the last two decades.

Methanol is seen as the most researched and promising alternative fuel as an alternative to gasoline. There are some drawbacks to the use of methanol in internal combustion engines. Methanol is an alcohol that has a very corrosive effect on metal parts. In addition, methanol wastes react with CO_2 and H_2O in the exhaust to form formic acid (CH_2O_2) and carbonic acid (H_2CO_3). Therefore, fuel mixtures of E85 (85% ethanol, 15% gasoline) and E10 (10% ethanol, 90% gasoline) of ethanol are commonly used instead of methanol. E10 fuel mixture is also known as gasohol.

The highest cost in biodiesel production is constituted approximately 70% by the raw material source. Therefore, the most important factor in reducing the cost of biodiesel is the cheap and sustainable raw material source. In this respect, the transesterification method emerges as the most suitable biodiesel production method in terms of fuel production efficiency, preservation of fuel properties, and production cost. Furthermore, the produced biodiesel fuels must provide the certain standards so that it does not damage the engine and engine parts, and does not cause both economic losses and environmental damages. Thus, adding alcohol to biodiesel is one of the best options. In this respect, butanol fuel appears a promising biodiesel additive in terms of both its fuel properties and good mixing characteristics with biodiesel. It can be concluded that one of the best ways to reduce the sulfur content of diesel fuel without losing its lubricating property and to reach Euro standards is to mix biodiesel-butanol into diesel fuel.

The author recommends that alternative fuels should be investigated in terms of the long-term effects of alternative fuels based on lifecycle analysis, environmental and enviro-economic effects for future study.

Acknowledgements I would like to express my thanks to Amasya University for its supports.

Declaration of Competing Interest The authors declare that they have no known competing financial interests or personal relationships that could have appeared to influence the work reported in this paper.

References

Alptekin E (2013) Hayvansal Atık Yağlardan Biyodizel Üretimi ve Bir Dizel Motorda Kullanımının İncelenmesi. Kocaeli Üniveristesi Kocaeli

Atmanli A (2016) Comparative analyses of diesel-waste oil biodiesel and propanol, n-butanol or 1-pentanol blends in a diesel engine. Fuel 176:209–215. https://doi.org/10.1016/j.fuel.2016.02.076

Automobile Technology (2021) European emission standards and emission values. https://otomobilteknoloji.blogspot.com/2019/05/avrupa-euro-emisyon-standartlari-ve-emisyon-degerleri.html. Accessed 22 July 2021

Beşergil B (2009) Yakıtlar Yağlar "fuels oils". Ege Üniversitesi Yayını, İzmir

Choi B, Jiang X, Kim YK, Jung G, Lee C, Choi I, Song CS (2015) Effect of diesel fuel blend with n-butanol on the emission of a turbocharged common rail direct injection diesel engine. Appl Energy 146:20–28. https://doi.org/10.1016/j.apenergy.2015.02.061

Fayyazbakhsh A, Pirouzfar V (2017) Comprehensive overview on diesel additives to reduce emissions, enhance fuel properties and improve engine performance. Renew Sustain Energy Rev 74:891–901. https://doi.org/10.1016/j.rser.2017.03.046

Hajba L, Eller Z, Nagy E, Hancsok J (2011) Properties of diesel-alcohol blends. Hung J Ind Chem 39(3):349–352

Ibrahim A (2016) Performance and combustion characteristics of a diesel engine fuelled by butanol–biodiesel–diesel blends. Appl Therm Eng 103:651–659. https://doi.org/10.1016/j.applthermaleng.2016.04.144

İleri E (2012) Biyodizele antioksidan eklenmesinin NO_x emisyonuna etkilerinin deneysel olarak Araştırılması. Ege Üniversitesi İzmir-Bornoca

Koncuk F (2008) Biyodizel Üretimi ve Dizel Motorunda Motor-Yanma Karakteristiklerinin İncelenmesi. Atatürk Üniversitesi Erzurum

Li Y, Meng L, Nithyanandan K, Lee TH, Lin Y, C-F L, Liao S (2016) Combustion, performance and emissions characteristics of a spark-ignition engine fueled with isopropanol-n-butanol-ethanol and gasoline blends. Fuel 184:864–872. https://doi.org/10.1016/j.fuel.2016.07.063

Math MC, Kumar SP, Chetty SV (2010) Technologies for biodiesel production from used cooking oil—a review. Energy Sustain Dev 14(4):339–345. https://doi.org/10.1016/j.esd.2010.08.001

McAllister S, Chen J-Y, Fernandez-Pello AC (2011) Fundamentals of combustion processes (Springer Science + Business Media, LLC, New York, 2011)

Misra RD, Murthy MS (2011) Blending of additives with biodiesels to improve the cold flow properties, combustion and emission performance in a compression ignition engine—a review. Renew Sustain Energy Rev 15(5):2413–2422. https://doi.org/10.1016/j.rser.2011.02.023

OECD-FOA Agricultural Outlook 2016-2025 BIOFUEL-OECD-FOA Agricultural Outlook 2016-2025 Turkey. https://stats.oecd.org/index.aspx?r=895962&lastaction=login_submit. Accessed 18th July 2021

OECDiLibrary. OECD-FAO-Biofuels. https://www.oecd-ilibrary.org/docserver/89d2ac54-en.pdf?expires=1632476749&id=id&accname=guest&checksum=7D153A74DB3B69CCA6F81ECCB1E68C14. Accessed 24th Sept 2021

OECDiLibrary. OECD-FAO commodity chapters, Biofuels. http://www.oecd-ilibrary.org/agriculture-and-food/oecd-fao-agricultural-outlook-2015/biofuels_agr_outlook-2015-13-en. Accessed 18 Oct 2017

Öğüt H, Oğuz H (2006) Üçüncü Milenyumun Yakıtı; Biyodizel 2. Baskı. Nobel Yayın Dağıtım, Ankara

Orsi F, Muratori M, Rocco M, Colombo E, Rizzoni G (2016) A multi-dimensional well-to-wheels analysis of passenger vehicles in different regions: primary energy consumption, CO_2 emissions, and economic cost. Appl Energy 169:197–209. https://doi.org/10.1016/j.apenergy.2016.02.039

Özsezen AN (2007) Atık Palmiye Yağından Üretilen Biyodizelin Motor Performans ve Emisyon Karakterleri Üzerine Etkisinin İncelenmesi. Doktora Tezi, Kocaeli Üniversitesi Kocaeli

Pulkrabek WW (2016) İçten Yanmalı Motorlar Mühendislik Temelleri. 1. Baskı. İzmir Güven Kitabevi, İzmir

Rajesh Kumar B, Saravanan S (2016) Use of higher alcohol biofuels in diesel engines: a review. Renew Sustain Energy Rev 60:84–115. https://doi.org/10.1016/j.rser.2016.01.085

Randelli F (2008) An integrated analysis of production costs and net energy balance of biofuels. Reg Environ Change 9(3):221–229. https://doi.org/10.1007/s10113-008-0055-7

Safgönül B, Ergeneman M, Arslan E, Soruşbay C (1995) İçten Yanmalı Motorlar. Birsen Yayınevi İstanbul

Şanlı H (2014) Atık Kızartma Yağlarının Karakterizasyonu ve Biyodizel Üretiminde Değerlendirilmesi. Kocaeli Üniversitesi, Kocaeli

Sarıkoç S (2019) Investigation of the effects of hydrogen and titanium dioxide addition on the performance and emission parameters of biodiesel-butanol-diesel blends. Erciyes University, Kayseri

Smerkowska B (2011) Biobutanol–production and application in diesel engines. Automot Ind Inst 65(6):549–556

Stone R (1992) Introduction to internal combustion engines, 2nd edn. The Machillan Press Ltd., London

Tüccar G, Özgür T, Aydın K (2014) Effect of diesel–microalgae biodiesel–butanol blends on performance and emissions of diesel engine. Fuel 132:47–52. https://doi.org/10.1016/j.fuel.2014.04.074

Wakil MA, Kalam MA, Masjuki HH, Atabani AE, Rizwanul Fattah IM (2015) Influence of biodiesel blending on physicochemical properties and importance of mathematical model for predicting the properties of biodiesel blend. Energy Convers Manage 94:51–67. https://doi.org/10.1016/j.enconman.2015.01.043

Yücel Y (2008) Bazı Enzimleri Kullanarak Biyodizel Üretimi ve Biyodizel Özelliklerinin Analitik Metotlarla Araştırılması. Uludağ Üniversitesi, Bursa
Zheng Z, Wang X, Zhong X, Hu B, Liu H, Yao M (2016) Experimental study on the combustion and emissions fueling biodiesel/n-butanol, biodiesel/ethanol and biodiesel/2,5-dimethylfuran on a diesel engine. Energy 115:539–549. https://doi.org/10.1016/j.energy.2016.09.054
Zhu L, Xiao Y, Cheung CS, Guan C, Huang Z (2016) Combustion, gaseous and particulate emission of a diesel engine fueled with n-pentanol (C5 alcohol) blended with waste cooking oil biodiesel. Appl Therm Eng 102:73–79. https://doi.org/10.1016/j.applthermaleng.2016.03.145

Chapter 3
Production of Bioethanol from Microalgal Feedstock: A Circular Biorefinery Approach

Sanjukta Banerjee, Debabrata Das, and Ananta K. Ghosh

Abstract The ever increasing energy demands have necessitated the development of alternative energy sources. Unbridled utilization of fossil fuels, coupled with public awareness has directed research focus towards the production of non-polluting and sustainable fuel sources. Algae have garnered tremendous research interest in the present time to meet the challenges faced by existing fuel sources. Ethanol has been widely used since the 1800s as a fuel of choice for internal combustion engines (ICE). Compared to conventional feedstock (corn/sugar beet), microalgae has a better potential in terms of ethanol yield: 40–140 ML ha^{-1} versus 3–6 ML ha^{-1} and can be a promising alternative for the future. Although various approaches have been presented to convert algal biomass to fuels, feed and chemicals; none of them are feasible from the economic point of view. Therefore, it is prudent to develop a biorefinery concept where multiple numbers of products can be recovered from the same algal biomass with a limited expense on unit operation. In order to address the present bio-economy challenges, it is pertinent to focus on the development of multiple products from the same biomass. In the present chapter the recent developments on algal refinery concept have been studied focussing on integrated bioethanol production under a biorefinery paradigm. The review also covers the technical challenge which inhibits the realization of algal biorefinery, along with providing insight on how to overcome the technical hurdles.

Keywords Microalgae · Biofuel · Bioproducts · Bioethanol · Circular biorefinery

---

All authors have contributed equally towards the chapter.

S. Banerjee (✉) · D. Das · A. K. Ghosh
Department of Biotechnology, Indian Institute of Technology Kharagpur, Kharagpur 721301, India

A. V. Agarwal and H. Valera (eds.), *Potential and Challenges of Low Carbon Fuels for Sustainable Transport*, Energy, Environment, and Sustainability,
https://doi.org/10.1007/978-981-16-8414-2_3

3.1 Introduction

Rapid industrialization and urbanization, population growth coupled with climate change are the major global concerns of the present millennium. As reported in 2017, the energy sector witnessed a 2.1% increase in the energy demand of which fossil fuel accounted for 70%. Due to technical feasibility and cheaper rates, fossil fuels are used indiscriminately leading to obnoxious release of greenhouse gases leading to climate change. Development of renewable fuel sources with synchronous greenhouse gas mitigation is the only feasible solution for the near future (BP 2018; Katiyar et al. 2017; Arun et al. 2020). Biomass based biofuels; especially third generation fuels derived from microalgae as the feedstock have garnered tremendous research interest in the present time. Microalgae as polyphyletic group of microorganism requires water, CO_2 and sunlight for its respiration and metabolic needs to produce biomass and can be cultivated under a wide range of condition (Banerjee et al. 2020). Microalgae can be cultivated and harvested all through the year and are not prone to seasonal variation. In addition to CO_2 sequestration, microalgae are also a good source of pigments like carotenoids, phycocyanin and phycobilliprotein which are of commercial and cosmetic importance. Microalgae can also store metabolites like carbohydrate and lipid which can be processed for the production of biofuels (Das 2015; Demirbas 2009). Depending on the carbon source, microalgae can be cultivated under three main categories: autotrophic, mixotrophic and heterotrophic growth condition. In autotrophic condition, the algal cells require light for their energy requirement and can sequester atmospheric CO_2 for its respirational need. Whereas, in mixotrophic and heterotrophic growth mode apart from CO_2, organic carbon is supplemented in the growth media to meet the energy requirement of the growing cells. In heterotrophic cultivation system the cells do not require light for its growth, and is unsuitable for outdoor cultivation. Compared to mixotrophic cultivation, for large scale algal production autotrophic growth mode is generally preferred as described in Table 3.1. Mass cultivation of algae is generally preferred in outdoor system where abundant light is available for algal growth and atmospheric CO_2 is

Table 3.1 Microalgal cultivation requirements

Parameters	Autotrophs	Mixotrophs	Heterotrophs
Light source	+	+	−
Organic carbon source	−	+	+
Inorganic carbon source	+	+	+
Maintenance cost	+	++	+++
Outdoor cultivation	+	+	−
Indoor cultivation	+	+	+

+Indicates positive response and −Indicates negative response

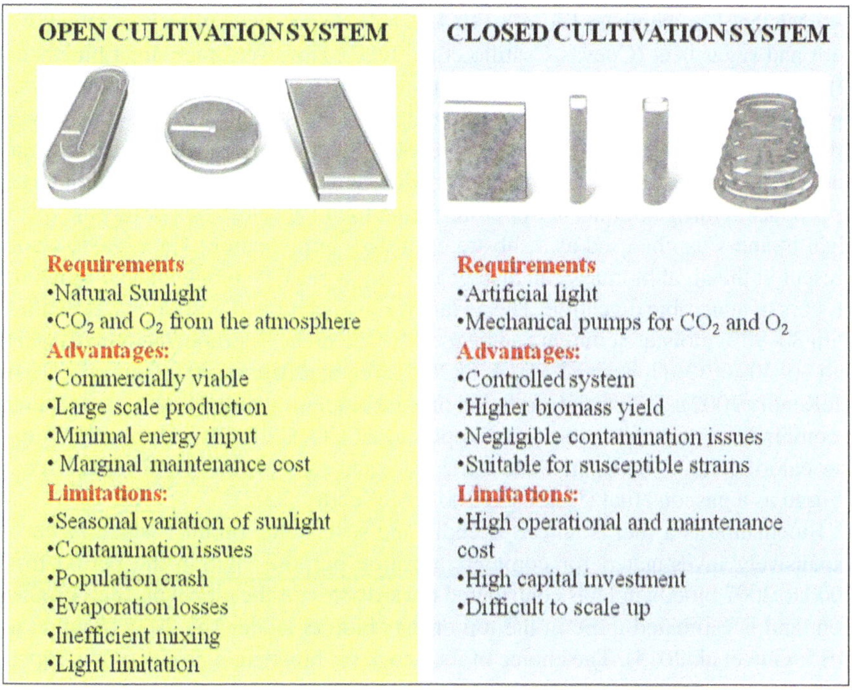

Fig. 3.1 Comparison of outdoor and indoor algal cultivation (Banerjee et al. 2020)

used as carbon source (Chisti 2007). Examples of outdoor cultivation system include raceway ponds and artificial ponds, which require minimal maintenance cost and energy input. In closed cultivation system, photobioreactors (PBR) which mimics the growth conditions of the microalgae are maintained to obtain high biomass yield with negligible contaminations. PBRs offer better control over physical and chemical parameter and can be efficiently used for cultivating mutant strains also (Fig. 3.1). However, the use of artificial lighting and mechanical pumps for air circulation, increase the operational and maintenance cost of these systems by almost 2.5 times (Davis et al. 2011).

The main liquid fuels which can be produced from microalgae are bioethanol, biodiesel and biobutanol. Gaseous fuels include biohydrogen and biomethane. Long chain hydrocarbons resembling crude oil and biogas are other fuels which can be produced using algal biomass as substrate. The transesterification of lipids produce biodiesel composed of monoalkyl esters of long chain fatty acid methyl esters. Algal biodiesel is renewable and non-toxic in nature and emits 78% less CO_2 as compared to petroleum, making it a sustainable fuel for the future (Chen et al. 2012). Due to the absence of high lignin and hemicelluloses on the algal cell wall they require mild pretreatment step compared to lignocellulosic biomass, easing the conversion of algal carbohydrates into fermentative biofuels like bioethanol, biohydrogen and biobutanol (Harun et al. 2010; Shaishav et al. 2013). Microalgal ethanol yield is

3–6 ML ha^{-1} as compared to 40–140 ML ha^{-1} from conventional feedstock like corn and sugar beet (Cuevas-Castillo et al. 2020). However, microalgal bioethanol production is still in its formative stage necessitating research for its further developments. Production of biobutanol using algal biomass as feedstock is a promising option. The bacteria involved for biobutanol production are saccharolytic in nature and does not require a separate saccharification step. The algal system is also adapted to produce hydrogen either by photofermentation or through biophotolysis; or the algal biomass can be used as a substrate for dark fermentation. The organic matter present in the algal biomass can also be utilized to produce biomethane through the process of anaerobic digestion. The advantage of anaerobic digestion is wet biomass with 80–90% moisture content can be used for the process, producing 60–70% CH_4 and 30–40% CO_2 (Jones and Mayfield 2012; Yen and Brune 2007; Demirbaş 2001; McKendry 2002). The partial oxidation of algal biomass at 800–1000 °C can produce a combination of combustible gas comprising of CO, CO_2, CH_4, H_2 and N_2 with a low calorific value (between 4 and 6 MJ m^{-3}) and can be either combusted directly or used as a gaseous fuel (Demirbaş 2001; McKendry 2002).

Bioethanol as a fuel is an eco-friendly and sustainable product which has been expansively investigated for commercialization purpose. Within the period from 2000 to 2007 bioethanol has contributed towards 85% of the global biofuels production; and is estimated to be in the top of the "biofuel ladder" by 2025 (Saini et al. 2015; Guo et al. 2015). The choice of feedstock for bioethanol production is dependent on several important factors like environmental, socio-economic and industrial. Microalgae with its abundant carbohydrate reserve stored in the form starch and sugar and the absence of lignin and hemicelluloses on its cell wall are an ideal candidate for bioethanol production. However, technical and economic inhibitions coupled with inefficient process designs have limited the use of microalgal biomass as a feedstock for bioethanol production (Banerjee et al. 2021).

Microalgae can serve as excellent bioethanol cell factories as they are composed of lipids, carbohydrates and proteins which can be efficiently converted to a plethora of fuels and other value added products. However, the main factor limiting the commercialization of algal ethanol is the cost of production (Agarwal et al. 2017). The cultivation, harvesting and drying of microalgae demands high energy input along with the requirement of sophisticated photobioreactors. To make algal biofuels a viable option research should be aimed towards the reduction of upstream and downstream processing steps. Adapting a biorefinery route of microalgal biofuels production where a single biomass is used as a feedstock for the production of numerous products is a plausible solution (Banu et al. 2020). The biofuels biorefinery concept arises from the petroleum refinery and involves a cascade of operations and processes to utilize the entire raw material for the production of maximum products and co-products without causing damage to any single product (Arun et al. 2020). The major consideration for a sustainable algal biorefinery process is the integration of upstream and downstream processes with green chemistry. These approaches focus on the maximum exploitation of the algal biomass while the energy input is minimal (Banu et al. 2020). Presently, circular biorefinery concept is gaining interest as a sustainable approach for achieving a closed loop algal technology process starting from algal cultivation and involving different channels of recycles of the waste stream back to the cultivation media. This chapter primarily explores the utilization of algal biomass for sustainable bioethanol production under a biorefinery paradigm. The review proposes different biorefinery routes which can be considered for algal bioethanol production to reduce the economic and carbon footprint of the process. Also, the role of techno economic analysis of the processes is evaluated to understand whether the circular biorefinery concept to produce bioethanol coupled with the recovery of value added products can be effectively implemented for commercialization.

3.2 Microalgae as a Sustainable and Renewable Feedstock for Energy Generation

Microalgae can be described as a diverse group of microscopic and photosynthetic organisms which may or may not be unicellular. In contrast to higher plants, microalgae are devoid of roots, stems, and leaves. Bestowed with high photosynthetic capacity, microalgae are important life forms as they approximately produce half the atmospheric oxygen with synchronous green house gas mitigation. The most abundant classes of microalgae are the diatoms (Bacillariophyceae) followed by the green algae (Chlorophyceae) and the golden algae (Chrysophyceae) (Li et al. 2014). Having the advantage of carbon neutrality, algal biomass represents an important raw material for the production of different by-products, ranging from food to fuel. Apart from biodiesel, bioethanol, biohydrogen, biobutanol and other fuels, algal biomass can also be utilized for the production of different, pigments (like lutein, astaxanthin, β-carotene, etc.), recombinant proteins, biomaterials and other bioproducts. The conversion of microalgal biomass to biofuels is mainly carried out using biochemical, thermochemical or catalytic conversion. The selection of the conversion technique is based on several factors which include feedstock composition, biofuel yield and productivity, reaction conditions and the energy efficiency of the system for sustainable biofuel production (Mishra et al. 2019). Among the different biofuels, bioethanol is the most commonly used fuel which is produced through the saccharification of carbohydrates followed by alcoholic fermentation using yeast or bacterial species. According to The United States Environmental Protection Agency report, renewable biofuel production is attaining worldwide attraction due to its advantage over fossil fuels, with bioethanol as the preferred fuel in the last decade compared to biodiesel and bio oil (Arun et al. 2020; Banerjee et al. 2021). During fermentation, the cellulose, starch and sugar content of the biomass is converted into ethanol through the metabolic action of *Saccharomyces cerevisiae*, which is commercially the dominant strain for fermentation (Suali and Sarbatly 2012). Ethanol is produced through the glycolytic pathway where sugars are converted into pyruvate, which ends up in acetaldehyde formation and CO_2 is released as by-product during the process. Further this acetaldehyde is reduced to form ethanol (Banerjee et al. 2020; Costa et al. 2015). According to the stoichiometry, one mole of glucose can produce two mols of ethanol and carbon dioxide each. The theoretical conversions efficiency of glucose to ethanol is around 51% (Banerjee et al. 2020, 2021). Microalgae with its high carbohydrate content (almost 60% of the dry cell weight (Nguyen et al. 2009)) and lacking the presence of lignin is a potential raw material for bioethanol production. In a study by Brennan and Owende (2010), *Chlorella vulgaris* with 37% starch content (per dry cell weight) had 65% ethanol conversion yield (Brennan and Owende 2010). In another study using *Chlamydomonas reinhardtii* biomass as feedstock, 0.235 g of bioethanol per gram of biomass was achieved when separate hydrolysis and fermentation (SHF) technique was used (Choi et al. 2010).

3.2.1 Microalgal Metabolites for Fuel and Feed

The buildup of different metabolites like lipid, carbohydrate and protein in algae are dependent on the environmental factors light, temperature, pH, salinity and CO_2 content) and the availability of nutrient (Fig. 3.2). Algal chloroplast plays a major role for the synthesis of fatty acid. The TAG (triacylglycerol) synthesis pathway inside the algal system can either be dependent or independent of acetyl CoA. Microalgal lipids (50–70% of the dry cell weight) are a promising feedstock for biodiesel production which has a calorific value of 39–41 MJ kg^{-1} and is comparable to petroleum-diesel

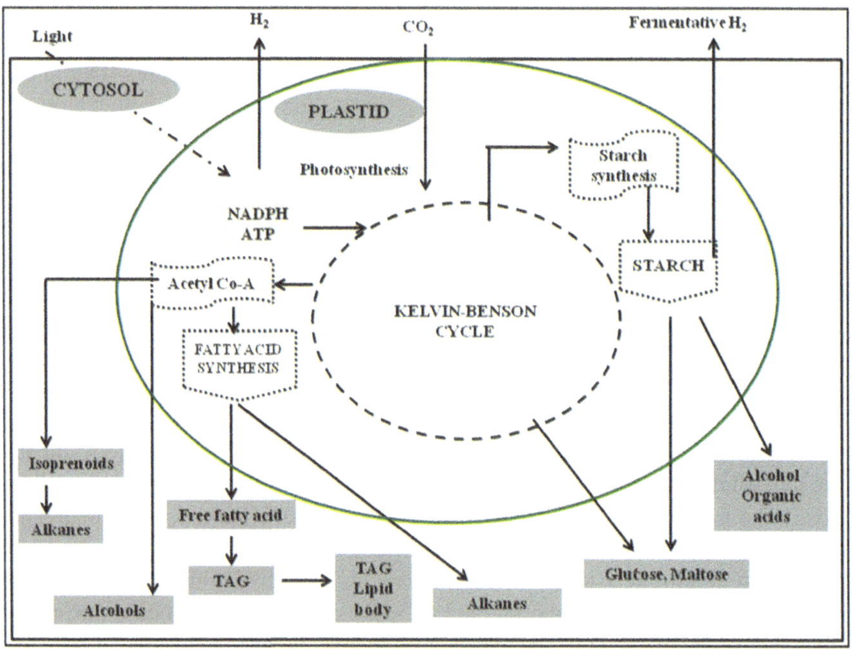

Fig. 3.2 Overview of metabolite production in microalgae (Banerjee et al. 2020)

(46 MJ kg^{-1}) (Demirbas 2010; Banerjee et al. 2019a). Based on the carbon numbers algal lipids can be either 12–20 carbon containing fatty acids which can be used for biodiesel production; and 20 carbon containing polyunsaturated fatty acids (PUFA) which are mainly used for food production. PUFAs are a major source of omega-3 fatty acids which have many nutritional and therapeutic properties. PUFAs are also known to have curative and protective activities against cardiovascular diseases (Liu et al. 2019). However, PUFAs can be synthesized by lower eukaryotes with few exceptions like *Chlorella minutissima* and *Crypthecodinium*, *Schizochytrium*, *Thraustochytrium* and *Ulkenia* (Chen et al. 2012).

Carbohydrate in microalgae is mainly stored in the form of starch in the plastid and cellulose, pectin and sulfated polysaccharides on the cell wall. Carbohydrate in microalgal system is synthesized during the photosynthesis process due to carbon fixation. During the Calvin-Benson cycle, the phosphoglycerate (a three-carbon compound) formed is the main precursor of the carbohydrate synthesis pathway (Fig. 3.3). The carbohydrate content of microalgae can be exploited for the production of many fuels like bioethanol and biohydrogen by photofermentation or the biomass can be used as a raw material for fuel production (Banerjee et al. 2020; Lam and Lee 2015). Biohydrogen shows potential as a carbon neutral and clean biofuel for the future, as on combustion it releases only water without any SO_x or NO_x emissions. Microalgae can produce hydrogen in the presence of water and sunlight by the process of photolysis, where the water molecule is broken down into one O_2 and

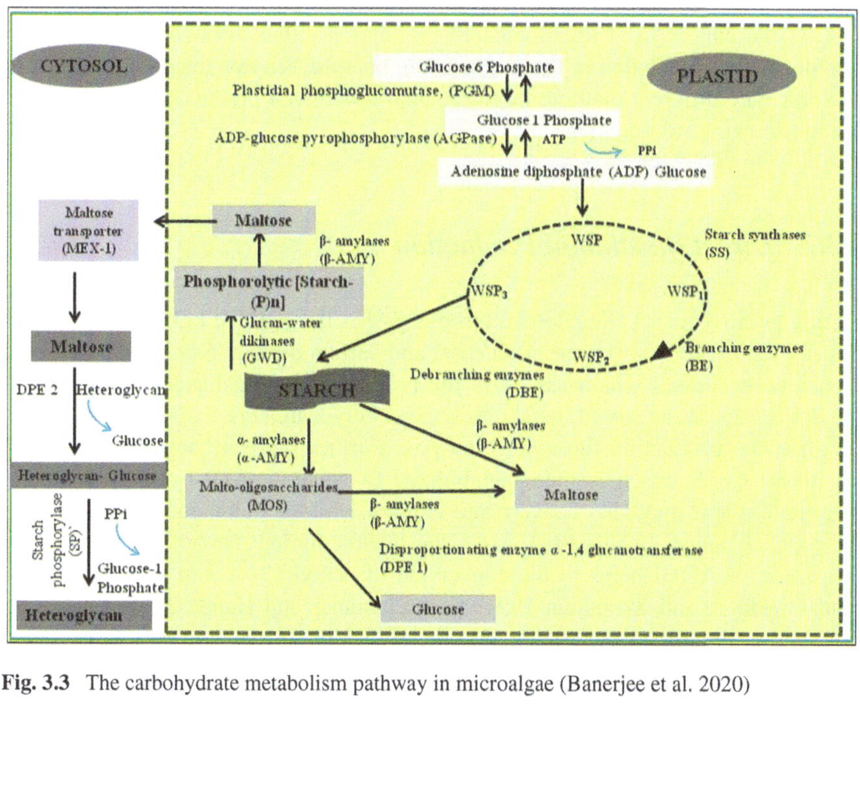

Fig. 3.3 The carbohydrate metabolism pathway in microalgae (Banerjee et al. 2020)

four H_2 ions during photosynthesis. Further, these hydrogen ions can be converted to hydrogen by the action of hydrogenase enzyme. But the inhibitory effect of oxygen on the hydrogenase enzyme and low hydrogen yield has guided the altered use of algal biomass as a substrate for dark fermentation to produce hydrogen under anaerobic condition (Singh and Das 2018; Kiran et al. 2014). Carbohydrate rich algal biomass can also be used for the production of biomethane by anaerobic digestion to produce energy in the range of 16,200–30,600 kJ m^{-3} depending on the nature of the biomass (Konur 2021). However, the high protein content in the biomass decreases the C/N ratio leading to more ammonia production which hinders the process of anaerobic digestion. This limitation can be trounced by using the microalgal biomass as a co-substrate with products having high C/N ratio (Kiran et al. 2014).

3.3 Bioethanol Production from Microalgae

In the current times production of bioethanol from first and second generation feedstock is more prevalent worldwide. However, there are many economical and technical problems associated with the upstream and downstream processes. Moreover, the food vs fuel debate and pretreatment of lignocellulosic biomass further limits the production of bioethanol from these feedstocks. Production of bioethanol from microalgae presents numerous advantages as has been discussed above. Starch, glycogen, and cellulose are the common carbohydrates available in microalgal system and can be used for bioethanol production either through dark fermentation or photofermentation or the biomass can be used as a raw material for yeast or bacteria. The different methods of microalgal bioethanol production are discussed in the following sub sections.

3.3.1 Direct Bioethanol Production

During photosynthesis the photoautotrophs passes through the light and dark reaction. In light reaction, oxygen production and carbon dioxide fixation are the main metabolic processes which results in the production of algal metabolites. But in the dark reaction the stored metabolites, especially starch are hydrolysed to sugars for pyruvate production through glycolysis. Further pyruvate acts as a substrate for acetyl Co-A which is then metabolized to generate ATP by the conversion of acetate. Or it may also convert into ethanol while simultaneously maintaining the redox balance. In this final stage, the bi-functioning alcohol/aldehyde dehydrogenase or ADHI helps in the conversion of acetyl Co-A into acetaldehyde or ethanol (Heyer and Krumbein 1991; Hemschemeier and Happe 2005; Mus et al. 2007). Traditionally, dark fermentation mainly addresses the conversion of organic substrates into biohydrogen, where the amylase activity helps in the hydrolysis of complex organic polymers into monomers. These monomers are further converted into a mixture of different lower molecular weight organic acids and alcohol. The end products generally differ among different species depending on the environmental conditions. However, acetic acid and ethanol are the most common products (de Farias Silva and Bertucco 2016). Microalgae can synthesize ethanol through fermentative metabolism during stress condition by directly utilizing their carbohydrate reserves during the dark reaction. Therefore, the accumulation of carbohydrates in the microalgal system is favourable for the production of ethanol. The different microalgae which are competent to produce ethanol (and release it through the cell wall) through intracellular mechanisms are *Chlamydomonas reinhardtii*, *Chlorella vulgaris*, *Chlorella moewusii*, *Chlorococcum littorale* and few other species of cyanobacteria (de Farias Silva and Bertucco 2016; Ueno et al. 1998). However the molecular yield of ethanol through dark fermentation reaction is very low and is not advantageous for the bioethanol production process (de Farias Silva and Bertucco 2016; Lakatos et al. 2019). Another method through which cyanobacteria and few microalgae can produce ethanol is photofermentation or the photonol route. The metabolic pathway for ethanol synthesis involves two stages: photosynthesis followed by fermentation. After the inorganic carbon is fixed in the Calvin-Benson cycle by utilizing the reducing power of photosynthesis, phosphoglycerate is formed which is then further converted into pyruvate by the action of two enzymes, namely pyruvate decarboxylase (PDC) and alcohol dehydrogenase (ADH). Acetaldehyde is formed as the intermediate product which is finally converted into ethanol by the action of dehydrogenase (ADH). This method requires genetically modified algal system where the PDC/ADH II cassette is expressed. However, the method requires further research as ethanol yield and production is very low in this system; and also ethanol concentration above a certain limit (2.5 and 4.5 g/L for *Synechococcus* sp. PCC 7942 and *Synechocystis* sp. PCC 6803, respectively) was found to have inhibitory effect on the algal system. For producing ethanol through this route use of genetically modified microalgal species are required which has high tolerance against ethanol with simultaneous high ethanol yield (de Farias Silva and Bertucco 2016; Lakatos et al. 2019).

Direct bioethanol production from microalgae is un-favourable in the present state because of the relatively toxic effect of the produced ethanol on the algal cells. But with progress in system biology and metabolic engineering, specific genes and pathways have been identified in the model cyanobacteria *Synechocystis* sp. PCC6803 which can be regulated for spiking up bioethanol production by algae. However, the development of these approaches in microalgae is currently limited to *C. reinhardtii* (Banerjee et al. 2020; Montagud et al. 2010; Noor-Mohammadi et al. 2012). Quantitative proteomics based approach with iTRAQ-LC–MS/MS has also helped in the identification of metabolic responses of genes which can be regulated during alcohol production in *C. reinhardtii* under different stresses Also the pathways which can help protect the microalgal cells against the lethal effect of organic solvents are identified in the present time (Jiang et al. 2017). An added advantage of *C. reinhardtii* playing an easy host for genetic modification for ethanol production is the known localization of the PTA and ACK genes and the presence of both the acetyl-CoA and acetaldehyde pathways (Atteia et al. 2006). Further, recent developments for improving the extraction of fermentable carbohydrate content in algae, the amylase gene from the hyper thermophilic bacterium *Thermotoga neapolitana* has been cloned in *C. reinhardtii* (Wang et al. 2015). Such transgenic algae which can express starch hydrolyzing enzyme can help in the reduction of additional pre-treatment costs which will help in improving the economics of algal bioethanol production.

3.3.2 Microalgal Biomass as a Feedstock for Bioethanol Production

Bioethanol has the major advantage of easy storage and distribution. This coupled with the easy blending characteristics of bioethanol with naturally occurring fossil fuel have propelled its usage as a suitable automobile fuel. Also, the levels of unburned hydrocarbon and carbon monoxide emission during bioethanol combustion are significantly less as compared to gasoline combustion (Banerjee et al. 2020; Yoon and Lee 2011). The quality of bioethanol produced through microbial fermentation relies majorly on the choice of biomass (sugarcane, cane molasses, cassava, microalgal biomass, etc.). The fermentation method comprises of three important steps which are saccharification, fermentation and separation and purification of the produced ethanol. It has been observed that a 10% blend of bioethanol with gasoline can help in 8–30% and 5–15% reduction of carbon monoxide and toxic gasses emissions, respectively on combustion (Tibaquirá et al. 2018). Compared to third generation feedstock for bioethanol fermentation, the use of first and second generation feedstock are more common. However, the environmental impact of first and generation biofuel production are harsh as compared to third generation biofuel. In a study by Cox et al. (2014), the environmental impact of three generation of aviation biofuel produced from sugarcane molasses, pongamia and microalgae were compared. From the study it was observed that the fossil energy ratio of sugarcane molasses is better (1.7 MJ output (MJ input)$^{-1}$) compared to second (1.1 MJ output (MJ input)$^{-1}$) and third (1.0 MJ output (MJ input)$^{-1}$) generation biomass (when compared with kerosene). However, the impact of land and water use pattern for microalgae based biofuel is lower than the other two generation of feedstock. A stark difference in result of global warming potential was observed, with third generation biofuel having the lowest impact followed by second and first generation biofuel (Cox et al. 2014). In another study it was observed that when bioethanol is produced under a biorefinery route, the economic return of the residual product is more for third generation feedstock (algal meal) as compared to first (sugarcane molasses) and second (*Pongamia*) generation feedstock. The nutritional value and toxicity of the residual meal were the major factor along with customer adoption in determining the cost (Klein-Marcuschamer et al. 2013). These properties have further necessitated research in the area of utilizing microalgal biomass as a source of carbohydrate for bioethanol production through the method of fermentation. During the light reactions of photosynthesis solar energy is used by the cell for the splitting of water into protons and electrons with the release of oxygen. These electrons and protons are further used for the generation of energy carriers like NADPH and ATP, to support the metabolic requirement of the cell. Whereas, during dark reaction the carbon dioxide is reduced to form carbohydrates using NADPH and ATP in the Calvin-Benson cycle. The major carbohydrate found in microalgae are polyglucans which are either accumulated in the plastids in the form of storage molecules (like starch and glycogen); or they function as the structural material of the cell wall (like cellulose and sulphate polysaccharides). The storage carbohydrate of microalgae varies among different strains and species, and can be optimized by altering the biochemical or metabolic pathway of the cell (Banerjee et al. 2020; Yoon and Lee 2011; Juneja et al. 2013).

The accumulation of carbohydrate in the microalgal system is synchronised with the availability of nutrient and physico-chemical conditions (like cultivation pH and temperature, level of irradiance, and CO_2 content). Air is the requisite and vital element required for microalgal growth along with maintaining an optimum level of pH, temperature and salinity. Macronutrients like magnesium, nitrogen, phosphorous, potassium, and sulphur are crucial for algal growth whereas, micronutrients like manganese and iron are required in small quantity between 2.5 and 30 ppm. Certain elements like boron, cobalt, copper, molybdenum and zinc are required in trace quantity between 2.5 and 4.5 ppm (Juneja et al. 2013). Changes in these physical or chemical parameters help in directly altering the flux distribution towards carbohydrate accumulation depending upon the species (Banerjee et al. 2020; Meher et al. 2006). The high-carbohydrate content of the holocellulose-based cell walls and starch-based cytoplasm of microalgae have made them as a suitable substrate for bioethanol fermentation. After cultivating the microalgae under optimum growth condition, the microalgae have to be harvested and dried before using it as a feedstock for bioethanol production. Further to the drying step, the microalgal biomass has to undergo the sequential processes of pre-treatment followed by saccharification and fermentation; and ultimately product recovery. The pre-treatment and saccharification condition of the microalgal biomass greatly influences the performance of fermentative bioethanol production. The optimum condition of pre-treatment and saccharification helps in the easy accessibility of the cell wall-bound carbohydrates for hydrolysis followed by conversion into simple and soluble sugars through the process of saccharification. Algal cell walls are mainly composed of cellulose. These are obtained from β-D-glucopyranose units, which condense through β-1, 4-glycosidic bonds forming a crystalline structure. The initial degree of crystallinity is an important factor which determines the hydrolysis rate as for randomly ordered crystalline or amorphous cellulose the hydrolysis rate is higher (Mittal et al. 2011). The main function of the pre-treatment step is to break down the microalgal cells so that the intracellular and complex carbohydrates become available for the saccharification process. Saccharification involves the splitting of the polysaccharides into monomers and it involves a two-stage process consisting of a pretreatment step prior to enzymatic hydrolysis. The general composition of algal carbohydrates is mainly a mixture of amino sugars, neutral sugars, and uronic acids; the composition of which varies according to the cultivation conditions and algal strains. The primary aim of an efficient pretreatment step is to increase the carbohydrate availability such that the ethanol fermentation rate is enhanced. However, if the method of pre-treatment is not suitable, then it leads to the degradation of the monomers into unwanted products like acetic acid, furanic compounds or formic acids; thereby hindering the bioethanol production process (Hargreaves et al. 2013; ye Lee et al. 2013).

3.3.3 Microalgal Bioethanol Production: Limitations

In the present time microalgae have emerged as the most potent substrate for the generation of biofuel and value added products. However, the productions of bioproducts from microalgal biomass have several shortcomings which restrict the commercialization of microalgal products. The chief limitation associated is the high production cost and energy demands during the cultivation, harvesting and drying processes. Post harvesting and drying, the cost associated with bioethanol fermentation from the biomass is also high. In addition, for the sustainable production of microalgal bioethanol, attention should also be focussed on the reduction and reuse of the liquid and solid fragments obtained after the upstream and downstream processes. Compared to terrestrial plants, algae have higher biofuels productivity but during large-scale algal cultivation a huge volume of water is required which directly increases the water footprint of the feedstock. Considering the life cycle analysis of the Yang et al. (2011); it has been shown that the reuse of water reduces the freshwater demand and nutrient usage by 84% and 55%, respectively. The biochemical composition of microalgae greatly influences the fuel property of the produced fuel. For example, the high moisture and ash content of microalgal feedstock limits its application for thermochemical conversions. Hence, pretreating the biomass before its application is necessary to reduce the ash content. In recent times, application of hydrothermal liquefaction (HTL) or wet torrefaction (WT) are useful pretreatment methods. Here the biomass is subjected to high temperature in the range of 180–260 °C which helps in improving the calorific and physical properties of the biomass to aid in biofuel production (Gan et al. 2020; Bach et al. 2017; Yu et al. 2020; Singh et al. 2015; Duan et al. 2013). In addition, the use of acid catalyst at optimum concentrations further helps in improving the hydrolysis of the microalgal biomass; thereby improving the algal biofuel yield (Gan et al. 2020). Further, some microalgae have high nitrogen and sulfur which directly leads to increase NO_x and SO_x emission (Obeid et al. 2019). In general, having low nitrogen content is favorable for biofuel production as it leads to enhanced C/N ratio. To reduce the NO_x and SO_x emission from microalgal biofuels approaches which can be adopted are; using metal oxides as catalyst coupled with co-combustion with oil shale (Obeid et al. 2019), K_2CO_3 and activated carbon promoted denitrogenation and desulphurization. Researchers have also suggested two stage HTL processes coupled with denitrogenation helps in reducing the nitrogen and sulfur content of the biomass along with decreasing the moisture and ash content (Chen et al. 2017).

Production of low value high volume products, like bioethanol from algal biomass is not feasible economically due to the high cost associates with downstream processing. This shortcoming can be mitigated by the generation of multiple products from the same microalgal biomass, envisaging a biorefinery concept. The biorefinery approach considers the sequential extraction of high value product like pigments followed by using the residual biomass for the production of biofuels. The spent media obtained after bioethanol production is usually rich in organic matters like vinnase and unhydrolyzed solid substrate. Therefore another approach can be the utilization of the spent wash of bioethanol fermentation for the production of biogas or biomethane (Harun et al. 2014; Zhang et al. 2014; Lu et al. 2020).The conversion of this heterogeneous substrate for producing biomethane will further help in addressing the twin problem of waste remediation and energy crisis. The reuse and recycle of both the solid and liquid residues which contains a considerable amount of biomolecules and nutrients for the production of value-added co-products and bioenergy predicts a circular bioeconomy model (Banerjee et al. 2021; Mohan et al. 2016). The microalgal biorefinery concept helps us to consider and compare different routes for the economical and sustainable microalgal biofuels production. The present chapter discusses the different methodology and routes which can be adapted for realising the algal biorefinery concept.

3.4 The Biorefinery Approach of Bioethanol Production from Microalgae

Developing microalgal based technology for the production of biofuels have several advantages such as high photosynthetic efficiency coupled with CO_2 sequestration, non competition with farmlands, high metabolite content with no seasonal variation. However, the main focus point of microalgae is its versatile range of application from fuel to feed. The concept of biofuel production from microalgae is not new; nevertheless there are certain technical limitations of microalgal biofuel production which hinders the technical exploitation of the processes. Capital cost is one of the major problems and can account for 50% of the total cost followed by harvesting and drying of the algal biomass (accounting for 30% of the total cost). Extensive research and development for system integration with detailed cost analysis can be one of the plausible solutions. For the maximum utilization of microalgal feedstock research focus has been shifted towards the integrating the different microalgal metabolites under a biorefinery approach. For example, after thermochemical conversion of microalgal lipids technologies like hydrothermal conversion (HTC) can transfer around 20–50% of the remaining organic material into the aqueous phase (HTC-AP) during the process. If the effluent is rich in monosaccharide content, then to reduce the COD load it can be subjected to anaerobic digestion or ethanol fermentation (Francavilla et al. 2015). The fermentation residues are usually rich in nitrogen and mineral content and can be recycled back for microalgal growth. To improve the energy recovery from microalgal biomass, two different downstream approaches can be adopted, one where the biomass is processed for the extraction of value added products prior to subjecting it for biofuel production. Or, the wet biomass can be used for the production of bioproducts like in hydrothermal liquefaction (HTL). Both these approaches have been successfully applied for biodiesel and bio-oil production from microalgae. Microalgal lipids have been widely subjected to biofuel production under a biorefinery approach (Francavilla et al. 2015; Muñoz et al. 2015; Chiaramonti et al. 2017). Authors have utilized biomass from *Botryococcus braunii, Nannochloropsis gaditana, Dunaliella tertiolecta*; for the extraction of value added products (like pigments) prior to using the depigmented biomass for lipid extraction and biodiesel production. Further, the residual biomass after the transesterification reaction was utilized for bio-oil and bio-char production through the process of pyrolysis. The bio-oil obtained had neutral pH unlike lignocellulosic oil; eliminating the problem of corrosion in combustible engines (Chiaramonti et al. 2017). Alternatively, researchers have also utilized the biomass after lipid transesterification for the production of synthesis gas (syngas) in addition to CO_2, CH_4 and tar as by-products through the process of gasification (Raheem et al. 2021; Hu et al. 2020). In a study by Kim and group (Kim et al. 2015), the biomass from *Dunaliella tertiolecta*, was used for lipid extraction followed by saccharification. The leftover biomass was used for the production of bio-oil through pyrolysis. The biorefinery approach for microalgal biomass utilization helped in increasing the activation energy from 163.12 to 670.24 kJ mol^{-1} (Kim et al. 2015). Being a complete amalgamation of multifunctional processes, biorefinery aims at the sustainable and concomitant generation of diverse spectrum of products and intermediates including feed and fuel as discussed in the following sections.

3.4.1 Biorefinery Route 1: Bioethanol Production Coupled with Other Biofuel Production

The eco-friendly nature of bioethanol makes it a sustainable option of liquid biofuel production from microalgae which have been widely investigated. However, commercialization of microalgal bioethanol is hindered due to its high production cost. Therefore, exploring integrative strategies with the available resources under a decentralized approach is a rational approach. Utilizing the microalgal biomass for different fuel production apart from bioethanol fermentation is a plausible option. One scheme can be extraction and transesterification of microalgal lipids before subjecting the biomass for bioethanol fermentation (Kumar et al. 2020; Sivaramakrishnan and Incharoensakdi 2018). Biodiesel is presently the most researched microalgal fuel and can be an efficient alternative for depleting fossil fuels. The biomass obtained after lipid extraction is termed as deoiled or deffated algal biomass (DAB) and contains significant amount of usable carbon for bioethanol production. This is the most investigated approach for improving the economics of microalgal ethanol production at laboratory scale. Before utilizing DAB for bioethanol fermentation the biomass has to be subjected to suitable pretreatment method for the conversion of complex carbohydrates into simple sugars which can be readily fermented by the microorganisms (Fig. 3.4a). In a study by Sivaramakrishnan and Incharoensakdi, transesterification followed by ethanol fermentation under an integrated process helped in improving the theoretical ethanol yield from 86 to 93% when the hydrolysate was prepared from the lipid extracted biomass (Sivaramakrishnan and Incharoensakdi 2018). In another study by Kumar et al. (2020), DAB was utilized for bioethanol production followed by polyhydroxybutyrate production yielding 0.145 ± 0.008 g ethanol (g DAB)$^{-1}$ and 0.43 ± 0.20 g_{PHB} (g DCW)$^{-1}$, respectively (Kumar et al. 2020). Table 3.2 describes different studies where DAB was utilized for the production of bioethanol under a biorefinery approach.

The integrative biorefinery concept aims to improve the process inputs and outputs with simultaneous reduction of the production cost and associated environmental impacts. It increases the viability of the microalgal biomass by optimizing the process economics for large scale production of microalgal biofuels. Production of biomethane through the process of anaerobic digestion is another sustainable replacement of fossil fuels which can be produced after ethanol production under a biorefinery paradigm. A heterogeneous spent is obtained after the fermentation and distillation of ethanol containing residual or unfermented sugar of the microalgal

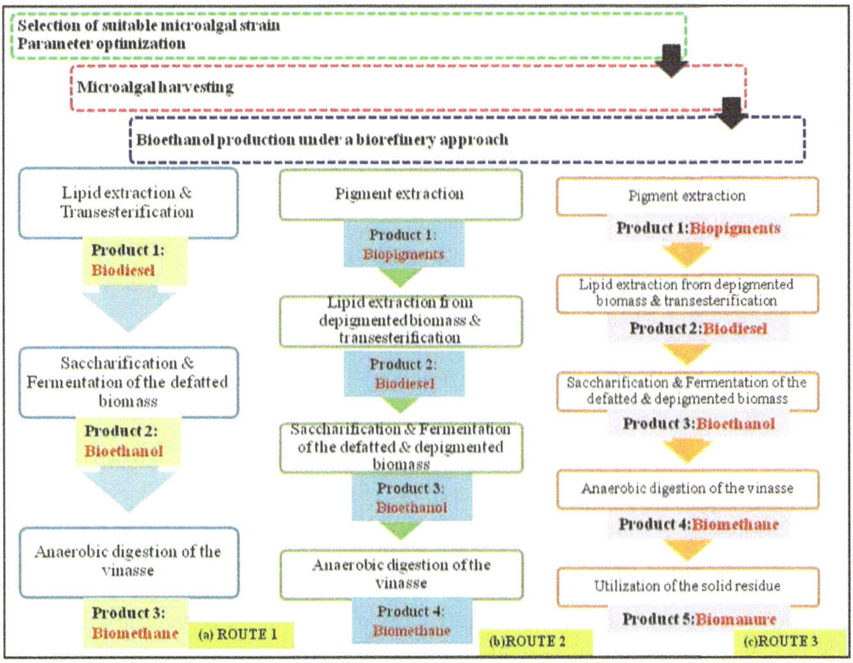

Fig. 3.4 **a–c** The different routes of bioethanol production from microalgae under a biorefinery approach

Table 3.2 Application of lipid extracted (deoiled) microalgal biomass for bioethanol production

Algal species	Application	Production/yield	References
Pseudochoricystis ellipsoidea	Lactic acid, Bioethanol	Utilization of the hydrolysate for replacing peptone and sugar in the fermentation media	Gao et al. (2012)
Dunaliella tertiolecta	Bioethanol	80.9% (w/w) sacharification yield based on total carbohydrate	Lee et al. (2013)
Scenedesmus dimorphus	Bioethanol	0.26 g bioethanol per gram was achieved, absence of additional pretreatment	Chng et al. (2016)
Scenedesmus obliquus	Protein and reducing sugars	Ultrasonication yielded better protein when deoiled biomass was used	Ansari et al. (2015)
Scenedesmus sp.	Saccharification	37.87% (w/w) and 43.44% saccharification yield when HCl and viscozyme L were used to pretreat the deoiled biomass	Pancha et al. (2016)
Chlorella vulgaris	Maltodextrin	90% malto-dextrin yield was achieved under 3 vol% of H_2SO_4 at 90 °C for 1 h	Lam et al. (2014)
C. reinhardtii UTEX 90	Bioethanol and biomethane	0.49 g ethanol/g of soluble carbohydrate, 1.5 L methane/L of spent media	Banerjee et al. (2021)
C. reinhardtii CC 2656	Bioethanaol and biomathane	0.46 g ethanol/g of soluble carbohydrate, 1.9 L methane/L of spent media	Banerjee et al. (2021)

biomass, solid residue of the saccharified biomass along with CO_2 and the yeast biomass known as vinnase (Yahmed et al. 2016). The traditional process of anaerobic digestion can be applied for the conversion of this complex organic residue for the production of methane by anaerobic bacteria and archaea bacteria addressing the issue of waste remediation through biofuel production (Banerjee et al. 2021; Yahmed et al. 2016). Although the application of this strategy is presently limited to macroalgae and duckweeds, few literatures are available for microalgae (Table 3.2). The general composition of microalgal biomass is $CH_{1.7}O_{0.4}N_{0.15}P_{0.0094}$ (Borowitzka and Borowitzka 1988) but can vary among different species and strains (Banerjee et al. 2021; Borowitzka and Borowitzka 1988). Majorly all microalgae can store different metabolites like lipid, carbohydrates and protein, which can be utilized sequentially for the production of different biofuels from the same biomass. The lipid content of the microalgal biomass can be refined to produce biodiesel; while the carbohydrate content (including the starch and cellulose) can be fermented to produce bioethanol. The carbohydrates, proteins and fats of the microalgae tissues present in the residual biomass can be further converted into methane through anaerobic digestion (Yahmed et al. 2016; Zhu 2014). The production of multiple fuels from the same biomass increases the total calorific value by approximately ten times, as reported by Zhu (2014), which directly improves the substrate energy recovery from the process. A similar study by Banerjee et al., showed that the substrate energy recovery increased by 12.73–17.05% when biomass of *C. reinhardtii* was used as a substrate for the sequential production of biodiesel, bioethanol and biomethane as opposed to only biodiesel or bioethanol production (Banerjee et al. 2021).

3.4.2 Biorefinery Route 2: Value Added Product Extraction Prior to Bioethanol Fermentation

Producing biofuels from microalgal feedstock as a replacement of fossil fuels is the present topic of research worldwide. However, technical limitations make microalgal biofuel economically uncompetitive. As discussed above, this anomaly can be overcome by producing different biofuels and byproducts from the microalgal biomass under a biorefinery paradigm. One more strategy to achieve this goal can be through the extraction of high value products before the simultaneous conversion of the algal metabolites into other products of interest under an integrated process which will help in increasing the economic value of the biofuel production process. Integrated or biorefinery system of biofuel productions helps in making the process economically sustainable and competitive in the fuel market (Da Silva et al. 2014; Huang et al. 2020). Microalgae contain many pigments of commercial importance like β-carotene or lutein which are utilized as pharmaceutical supplements and food-additives. Therefore, to improve the economic sustainability of microalgal bioethanol fermentation, extraction of value added co-products prior to biofuel production is a rational approach (Fig. 3.4b). In literature maximum biorefinery studies are focused on biodiesel production coupled with other fuels or value added production. Few references are available where simultaneous bioethanol fermentation with other products is reported. Nevertheless, these studies may act like base cases to design bioethanol biorefinery processes. Authors have cultivated microalgae in wastewater for the concomitant accumulation of carotenoids and lipids, addressing the combined problem of wastewater treatment and biofuel production (Liu et al. 2012; Mostafa et al. 2012). Co-extraction of pigments and oils followed by using the depigmented and defatted biomass for biohydrogen production has been reported in the marine microalgae *Nannochloropsis* sp. (Nobre et al. 2013; Ferreira et al. 2013). The authors employed supercritical CO_2 doped with ethanol for the extraction of pigments and lipids. However, in none of the studies the efficiency of the process or the energy recovery under the biorefinery paradigm as compared to the control was mentioned.

In a study by Huang and group (Huang et al. 2020), a consolidated bioprocessing (CSP) system was developed using a recombinant yeast species for the production of bioethanol from pigment extracted biomass. The genetically modified *Saccharomyces cerevisiae* had cellulases/amylases on its cell surface which could simultaneously hydrolyze as well as ferment the algal sugars into ethanol. *C. reinhardtii* JSC4 was the microalgae of choice which was cultivated under its optimal condition to enhance its carbohydrate and lutein content. The depigmented *C. reinhardtii* JSC4 biomass improved the overall bioethanol production by 10.73% and 31.6% as compared to raw starch (positive control) and pigment containing *C. reinhardtii* JSC4 cells, respectively. The life cycle analysis of the process indicated a 2.7–10.7 time's lower total environmental impact of the CBP process as compared to traditional ethanol production process. The total economic output from the process was estimated at $60.86, with a production of 2.43 kg ethanol and 5 g lutein from 1 kg of algal biomass (Huang et al. 2020). In another study by Banerjee et al., a similar approach was adopted where the biomass of *C. reinhardtii* was extracted using methanol prior to using the depigmented biomass linearly for biodiesel and bioethanol production. This was followed by using the spent media after bioethanol fermentation for biomethane production. The authors reported an overall energy recovery of 41% considering substrate energy and ignoring process energy (Banerjee et al. 2021).

3.4.3 Biorefinery Route 3: Bioethanol Production Followed by Utilization of the Solid Residue as Biomanure

As discussed in the previous sections a successful microalgal biorefinery depends on the optimum generation of different product and co-products. A strategy to integrate microalgal biorefinery with wastewater treatment is a plausible step towards improving the economics of the process. Further for enhancing the process microalgal biomass can also be used as an environment compatible alternative to commercial fertilizers (Fig. 3.4c). In a study by Mukherjee and group (Mukherjee et al. 2016), microalgae cultivated in rice mill effluent were able to accumulate phosphorous, ammonia nitrogen and poly-p with simultaneous reduction in BOD and COD by 98.7% and 93.5%, respectively; establishing the potential of microalgae as a suitable biofertilizer with simultaneous phycoremediation (Mukherjee et al. 2016). In another study, Renuka and group (Renuka et al. 2016) were able to achieve 7.4–33% increase in the dry weight of wheat plant along with 10% increase in spike when different strain of unicellular and filamentous microalgae were used as co-fertilizer along with vermicompost (Renuka et al. 2016). Such studies help in establishing the judicious application of microalgae as biomanure or biofertilizer. A proposal by Zhu in the manuscript titled "Perspective: The combined production of ethanol and biogas from microalgal residuals to sustain microalgal biodiesel" aims at integrating the entire channel of microalgal biorefinery along with usage of residual algal matter as biomanure. From the elemental composition of microalgae it was observed that the yearly demand of nitrogen and phosphorous per unit surface area can be estimated at 8–16 tons N ha^{-1} $year^{-1}$ (Zhu 2014; Sialve et al. 2009) which may create a nutrient imbalance. After extraction of the products from the microalgal biomass in a biorefinery scheme the nutrients available in residual microalgal biomass may be further recycled as biomanure or soil conditioner, creating a close loop of the algal biorefinery cycle. This closed integration of the microalgal biorefinery concept extracts the utmost potential of microalgal biomass giving the maximum output in terms of energy generation and recovery (Zhu 2014; Khan et al. 2019). The practical implication of this concept has been barely reported in literature, however the authors are of the view that the application of this route will help to achieve the absolute benefit of third generation energy production from microalgal biomass.

3.5 Circular Bioeconomy for Third Generation Bioethanol Production

The concept of circular bioeconomy was coined as an antagonist concept to the linear economy model to enhance the positive impact of industrial procedures. The aim of the circular economy is to restructure the life-cycle of a product where the net environmental impacts are reduced with a simultaneous reduction of resource expenditure and waste production (Pearce and Turner 1990). It supports the idea of symbiosism by adapting product reusing and decreasing material input and waste generation. Therefore circular bioeconomy can be visualized as a concept of recycle, reuse and reduce which helps in improving the overall dynamics of the processes. As defined by European Commission (2018), bioeconomy is an umbrella term for developing various renewable and bio-products of high value ranging from fuel to feed through biochemical conversions (Strategy 2018). The aim of these concepts is climate change mitigation through renewable energy production. For the practical realization of these goals, the biorefinery approach is an essential component which can aid in the optimal conversion of biomass to bioenergy and bioproducts. Developing a closed circle loop in the circular bioeconomy framework improves the overall sustainability and economic viability of the bio-stream production processes. Four main conversion techniques form the basis of the biorefinery concept, which are chemical, mechanical, biological and physical methods. These methods help in converting the biomass into the desired product under optimum process parameters. In a typical circular biorefinery concept usually two types of products are generated, primary and secondary. Primary products are the crude bioproducts like fuel and fertilizers and are rated as low value high volume products. Whereas secondary products are produced in a lower volume but have a high value mainly due to the refinement processes involved in them, like neutraceutical and cosmetic by products (Arun et al. 2020; Banerjee et al. 2021; Ubando et al. 2020).

The application of circular biorefinery concept for bioethanol production has not yet been realized practically. As discussed in the above sections, bioethanol biorefinery has limited literature available. Most of the available literatures are on proposals as to how best apply the algal biorefinery concept. However, applying closed loop biorefinery concept for microalgal bioethanol production is largely missing. In this section the authors have tried to propose a closed loop circular biorefinery concept for bioethanol production coupled with other value added products. The schematic representation of the concept is given in Fig. 3.5. As discussed in the previous sections, microalgae can synthesize and store different metabolites like carbohydrate, lipid, pigments and proteins; the concentrations of which are species dependent and can be altered under different physiological conditions. Literature studies have suggested the application of mixotrophic cultivation coupled with a suitable organic carbon source and optimum CO_2 dosage is beneficial for algal growth. Microalgae are naturally adept in sequestering high concentration of CO_2 during their growth which is then fixed in the form of biomass and storage molecules mediated by RuBisCo enzyme. This helps in addressing the twin problem of CO_2 mitigation and biomass/metabolite production. Apart from the carbon source, the concentration of nitrogen in the growth media directly affects the nucleic acid composition of algae (Mantzorou and Ververidis 2019). Therefore, under nitrogen starvation, algal cells

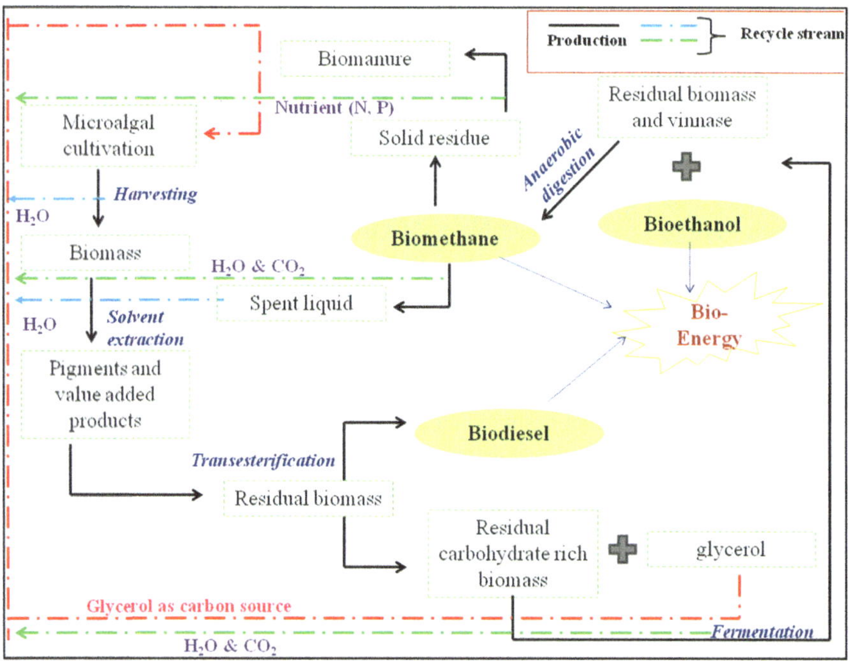

Fig. 3.5 The circular biorefinery concept for bioethanol production

are known to retain starch or lipid as its storage molecule. Other physical and chemical parameters like pH, salinity, temperature, trace metals etc. are also known to influence the accumulation of metabolites and biomass concentration of microalgae. Therefore, the primary step in the circular biorefinery cycle is the selection of suitable microalgal species and to optimize its growth condition such that its maximum potential in terms of biomass and metabolite concentration can be achieved (Banerjee et al. 2020, 2021; Mantzorou and Ververidis 2019; Zhan et al. 2017; Lal et al. 2021). The next unit operation involved is the separation of the microalgal cells from the cultivation media. The dilute nature of microalgal suspensions makes the microalgal harvesting a cost intensive step, accounting for 30% of the total production cost. Also, water is a crucial input which is necessary for the survival of microalgae due to its role as a thermal regulator and nutrient provider in the culture (Lal et al. 2021; Ray et al. 2021). Literatures have reported that 3726 kg of water is required for the production of 1.0 kg of microalgal biodiesel, thereby increasing the overall water footprint of the process (Yang et al. 2011; Farooq et al. 2015). Therefore, development of efficient harvesting techniques which is energy and cost-effective, environmentally friendly as well as allows water or media recycle should be developed. The different methods for microalgal harvesting are centrifugation, flocculation, filtration and sedimentation. Sedimentation is the most preferred method for solid liquid separation but is time consuming and sluggish. Centrifugation and filtration have high efficiency in terms of biomass recovery but is cost intensive, therefore limiting their application for large scale microalgal harvesting for biofuel filtration. Flocculation through chemical flocculant addition or autoflocculation is cost effective and can be applied for commercial scale but has the major problem of biomass contamination. Biological flocculation is an ecofriendly and cheap method which can be employed for biomass harvesting but is time consuming and can alter the biochemical composition of the biomass. In recent time electroflocculation is considered as an efficient method for biomass recovery and has low maintenance with minimum operational requirement and can also be carried in continuous mode. Biomass recovery efficiency as high as 98% have been reported by authors when electrocoagulation flocculation was employed as the harvesting method (Ray et al. 2021; Pandey et al. 2020; Demirbas and Kobya 2017; Bayramoglu et al. 2004). Different studies have shown that when the liquid media obtained after the electroflocculation method was recycled back for microalgal cultivation; it did not affect the biomass quality in terms of its biochemical composition (Pandey et al. 2020; Koley et al. 2017). A study by Loftus and Johnson to understand the influence of different factors affecting microalgal growth in recycled medium by surveying over 86 different literatures suggested that rather than the harvesting method, it is the microalgal taxa which dictates the growth outcomes during media recycle. Nevertheless, electroflocculation was found to be the most preferred method for microalgal harvesting followed by media recycles (Loftus and Johnson 2017).

Post harvesting, the biomass has to be dried and processed for further product development. Utilizing the wet biomass directly for biofuel production further reduces the economics of the process (Banerjee et al. 2019b). Apart from the energy content of microalgal biomass, these photosynthetic microorganisms also have the potential to synthesize certain bioactive compounds such as pigments, anti oxidants and anti inflammatory products, etc. which have an wide array of application as food or feed, neutraceutical and pharmaceutical usage, biomaterials, etc. (González-González and de-Bashan 2021; Bai et al. 2011).These bioactive compounds from microalgae are considered as high-value products which can be co-generated in the biorefinery cycle. Extraction and purification of these compounds are effective economically and can help to balance the high cost incurred during microalgal cultivation, harvesting and further processing. This approach can make microalgal biorefinery competitive in terms of energy and cost recovery (Banerjee et al. 2021; Bai et al. 2011). All these applications are in addition to the original energy purposes. The circular biorefinery paradigm consists of wide range production of different biofuels and by-products. Various technologies are involved in realizing the biorefinery scheme which mainly aims at production of multiple products from the same biomass under a cost feasible and environmentally sustainable manner. The concept is influenced by the petroleum refinery, aiming at economically feasible microalgal biofuel production (Hosseini et al. 2020). The two major carbon partitioning in the algal system are through lipids and carbohydrate accumulation, the storage of which are antagonistic to each other and have a dynamic flux metabolism. Post pigment extraction, the depigmented biomass can be used for lipid extraction followed by biodiesel production. Removing the pigment from the biomass makes the extraction of lipid easier compared to intact biomass. The extraction of lipid followed by transesterification requires the addition of different solvents, mainly methanol, hexane and chloroform. After lipid extraction it is converted into fatty acid methyl esters (FAME) in the presence of alcohol through the process of transesterification. A byproduct of the reaction is glycerol. It is estimated that after lipids transesterification, 10 kg of waste glycerol is generated for 100 kg of biodiesel produced (Leoneti et al. 2012; Xu et al. 2019). A high capital cost is involved for the purification of the waste glycerol and is typically avoided by industries, therefore making it an additional waste product causing disposal issues. Therefore strategies to recycle and supplement this glycerol as an organic carbon for microalgal cultivation can be an effective solution. Many studies have reported the successful recycle of waste glycerol back in the cultivation media without affecting the biochemical composition of microalgae (Abomohra et al. 2018; Lu et al. 2018). Microalgal biorefinery aims at the complete conversion of the different cellular components into valuable products and byproducts. Lipid extraction is an efficient pretreatment step for concentrating the carbohydrate content of biomass, such that the defatted biomass can be utilized for further fuel production. Utilizing the depigmented and defatted biomass for bioethanol production improves the overall energy recovery of the process as was discussed in the previous sections. For the economic production of bioethanol, efficient and low-cost pretreatment coupled with a fast saccharification and fermentation process is a pre-requisite. Extraction of lipid prior to bioethanol fermentation leads to an accelerated rate of sugar utilization which enhances the production and productivity of ethanol. Also, it helps in overcoming the additional energy requirements of thermal or mechanical pre-treatment of the microalgal biomass, indirectly improving the carbon footprint of the process. Many authors have reported increase in bioethanol yield and productivity using defatted microalgal biomass as was discussed in Table 3.1. Therefore it can be rationalized that for cost effective bioethanol production form microalgal biomass using depigmented and defatted biomass is a convenient approach (Banerjee et al. 2019b, 2021; Zhu 2014; Xu et al. 2019). After bioethanol production and separation, the spent media has a heterogeneous combination of organic compounds which include volatile fatty acids and unutilized carbohydrate content of the algal biomass; and vinasse, which is the yeast biomass. This residual content has a high protein content which decreases the overall C/N ratio of the substrate, thus reducing the product yield. However, addition of organic carbon sources like wood straw or waste from kitchen, etc. can help to augment the C/N ratio. This mixture can be then subjected to anaerobic digestion for the production of biomethane. The cost of metabolite extraction, transesterification followed by fermentation is more than the energy recovered from biodiesel and bioethanol production. Therefore, coupling anaerobic digestion with the circular bioethanol biorefinery may help to boost the energy recovered from the process (Banerjee et al. 2021; Zhu 2014). After anaerobic digestion and gaseous product recovery, the solid and liquid part of the digestion mixture can be separated for further characterization. This liquid digestate can be further divided into a stream containing high solids content called as the cake and a liquid stream also known as anaerobic centrate. Different dewatering techniques can be applied for achieving this. The centrate is chiefly composed of low-density solids together with a high concentration of nutrients, majorly nitrogen and phosphorous. After characterization and suitable dilution this centrate can be recycled for microalgal cultivation. Also, the CO_2 released during the process of anaerobic digestion can assist in microalgal growth by acting as an inorganic carbon supplement, thereby improving biomass production (Lu et al. 2018). Promising results have been obtained by authors when microalgae were cultivated in anaerobic centrate obtained after the anaerobic digestion process (Lu et al. 2018; Uggetti et al. 2014; del Mar Morales-Amaral et al. 2015). After the biochemical characterization of the solid cake it can be successfully used as a biomanure or biofertilizer which can replace the synthetic fertilizer which are commercially available, as discussed in the previous section. The concept proposed in this section is a theoretical consideration and has been not realized practically. However parts of the cycle has been implemented and studied by different authors at laboratory scale, and needs validation at commercial level. Therefore, for the successful implementation of the circular bioethanol biorefinery concept, techno economic assessment is necessary and has been discussed in the following section.

3.6 Techno Economic Evaluation of the Biorefinery Routes

The concept of microalgal biorefinery helps to compare different routes which can be adopted for maximizing the energy output from the system. However, even they have certain limitations which need to be addressed before implementing them on large scale. Techno economic assessment (TEA) and life cycle assessment (LCA) are two powerful tools which helps us to predict the performance of a system based on acquired data. TEA helps in analyzing the economic performance of a process, product, or service using software which can estimate the different cost associated with the process and the revenue it can generate based on financial and technical parameters. Whereas, LCA analyses the environmental impact of a process, product or service and typically considers the different stages of the process life like cradle to grave or cradle to cradle approach. Techno-economic analysis (TEA) should be performed prior to the procurement and construction of engineering designs to evaluate the economic feasibility of the process (Banu et al. 2020; Li et al. 2014; Mishra et al. 2019). It helps in enhancing the cost dynamic of the process by identifying and eliminating the weak steps involved in the cycle and provides alternatives for the optimum functioning of the process. In general, the steps involved in TEA are:

(a) Identifying the system and its boundaries
(b) Developing the process flow design
(c) Energy and mass balance analysis
(d) Calculating the capital cost involved along with estimating the cost involved in different unit operations involved.

The initial step mainly focuses on the acquiring details of the market to predict the market outlook and assess the different factors which may affect the process commercialization. Designing the process flow diagram is crucial for the TEA process as it illustrates the different steps and sub processes associated with the main process. It helps to identify the various biorefinery routes which are involved and their integration to obtain the maximum product recovery from the process. Based on these data, it becomes easy to understand which biorefinery route is sustainable in terms of economic investments by considering the net value of the process. Economic profitability is always higher when integrated circular biorefinery processes are used for biofuel and byproduct generation. According to Solis et al., in an integrated algal biorefinery if we consider i process units involving j number of raw material and energy streams, then the process matrix is represented by Aij wherein i indicates the process units and the material and energy streams is represented by j. The negative and positive values of the matrix Aij indicate whether the stream is an input or output of the process unit, respectively. The product output demand for every j is represented by Dj. The variable costs for each of the raw materials involved and the energy stream is represented by VCj whereas, P is the selling cost of the for each product in the stream. To evaluate the environmental impact of the process, impact classes are designated as k and Eik is the representation of the different impacts associated with each step in the process. Weights (Wk) are allotted to each impact category for developing one single score for evaluating the complete process and are represented by Si (Pandey et al. 2020). The above matrix designing system forms the basis of different softwares for assessing the TEA of a biorefinery process. Aspen Plus, Aspen HYSYS, DYNSIM, CHEMCAD, DESIGN II, SuperPro Designer PRO/II, UniSim Design are few of the available softwares which are used to estimate the energy and mass balance for a biorefinery process by developing a process a flow diagram (da Silva et al. 2020; Solis et al. 2020).

Life cycle assessment or LCA helps to calculate whether microalgal biofuel production is feasible in terms of its environmental impact. This tool starts evaluating the biorefinery cycle right from the microalgal cultivation and ends with product generation and byproduct recovery. It basically assesses the environmental footprint of a product during its entire lifetime. Similar to TEA, for the LCA procedure defining the system boundaries and inventories involved is also essential to categorize the different input and outputs for comparing the sustainability of the biorefinery process with conventional fuel production processes. A throughput analysis of the inputs and outputs of material and energy within the life-cycle inventories will help in determining the sustainability of the bio-energy products against the conventional fuels. A LCA study performed in 2015 estimated that during the production of bio-energy by hydrothermal liquefaction and pyrolysis, the net energy recovery and the green house gas emission from the process were 1.23 and −11.4 and 2.27 and 210 g $CO_{2\text{-eq}}$ respectively (He et al. 2015). Whereas in another study four different bioenergy production processes were evaluated (hydrothermal liquefaction, pyrolysis, trans-esterification and anaerobic digestion with and without additional pretreatment). It was observed that among these processes anaerobic digestion with pretreatment was better compared to others in terms of the net energy recovery and green house gas emission (He et al. 2015; Shah et al. 2016; Tesfa et al. 2013).The generally used softwares for conducting LCA studies are BEES 4.0, CMLCA, Enviance, EarthSmart, Economic input output LCA, GaBi, GREET Model, GEMIS, SimaPro, Sustainable Minds and Umberto (Banu et al. 2020; Mishra et al. 2019; Banerjee et al. 2019b).

For the commercial scale production of microalgal biofuel with a positive TEA and LCA, the following points can be considered by researchers and investors:

(a) Identification of microalgae strain which has high biomass productivity and metabolite accumulating potential
(b) Cost effective and efficient harvesting techniques for maximum biomass recovery
(c) Evaluating the effect of genetic engineering methods for strain alteration such as to improve its environmental tolerance
(d) Microalgal engineering developing strains with high lipid or carbohydrate content
(e) Improving metabolite extraction processes such that the presence of toxic elements in the waste stream can be reduced such that it can be recycled back in the biorefinery cycle
(f) Techno economic analysis and life cycle inventories should be performed to identify the strongest and weakest steps involved in the process such that actions can be taken for improving the sustainability of the process (Banu et al. 2020; Mishra et al. 2019; He et al. 2015).

3.7 Practical Implication of the Study

Microalgal biofuel is a promising alternative to fossil fuels as compared to first and second generation energy crops. However, due to lack of economic viability microalgal biofuel production has not yet reached commercial stage. Researchers' around the world have proposed various strategies for enhancing the economic feasibility of the biofuel production from microalgae; one among them is adopting a circular biorefinery approach. In the present chapter the authors have tried to present different approaches for integrating diverse bioprocesses for bioethanol production. It was observed that for utmost energy recovery adopting a circular biorefinery approach for maximum product generation from the same microalgal biomass is the most promising option. Biofuel production preceded by value added product extraction and followed by residual utilization is a judicious solution. To balance the input and output energy ranges during the microalgal biofuel production; depending on the cultivation method adopted energy savings can be obtained by generating different by-products like glycerol, or anaerobic centrate, etc. therefore it is very essential to estimate the net energy recovery of the entire process especially when the overall metabolite content of the microalgae is low. The main aim of the circular biorefinery process is to achieve a net positive energy recovery from the entire process. Similarly, recycling of residual water or produced CO_2 back for algal cultivation is a rational step as it helps in reducing the input cost for the next cycle. Cascading production usually increases the overall efficiency of the process with judicious use of resources; however the complexity arises when the cycle is connected with decreasing greenhouse gas emissions. It is therefore necessary to identify the main purpose of the cycle, for example if addressing climate change is the main concern, then cascading circular chains which has long term carbon sequestration should be implemented. Considering biomass production, geographic aspects has an important role in influencing the biochemical property of the biomass and needs to be judiciously addressed. Most studies reviewed in this chapter are limited to lab scale studies where the biomass was cultivated in photobioreactors. In practice, in open ponds or tanks biomass productions is prone to outdoor climatic patterns and have to be accounted for during techno economic analysis. Apart from biomass production and utilization, transport and storage of input raw materials and export of products and byproducts should be taken into consideration while assessing the impact of the biofuel production process. Currently, commercial establishments for microalgal biofuels are limited to single product generation with waste recycle. In practice a big gap exists between the expected results of circular refinery process and the actual performance these biorefineries. The various stakeholders involved both at local and international level, private and public players as well as academic community have a huge role to play. Also, the ever changing policy frameworks impact the realization of these biorefinery schemes. Also, the sustainability assessment model should be developed aimed towards utilization of the actual data such that uncertainties and risks can be minimized with simultaneous improvement in the net energy recovery. This gap might be minimized when the starting point of the biorefinery cycle has sufficient input in term of microalgal biomass productivity and metabolite concentration. Shifting from linear to circular approach for biomass utilization together with techno economic analysis and life cycle assessment studies makes the microalgal bioethanol production looks promising. Nevertheless, research at laboratory scale should be implemented at pilot scale to get a real picture of the figures when this concept is commercialized. For this, academic and industrial collaborations with positive response from government agencies and policymakers are crucial for the application of biocircular bioeconomy in real time.

3.8 Conclusion

The thinning supply of fossil fuels together with accelerating green house gas emissions have given rise to many environmental concerns associated with the present energy use pattern. This has further necessitated the development of renewable and sustainable energy sources. The present chapter discussed the merits of third generation bioethanol production from microalgae along with the associated technical limitations. Economic challenge is the major issue faced by microalgal bioethanol production when compared with fossil fuels which are relatively inexpensive. This chapter discusses the different strategies for integrative bioethanol production. This is in accordance with the concept of judicious utilization of the microalgal biomass over the entire production chain. Integrative microalgal biorefinery aims at producing biofuels with concomitant byproduct development. Biofuels are considered as high volume low value product, whereas byproducts are classified as low volume and high value commodities with differential application in health and food industry. The amalgamations of biorefineries have the potential for products generation as well as decrease environmental impacts with simultaneous enhancement of societal benefits. Different reviews have suggested that instead of a linear biorefinery concept, developing a closed loop circular biorefinery is economically and environmentally more prudent. The circular economy joins various unit operations in a single cycle by linking the flow of material, associating production with recycling, extraction with reuse, to maximize the output from the same input material to achieve a positive net energy recovery. However, it is impertinent to combine circular biorefinery with circular economy such that an overall positive output is balanced in terms of economy and environment. In conclusion, optimizing each steps of the biorefinery cycle at laboratory scale, with in-depth techno economic analysis and life cycle assessment of the scale up procedure is necessary for the practical implementation of the proposed concept.

References

Abomohra AEF, Eladel H, El-Esawi M, Wang S, Wang Q, He Z, Feng Y, Shang H, Hanelt D (2018) Effect of lipid-free microalgal biomass and waste glycerol on growth and lipid production of Scenedesmus obliquus: innovative waste recycling for extraordinary lipid production. Bioresour Technol 249:992–999

Agarwal AK, Agarwal RA, Gupta T, Gurjar BR (2017) Introduction to biofuels. In: Biofuels. Springer, Singapore, pp 3–6

Ansari FA, Shriwastav A, Gupta SK, Rawat I, Guldhe A, Bux F (2015) Lipid extracted algae as a source for protein and reduced sugar: a step closer to the biorefinery. Biores Technol 179:559–564

Arun J, Gopinath KP, SundarRajan P, Felix V, JoselynMonica M, Malolan R (2020) A conceptual review on microalgae biorefinery through thermochemical and biological pathways: bio-circular approach on carbon capture and wastewater treatment. Bioresour Technol Rep 11:100477

Atteia A, van Lis R, Gelius-Dietrich G, Adrait A, Garin J, Joyard J, Rolland N, Martin W (2006) Pyruvate formate-lyase and a novel route of eukaryotic ATP synthesis in *Chlamydomonas mitochondria*. J Biol Chem 281(15):9909-9918

Bach QV, Chen WH, Lin SC, Sheen HK, Chang JS (2017) Wet torrefaction of microalga Chlorella vulgaris ESP-31 with microwave-assisted heating. Energy Convers Manage 141:163–170

Bai MD, Cheng CH, Wan HM, Lin YH (2011) Microalgal pigments potential as byproducts in lipid production. J Taiwan Inst Chem Eng 42(5):783–786

Banerjee S, Singh H, Das D, Atta A (2019a) Process optimization for enhanced biodiesel production by Neochloris oleoabundans UTEX 1185 with concomitant CO_2 sequestration. Ind Eng Chem Res 58(35):15760–15771

Banerjee S, Rout S, Banerjee S, Atta A, Das D (2019b) Fe_2O_3 nanocatalyst aided transesterification for biodiesel production from lipid-intact wet microalgal biomass: a biorefinery approach. Energy Convers Manage 195:844–853

Banerjee S, Banerjee S, Ghosh AK, Das D (2020) Maneuvering the genetic and metabolic pathway for improving biofuel production in algae: present status and future prospective. Renew Sustain Energy Rev 133:110155

Banerjee S, Ray A, Das D (2021) Optimization of *Chlamydomonas reinhardtii* cultivation with simultaneous CO_2 sequestration and biofuels production in a biorefinery framework. Sci Total Environ 762:143080

Banu JR, Kavitha S, Gunasekaran M, Kumar G (2020) Microalgae based biorefinery promoting circular bioeconomy-techno economic and life-cycle analysis. Bioresour Technol 302:122822

Bayramoglu M, Kobya M, Can OT, Sozbir M (2004) Operating cost analysis of electrocoagulation of textile dye wastewater. Sep Purif Technol 37(2):117–125

Borowitzka MA, Borowitzka LJ (1988) Micro-algal biotechnology. Cambridge University Press

BP statistical review of world energy, 67th edn (2018) http://www.indiaenvironmentportal.org.in/files/file/bp-stats-review-2018-full-report.pdf. Accessed 22 June 2021

Brennan L, Owende P (2010) Biofuels from microalgae—a review of technologies for production, processing, and extractions of biofuels and co-products. Renew Sustain Energy Rev 14(2):557–577

Chen R, Yue Z, Deitz L, Liu Y, Mulbry W, Liao W (2012) Use of an algal hydrolysate to improve enzymatic hydrolysis of lignocellulose. Biores Technol 108:149–154

Chen C, Chen F, Cheng Z, Chan QN, Kook S, Yeoh GH (2017) Emissions characteristics of NO_x and SO_2 in the combustion of microalgae biomass using a tube furnace. J Energy Inst 90(5):806–812

Chiaramonti D, Prussi M, Buffi M, Rizzo AM, Pari L (2017) Review and experimental study on pyrolysis and hydrothermal liquefaction of microalgae for biofuel production. Appl Energy 185:963–972

Chisti Y (2007) Biodiesel from microalgae. Biotechnol Adv 25(3):294–306

Chng LM, Chan DJ, Lee KT (2016) Sustainable production of bioethanol using lipid-extracted biomass from Scenedesmus dimorphus. J Clean Prod 130:68–73

Choi SP, Nguyen MT, Sim SJ (2010) Enzymatic pretreatment of *Chlamydomonas reinhardtii* biomass for ethanol production. Biores Technol 101(14):5330–5336
Costa RL, Oliveira TV, de Souza Ferreira J, Cardoso VL, Batista FRX (2015) Prospective technology on bioethanol production from photo fermentation. Biores Technol 181:330–337
Cox K, Renouf M, Dargan A, Turner C, Klein-Marcuschamer D (2014) Environmental life cycle assessment (LCA) of aviation biofuel from microalgae, Pongamia pinnata, and sugarcane molasses. Biofuels Bioprod Biorefin 8(4):579–593
Cuevas-Castillo GA, Navarro-Pineda FS, Rodríguez SAB, Rivero JCS (2020) Advances on the processing of microalgal biomass for energy-driven biorefineries. Renew Sustain Energy Rev 125:109606
Da Silva TL, Gouveia L, Reis A (2014) Integrated microbial processes for biofuels and high value-added products: the way to improve the cost effectiveness of biofuel production. Appl Microbiol Biotechnol 98(3):1043–1053
da Silva AF, Brazinha C, Costa L, Caetano NS (2020) Techno-economic assessment of a Synechocystis based biorefinery through process optimization. Energy Rep 6:509–514
Das D (ed) (2015) Algal biorefinery: an integrated approach. Springer
Davis R, Aden A, Pienkos PT (2011) Techno-economic analysis of autotrophic microalgae for fuel production. Appl Energy 88(10):3524–3531
de Farias Silva CE, Bertucco A (2016) Bioethanol from microalgae and cyanobacteria: a review and technological outlook. Process Biochem 51(11):1833–1842
del Mar Morales-Amaral M, Gómez-Serrano C, Acién FG, Fernández-Sevilla JM, Molina-Grima E (2015) Production of microalgae using centrate from anaerobic digestion as the nutrient source. Algal Res 9:297–305
Demirbaş A (2001) Biomass resource facilities and biomass conversion processing for fuels and chemicals. Energy Convers Manage 42(11):1357–1378
Demirbas MF (2009) Biorefineries for biofuel upgrading: a critical review. Appl Energy 86:S151–S161
Demirbas A (2010) Use of algae as biofuel sources. Energy Convers Manage 51(12):2738–2749
Demirbas E, Kobya M (2017) Operating cost and treatment of metalworking fluid wastewater by chemical coagulation and electrocoagulation processes. Process Saf Environ Prot 105:79–90
Duan P, Chang Z, Xu Y, Bai X, Wang F, Zhang L (2013) Hydrothermal processing of duckweed: effect of reaction conditions on product distribution and composition. Biores Technol 135:710–719
Farooq W, Suh WI, Park MS, Yang JW (2015) Water use and its recycling in microalgae cultivation for biofuel application. Biores Technol 184:73–81
Ferreira AF, Ribeiro LA, Batista AP, Marques PA, Nobre BP, Palavra AM, da Silva PP, Silva C (2013) A biorefinery from *Nannochloropsis* sp. microalga–energy and CO_2 emission and economic analyses. Bioresour Technol 138:235–244
Francavilla M, Kamaterou P, Intini S, Monteleone M, Zabaniotou A (2015) Cascading microalgae biorefinery: fast pyrolysis of Dunaliella tertiolecta lipid extracted-residue. Algal Res 11:184–193
Gan YY, Ong HC, Chen WH, Sheen HK, Chang JS, Chong CT, Ling TC (2020) Microwave-assisted wet torrefaction of microalgae under various acids for coproduction of biochar and sugar. J Clean Product 253, 119944
Gao MT, Shimamura T, Ishida N, Takahashi H (2012) Investigation of utilization of the algal biomass residue after oil extraction to lower the total production cost of biodiesel. J Biosci Bioeng 114(3):330–333
González-González LM, de-Bashan LE (2021) Toward the enhancement of microalgal metabolite production through microalgae–bacteria consortia. Biology 10(4):282
Guo M, Song W, Buhain J (2015) Bioenergy and biofuels: history, status, and perspective. Renew Sustain Energy Rev 42:712–725
Hargreaves PI, Barcelos CA, da Costa ACA, Pereira N Jr (2013) Production of ethanol 3G from Kappaphycus alvarezii: evaluation of different process strategies. Biores Technol 134:257–263

Harun R, Danquah MK, Forde GM (2010) Microalgal biomass as a fermentation feedstock for bioethanol production. J Chem Technol Biotechnol 85(2):199–203
Harun R, Yip JW, Thiruvenkadam S, Ghani WA, Cherrington T, Danquah MK (2014) Algal biomass conversion to bioethanol–a step-by-step assessment. Biotechnol J 9(1):73–86
He W, Zhang G, Zhang X, Ji J, Li G, Zhao X (2015) Recent development and application of thermoelectric generator and cooler. Appl Energy 143:1–25
Hemschemeier A, Happe T (2005) The exceptional photofermentative hydrogen metabolism of the green alga *Chlamydomonas reinhardtii*. Biochem Soc Trans 33(1):39–41. https://doi.org/10.1042/BST0330039
Heyer H, Krumbein WE (1991) Excretion of fermentation products in dark and anaerobically incubated cyanobacteria. Arch Microbiol 155(3):284–287
Hosseini A, Jazini M, Mahdieh M, Karimi K (2020) Efficient super antioxidant and biofuel production from microalga *Haematococcus pluvialis* via a biorefinery approach. Bioresour Technol 306:123100
Hu Y, Qi L, Rao KTV, Zhao B, Li H, Zeng Y, Xu CC (2020) Supercritical water gasification of biocrude oil from low-temperature liquefaction of algal lipid extraction residue. Fuel 276:118017
Huang X, Bai S, Liu Z, Hasunuma T, Kondo A, Ho SH (2020) Fermentation of pigment-extracted microalgal residue using yeast cell-surface display: direct high-density ethanol production with competitive life cycle impacts. Green Chem 22(1):153–162
Jiang Y, Xiao P, Shao Q, Qin H, Hu Z, Lei A, Wang J (2017) Metabolic responses to ethanol and butanol in *Chlamydomonas reinhardtii*. Biotechnol Biofuels 10(1):1–16
Jones CS, Mayfield SP (2012) Algae biofuels: versatility for the future of bioenergy. Curr Opin Biotechnol 23(3):346–351
Juneja A, Ceballos RM, Murthy GS (2013) Effects of environmental factors and nutrient availability on the biochemical composition of algae for biofuels production: a review. Energies 6(9):4607–4638
Katiyar R, Kumar A, Gurjar BR (2017) Microalgae based biofuel: challenges and opportunities. In: Biofuels. Springer, Singapore, pp 157–175
Khan SA, Sharma GK, Malla FA, Kumar A, Gupta N (2019) Microalgae based biofertilizers: a biorefinery approach to phycoremediate wastewater and harvest biodiesel and manure. J Clean Prod 211:1412–1419
Kim SS, Ly HV, Kim J, Lee EY, Woo HC (2015) Pyrolysis of microalgae residual biomass derived from *Dunaliella tertiolecta* after lipid extraction and carbohydrate saccharification. Chem Eng J 263:194–199
Kiran B, Kumar R, Deshmukh D (2014) Perspectives of microalgal biofuels as a renewable source of energy. Energy Convers Manage 88:1228–1244
Klein-Marcuschamer D, Turner C, Allen M, Gray P, Dietzgen RG, Gresshoff PM, Hankamer B, Heimann K, Scott PT, Stephens E, Speight R (2013) Technoeconomic analysis of renewable aviation fuel from microalgae, Pongamia pinnata, and sugarcane. Biofuels Bioprod Biorefin 7(4):416–428
Koley S, Prasad S, Bagchi SK, Mallick N (2017) Development of a harvesting technique for large-scale microalgal harvesting for biodiesel production. RSC Adv 7(12):7227–7237
Konur O (2021) Biodiesel fuels: a scientometric review of the research. In: Biodiesel fuels. CRC Press, pp 225–248
Kumar AN, Chatterjee S, Hemalatha M, Althuri A, Min B, Kim SH, Mohan SV (2020) Deoiled algal biomass derived renewable sugars for bioethanol and biopolymer production in biorefinery framework. Bioresour Technol 296:122315
Lakatos GE, Ranglová K, Manoel JC, Grivalský T, Kopecký J, Masojídek J (2019) Bioethanol production from microalgae polysaccharides. Folia Microbiol 64(5):627–644
Lal A, Banerjee S, Das D (2021) *Aspergillus* sp. assisted bioflocculation of Chlorella MJ 11/11 for the production of biofuel from the algal-fungal co-pellet. Sep. Purif. Technol. 272:118320
Lam MK, Lee KT (2015) Bioethanol production from microalgae. In: Handbook of marine microalgae, pp 197–208

Lam MK, Tan IS, Lee KT (2014) Utilizing lipid-extracted microalgae biomass residues for maltodextrin production. Chem Eng J 235:224–230

Lee OK, Kim AL, Seong DH, Lee CG, Jung YT, Lee JW, Lee EY (2013) Chemo-enzymatic saccharification and bioethanol fermentation of lipid-extracted residual biomass of the microalga, Dunaliella tertiolecta. Biores Technol 132:197–201

Leoneti AB, Aragão-Leoneti V, De Oliveira SVWB (2012) Glycerol as a by-product of biodiesel production in Brazil: alternatives for the use of unrefined glycerol. Renew Energy 45:138–145

Li K, Liu S, Liu X (2014) An overview of algae bioethanol production. Int J Energy Res 38(8):965–977

Liu J, Huang J, Jiang Y, Chen F (2012) Molasses-based growth and production of oil and astaxanthin by Chlorella zofingiensis. Biores Technol 107:393–398

Liu Y, Lai YJS, Barbosa TS, Chandra R, Parameswaran P, Rittmann BE (2019) Electro-selective fermentation enhances lipid extraction and biohydrogenation of Scenedesmus acutus biomass. Algal Res 38:101397

Loftus SE, Johnson ZI (2017) Cross-study analysis of factors affecting algae cultivation in recycled medium for biofuel production. Algal Res 24:154–166

Lu D, Zhang XJ, Liu X, Zhang L, Hines M (2018) Sustainable microalgae cultivation by using anaerobic centrate and biogas from anaerobic digestion. Algal Res 35:115–124

Lu Z, Loftus S, Sha J, Wang W, Park MS, Zhang X, Johnson ZI, Hu Q (2020) Water reuse for sustainable microalgae cultivation: current knowledge and future directions. Resour Conserv Recycl 161:104975

Mantzorou A, Ververidis F (2019) Microalgal biofilms: a further step over current microalgal cultivation techniques. Sci Total Environ 651:3187–3201

McKendry P (2002) Energy production from biomass (part 2): conversion technologies. Biores Technol 83(1):47–54

Meher LC, Sagar DV, Naik SN (2006) Technical aspects of biodiesel production by transesterification—a review. Renew Sustain Energy Rev 10(3):248–268

Mishra S, Roy M, Mohanty K (2019) Microalgal bioenergy production under zero-waste biorefinery approach: recent advances and future perspectives. Bioresour Technol 292:122008

Mittal A, Katahira R, Himmel ME, Johnson DK (2011) Effects of alkaline or liquid-ammonia treatment on crystalline cellulose: changes in crystalline structure and effects on enzymatic digestibility. Biotechnol Biofuels 4(1):1–16

Mohan SV, Nikhil GN, Chiranjeevi P, Reddy CN, Rohit MV, Kumar AN, Sarkar O (2016) Waste biorefinery models towards sustainable circular bioeconomy: critical review and future perspectives. Biores Technol 215:2–12

Montagud A, Navarro E, de Córdoba PF, Urchueguía JF, Patil KR (2010) Reconstruction and analysis of genome-scale metabolic model of a photosynthetic bacterium. BMC Syst Biol 4(1):1–16

Mostafa SS, Shalaby EA, Mahmoud GI (2012) Cultivating microalgae in domestic wastewater for biodiesel production. Notulae Sci Biol 4(1):56–65

Mukherjee C, Chowdhury R, Sutradhar T, Begam M, Ghosh SM, Basak SK, Ray K (2016) Parboiled rice effluent: A wastewater niche for microalgae and cyanobacteria with growth coupled to comprehensive remediation and phosphorus biofertilization. Algal Res 19:225–236

Muñoz R, Navia R, Ciudad G, Tessini C, Jeison D, Mella R, Rabert C, Azócar L (2015) Preliminary biorefinery process proposal for protein and biofuels recovery from microalgae. Fuel 150:425–433

Mus F, Dubini A, Seibert M, Posewitz MC, Grossman AR (2007) Anaerobic acclimation in *Chlamydomonas reinhardtii*: anoxic gene expression, hydrogenase induction, and metabolic pathways. J Biol Chem 282(35):25475–25486

Nguyen MT, Choi SP, Lee JW, Lee JH, Sim SJ (2009) Hydrothermal acid pretreatment of *Chlamydomonas reinhardtii* biomass for ethanol production. J Microbiol Biotechnol 19(2):161–166

Nobre BP, Villalobos F, Barragan BE, Oliveira AC, Batista AP, Marques PASS, Mendes RL, Sovová H, Palavra AF, Gouveia L (2013) A biorefinery from *Nannochloropsis* sp. microalga–extraction of oils and pigments. Production of biohydrogen from the leftover biomass. Bioresour Technol 135:128–136

Noor-Mohammadi S, Pourmir A, Johannes TW (2012) Method to assemble and integrate biochemical pathways into the chloroplast genome of *Chlamydomonas reinhardtii*. Biotechnol Bioeng 109(11):2896–2903

Obeid F, Van TC, Brown R, Rainey T (2019) Nitrogen and sulphur in algal biocrude: A review of the HTL process, upgrading, engine performance and emissions. Energy Convers Manage 181:105–119

Pancha I, Chokshi K, Maurya R, Bhattacharya S, Bachani P, Mishra S (2016) Comparative evaluation of chemical and enzymatic saccharification of mixotrophically grown de-oiled microalgal biomass for reducing sugar production. Biores Technol 204:9–16

Pandey A, Shah R, Yadav P, Verma R, Srivastava S (2020) Harvesting of freshwater microalgae *Scenedesmus* sp. by electro–coagulation–flocculation for biofuel production: effects on spent medium recycling and lipid extraction. Environ Sci Pollut Res 27(3):3497–3507

Pearce DW, Turner RK (1990) Economics of natural resources and the environment. JHU Press

Raheem A, Abbasi SA, Mangi FH, Ahmed S, He Q, Ding L, Memon AA, Zhao M, Yu G (2021). Gasification of algal residue for synthesis gas production. Algal Res 58:102411

Ray A, Banerjee S, Das D (2021) Microalgal bio-flocculation: present scenario and prospects for commercialization. Environ Sci Pollut Res 1–19

Renuka N, Prasanna R, Sood A, Ahluwalia AS, Bansal R, Babu S, Singh R, Shivay YS, Nain L (2016) Exploring the efficacy of wastewater-grown microalgal biomass as a biofertilizer for wheat. Environ Sci Pollut Res 23(7):6608–6620

Saini JK, Saini R, Tewari L (2015) Lignocellulosic agriculture wastes as biomass feedstocks for second-generation bioethanol production: concepts and recent developments. 3 Biotech 5(4):337–353

Shah A, Baral NR, Manandhar A (2016) Technoeconomic analysis and life cycle assessment of bioenergy systems. In: Advances in bioenergy, vol 1. Elsevier, pp 189–247

Shaishav S, Singh RN, Satyendra T (2013) Biohydrogen from algae: fuel of the future. Int Res J Environ Sci 2(4):44–47

Sialve B, Bernet N, Bernard O (2009) Anaerobic digestion of microalgae as a necessary step to make microalgal biodiesel sustainable. Biotechnol Adv 27(4):409–416

Singh R, Balagurumurthy B, Bhaskar T (2015) Hydrothermal liquefaction of macro algae: effect of feedstock composition. Fuel 146:69–74

Singh H, Das D (2018) Biofuels from microalgae: biohydrogen. In Energy from microalgae. Springer, Cham, pp 201–228

Sivaramakrishnan R, Incharoensakdi A (2018) Utilization of microalgae feedstock for concomitant production of bioethanol and biodiesel. Fuel 217:458–466

Solis CA, Mayol AP, San Juan JG, Ubando AT, Culaba AB (2020) Multi-objective optimal synthesis of algal biorefineries toward a sustainable circular bioeconomy. IOP Conf Ser Earth Environ Sci 463(1):012051

Strategy UB (2018) A sustainable bioeconomy for Europe: strengthening the connection between economy, society and the environment. European Commission (2018). https://ec.europa.eu/research/bioeconomy/pdf/ec_bioeconomy_strategy_2018.pdf. дата обращения 25 Feb 2020

Suali E, Sarbatly R (2012) Conversion of microalgae to biofuel. Renew Sustain Energy Rev 16(6):4316–4342

Tesfa B, Mishra R, Zhang C, Gu F, Ball AD (2013) Combustion and performance characteristics of CI (compression ignition) engine running with biodiesel. Energy 51:101–115

Tibaquirá JE, Huertas JI, Ospina S, Quirama LF, Niño JE (2018) The effect of using ethanol-gasoline blends on the mechanical, energy and environmental performance of in-use vehicles. Energies 11(1):221

Ubando AT, Felix CB, Chen WH (2020) Biorefineries in circular bioeconomy: a comprehensive review. Bioresour Technol 299:122585

Ueno Y, Kurano N, Miyachi S (1998) Ethanol production by dark fermentation in the marine green alga, Chlorococcum littorale. J Ferment Bioeng 86(1):38–43

Uggetti E, Sialve B, Latrille E, Steyer JP (2014) Anaerobic digestate as substrate for microalgae culture: the role of ammonium concentration on the microalgae productivity. Biores Technol 152:437–443

Wang X, Ruan Z, Boileau D, Sears BB, Liu Y, Liao W (2015) Transgenic expression of a bacterial thermophilic amylase in the *Chlamydomonas reinhardtii* chloroplast to facilitate algal biofuel production. BioEnergy Res 8(2):527–536

Xu S, Elsayed M, Ismail GA, Li C, Wang S, Abomohra AEF (2019) Evaluation of bioethanol and biodiesel production from Scenedesmus obliquus grown in biodiesel waste glycerol: a sequential integrated route for enhanced energy recovery. Energy Convers Manage 197:111907

Yahmed NB, Jmel MA, Alaya MB, Bouallagui H, Marzouki MN, Smaali I (2016) A biorefinery concept using the green macroalgae Chaetomorpha linum for the coproduction of bioethanol and biogas. Energy Convers Manage 119:257–265

Yang J, Xu M, Zhang X, Hu Q, Sommerfeld M, Chen Y (2011) Life-cycle analysis on biodiesel production from microalgae: water footprint and nutrients balance. Biores Technol 102(1):159–165

ye Lee J, Li P, Lee J, Ryu HJ, Oh KK (2013) Ethanol production from *Saccharina japonica* using an optimized extremely low acid pretreatment followed by simultaneous saccharification and fermentation. Bioresour Technol 127:119–125

Yen HW, Brune DE (2007) Anaerobic co-digestion of algal sludge and waste paper to produce methane. Biores Technol 98(1):130–134

Yoon SH, Lee CS (2011) Lean combustion and emission characteristics of bioethanol and its blends in a spark ignition (SI) engine. Energy Fuels 25(8):3484–3492

Yu KL, Chen WH, Sheen HK, Chang JS, Lin CS, Ong HC, Show PL, Ng EP, Ling TC (2020) Production of microalgal biochar and reducing sugar using wet torrefaction with microwave-assisted heating and acid hydrolysis pretreatment. Renew Energy 156:349–360

Zhan J, Rong J, Wang Q (2017) Mixotrophic cultivation, a preferable microalgae cultivation mode for biomass/bioenergy production, and bioremediation, advances and prospect. Int J Hydrogen Energy 42(12):8505–8517

Zhang T, Xie X, Huang Z (2014) Life cycle water footprints of nonfood biomass fuels in China. Environ Sci Technol 48(7):4137–4144

Zhu L (2014) The combined production of ethanol and biogas from microalgal residuals to sustain microalgal biodiesel: a theoretical evaluation. Biofuels, Bioprod Biorefin 8(1):7–15

Chapter 4
Fuel Delivery System for Alternative Fuel Engines: A Review

Yuanxian Zhu and Liyun Fan

Abstract Application of low carbon alternative fuels for engines has been well-known as an attractive approach to reduce greenhouse gas emissions. In recent years, additional efforts have been made on the research and development of alternative fuel engines that run on natural gas, methanol, hydrogen and ammonia, because they provide more potential to achieve strict emission targets. However, a technical challenge for these alternative fuel engines is whether a qualified fuel delivery system is available. Blending is perhaps the most convenient method to burn alternative fuel. A popular blending fuel is the mixture of gasoline and 10% ethanol (E10) for use in passenger vehicles. In this case, no change is needed for the fuel delivery system. In other cases including using methanol as an alternative, the components of fuel delivery system require materials modification to prevent corrosion. High-pressure direct injection of methanol is preferred for heavy-duty engines because compression ignition exhibits high thermal efficiency. In this case, the conventional fuel delivery system needs more modifications. For example, a higher flow capacity is needed for increased flow rate due to the lower energy density. Additional cooling elements are needed to prevent vaporization. A special coating should also be applied to the surface of the pump plunger and control valve to improve the lubricity because those components are made of the anticorrosive material. For natural gas, an effective combustion mode is so called dual—fuel mode. There, natural gas introduced from the intake port and mixed with air homogenously is ignited by injecting a small amount of diesel fuel when the piston approaches the end of compression stroke. More recently, a more advanced fuel delivery system has been developed, which uses one injector for diesel fuel and another for natural gas (or alternatively a co-axial injector with inner nozzle hole for diesel fuel and outer nozzle holes for natural gas). It has been reported that diesel fuel injected for ignition can be reduced to 5% of total fuel with using this new system. This dual-fuel mode has been used for

Y. Zhu (✉)
Chengdu WIT Electronic Fuel System Co., Ltd., Chengdu, China
e-mail: zhu19821993@163.com

L. Fan
Harbin Engineering University, Harbin, China
e-mail: fanliyun@hrbeu.edu.cn

A. V. Agarwal and H. Valera (eds.), *Potential and Challenges of Low Carbon Fuels for Sustainable Transport*, Energy, Environment, and Sustainability,
https://doi.org/10.1007/978-981-16-8414-2_4

methanol and ammonia fuel, and could be the next main-stream for alternative fuel engines.

Keywords Fuel delivery system · Low carbon alternative fuels · Alternative fuel engine · Spark ignition engine · Compression ignition engine

Abbreviations

B5	5% Biodiesel
BMEP	Brake Mean Effective Pressure
CI	Compression Ignition
CNG	Compressed Natural Gas
ASTM	American Society of Testing Materials
CWS	Coal Water Slurry
DDC	Detroit Diesel Corporation
DLC	Diamond-like Carbon
E10	10% Ethanol
EUI	Electronic unit injector
EUP	Electronic unit pump
F-T	Fischer Tropsch
GDI	Gasoline direction injection
HCCI	Homogeneous Charge Compression Ignition
IC	Internal Combustion
LPG	Liquefied Petroleum Gas
LNG	Liquefied Natural Gas
MPI	Multi-point Port Injection
NVH	Noise Vibration Harshness
PTFE	Poly Tetra Fluoroethylene
R&D	Research and Development
SI	Spark Ignition
SOGAV	Solenoid Operated Gas Admission Valve
TDC	Top Dead Center

4.1 Introduction

Over the past few decades, a variety of alternative fuels has been successfully used in engines. (Darade and Dalu 2012; Verhelst et al. 2019; Varde and Frame 1984; Arcoumanis et al. 2008; Torres et al. 2021; National Renewable Energy Laboratory). Motivations for these applications include reducing pollutant emissions, lowering fuel cost, and conserving petroleum resources. More recently, with the renewed focus on climate change issues, world legislators have discussed a timetable for carbon neutralization. For the engine industry, the obvious big challenge is how to swiftly reduce greenhouse gas emissions. An attractive solution is to use low carbon or carbon-free fuels, which gives rise to new opportunities in engine technology advancement.

For a conventional engine, a quick and precise air–fuel mixing over its entire operating region is required to achieve a clean and efficient combustion; consequently, the fuel delivery system becomes a key element. For an alternative fuel engine, the availability of fuel delivery system is often a more challenging task due to following factors.

- Some alternative fuels often assumes a gas state at certain pressure and temperature, and the sealing and lubrication may become problems
- Some alternative fuels have much lower energy density than conventional gasoline and diesel fuel, the capacity of fuel flow passage should be increased greatly to keep engine power-output
- Some alternative fuels exhibit strong corrosivity to both metallic and non-metallic materials, which may cause the failure of components
- Major low carbon fuels, such natural gas, methanol and ammonia, may require to be applied in a special operating mode—dual fuel mode to achieve desired results. In this case, the conventional fuel delivery system is hard to play a qualified role.

This chapter reviews the progress of fuel delivery technologies for alternative fuel engines. Special focus is paid on those new emerging fuel delivery technologies for low carbon or carbon neutral fuels used in the last decade.

4.2 Fuel Delivery System of Conventional IC Engines

Conventional internal combustion (IC) engines generally operate on either the Otto cycle or the Diesel cycle modes. The former is also called the spark ignition (SI) engine, and the latter called the compression ignition (CI) engine. Both SI and CI engines have become the dominate power plant in the transportation field for more than a century.

4.2.1 Fuel Delivery System of SI Engine

The early fuel delivery system for the SI engine is essentially a carburetor, as shown in Fig. 4.1 (Heywood 1988). This device is used to control the fuel flow into the intake manifold and distribute the fuel across the air stream. It can provide a proper air—fuel ratio at different operating conditions, and then the homogenously formed mixture is ignited by a spark plug when the piston is near the position of TDC.

Due to more stringent emissions legislation and advancements in electronics technology, the fuel injection system gradually replaced the carburetor in the 1980s. The typical multi-point port injection system (MPI) operates when the fuel is injected into the intake port of each engine cylinder (Automotive Handbook 2004). Compared to

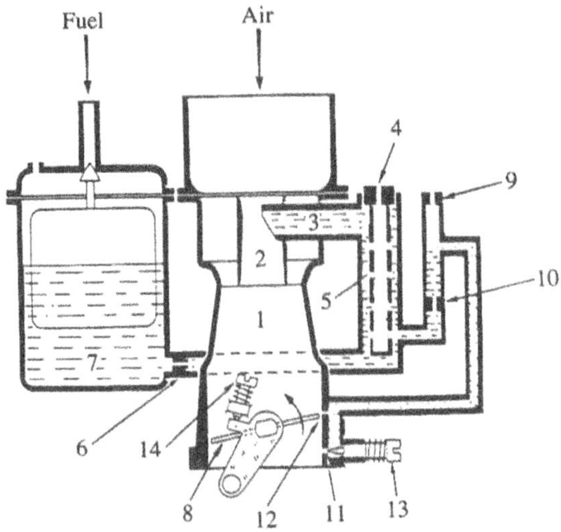

1 Main venturi, 2 Boost venturi, 3 Main metering spray tube or nozzle, 4 Air-bleed orifice, 5 Emulsion tube or well, 6 Main fuel-metering orifice, 7 Float chamber, 8 Throttle plate, 9 Idle air-bleed orifice, 10 Idle fuel orifice, 11 Idle mixture orifice, 12 Transition orifice, 13 Idle mixture adjusting screw, 14 Idle throttle setting adjusting screw

Fig. 4.1 Schematic of modern carburetor (Heywood 1988)

the carburetor, a port fuel injection system has the advantage of increased power and torque as a result of improved volumetric efficiency, more uniform fuel distribution, and faster engine response to throttle position changes. Moreover, electronic engine management calculates the required injection quantity based on lambda closed-loop control and facilitates three-way catalyst application to reduce main pollutant emissions by more than 90% versus prior techniques.

In the early 2000s, fuel injection technology for the SI engine moved from multi-point port injection to in-cylinder injection to achieve higher power density and better fuel economy. This more advanced system is commonly called gasoline direction injection (GDI) (Reif 2014a). As shown in Fig. 4.3, the GDI system consists of a high-pressure pump, a fuel rail, and several high-pressure injectors. The system operates in two modes—a homogenous mode where the fuel is injected during the induction stroke and a stratified-charge mode where the fuel is injected during the compression stroke. With using GDI, a turbocharged SI engine can adopt a higher compression ratio as in-cylinder fuel vaporization reduces charge temperature. Another advantage with GDI is increased volumetric efficiency as only fresh air gets into cylinder during intake stroke (Fig. 4.2).

The engine as a result exhibits lower fuel consumption while outputting more power. Furthermore, the stratified-charge mode can provide improved fuel economy. Today, the power density for a turbocharged GDI SI engine could be as high as 100 kW/L.

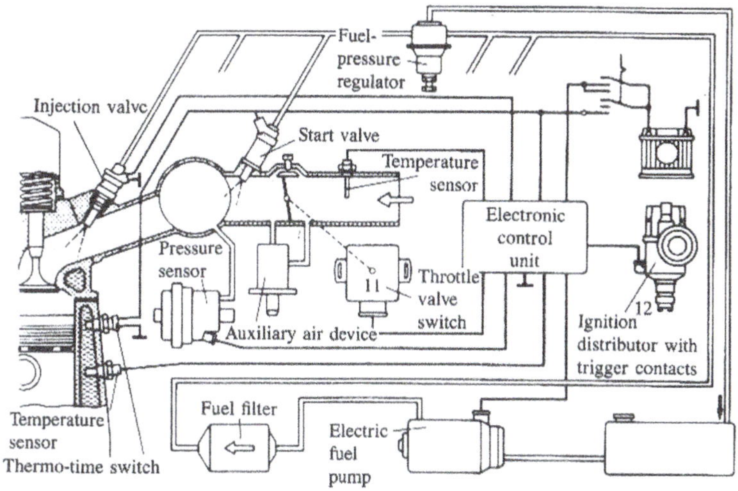

Fig. 4.2 Schematic diagram of MPI system (Heywood 1988)

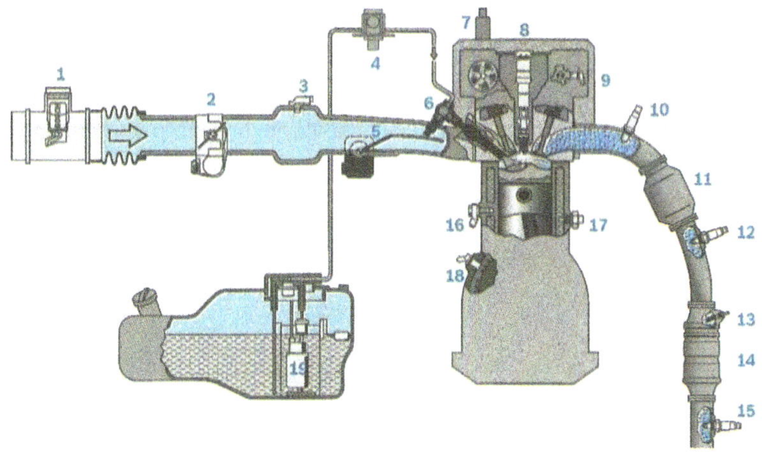

1 Hot film air mass meter, 2 Intake throttle,3 Intake pressure sensors,
4 high pressure pump, 5 Flow control valve,
6 Oil rail with high pressure injection nozzle,
7 Cam adjuster, 8 Ignition coil with spark plug, 9 Cam phase sensor,
10 Lambda oxygen sensor, 11 Primary catalytic converter,
12 Lambda oxygen sensor, 13 exhaust gas temperature sensor
14 NO_x catalytic converter, 15 Lambda oxygen sensor, 16 Knock sensor,
17 Temperature sensor, 18 Speed sensor
19 Gasoline supply module with electric oil pump

Fig. 4.3 Schematic diagram of typical GDI system (Reif 2014a)

4.2.2 Fuel Delivery System of CI Engine

When he was granted the first patent for a CI engine in 1893, Rudolf Diesel did not solve the fuel delivery problem. It was not until five years later in 1897, after creating a compressed air assistant fuel injection device (see Fig. 4.4), that he introduced a working prototype of the first CI engine (Reif 2014b).

Compared with the SI engine, the fuel delivery system for CI engine presents a more difficult task for design engineers because the diesel fuel needs to be injected into the cylinder and atomized under high pressure, and this obstacle confounded the early development of CI engines. In 1923, BOSCH invented a high injection pump and thus paved the way for using CI engines for transport.

Figure 4.5 illustrates an early-stage design for a high-pressure pump, whereby a plug-in pump is driven by a camshaft. With the camshaft rotating, the plunger in the pump barrel moves up and pushes the fuel out through a check valve. An elaborate designed mechanical device can adjust the fuel flow rate based on the engine requirement. This type of high-pressure injection pump has been the dominated fuel delivery system in CI engines for more than 60 years.

In the last decade of the twentieth century, the development of fuel delivery system for CI engines shifted into high gear as pollutants emission legislation tightened. Electronic unit injector (EUI), electronic unit pump (EUP), electronic in-line pump (Fan et al. 2008, 2010) and electronic distribution pump were several successful examples. However, the greatest technical breakthrough was the introduction of the common rail fuel injection system (Reif 2014b).

As shown in Fig. 4.6, a common rail fuel injection system has mechanical

Fig. 4.4 Compressed air assistant injector (Reif 2014b)

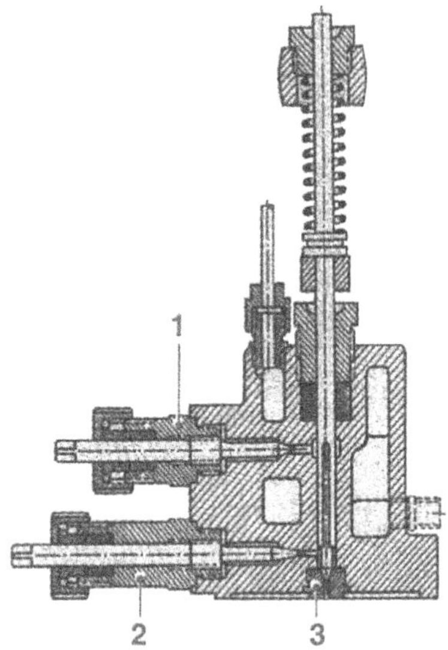

1 Compressed air supply, 2 Fuel supply, 3 Nozzle

1 Camshaft, 2 Roller tappet, 3 Control sleeve, 4 Control Rack, 5 Fuel inlet, 6 Pump cylinder, 7 Control sleeve, 8 Fuel line connector, 9 Delivery valve, 10 Oil dipstick, 11 Plunger

Fig. 4.5 Injection pump for CI engine at early stage (Reif 2014b)

electrical integration, and comprise of solenoid valve injectors, a high-pressure pump with inlet metering valve, a fuel accumulate, physical signal sensors, and an embedded controller. With this sophisticated system, the fuel injection process can be controlled more precisely, and allows for flexible injection pressure control independent of engine operating conditions, multiple injections per cycle, and lower pump driving power requirements, thereby resulting in a CI engine having significant improvements in pollutant emissions, power density, fuel economy, and NVH.

Table 4.1 lists all fuel delivery systems presented in this section.

4.3 Approaches to Develop Fuel Delivery System to Adapt Alternative Fuel

Using alternative fuels may have the benefits of either improving engine emissions, or thermal efficiency, or operating cost. What contemporary automotive engineers pursue is one of optimization: delivering specific alternative fuels for maximum benefit. Depending on the type of fuel being introduced into the engine (*i.e.*, pure or blend), the fuel is either spark-ignited or compression ignited. The injected fuel is then burned in a conventional combustion mode or a HCCI mode. These various

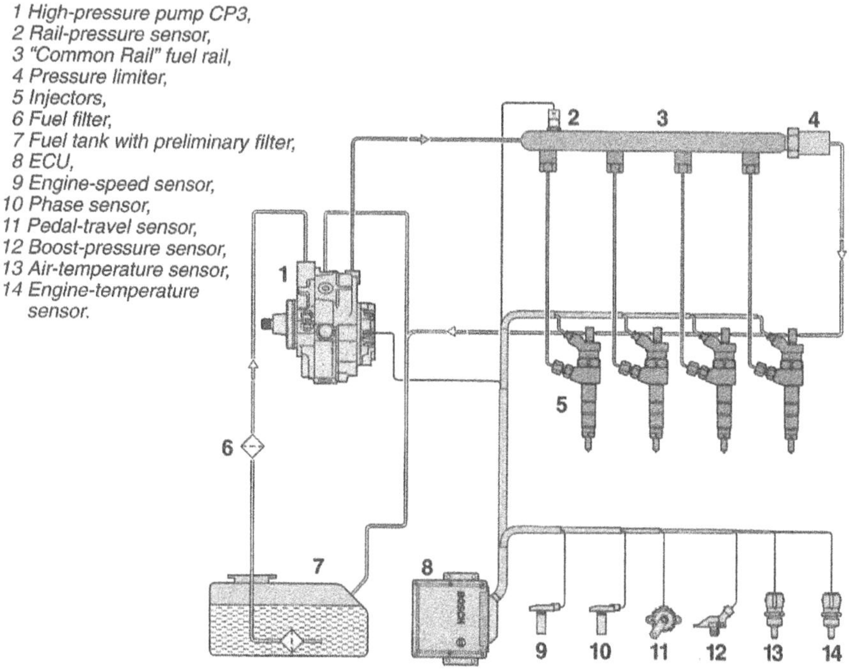

Fig. 4.6 Schematic diagram of common rail fuel injection system (Automotive Handbook 2004)

Table 4.1 Main fuel delivery system for SI engines and CI engines

Fuel delivery system	Engine	Main features	Starting era
Carburetor system	SI	Mechanically fuel quantity control	1870s
MPI system	SI	Port injection, Elec. control	1990s
GDI system	SI	Direct injection, Elec. control	2000s
PLN system	CI	Mechanically fuel quantity control	1920s
EUP/EUI system	CI	Elec. timing/fuel quantity control	1980s
CR system	CI	Elec. timing/pressure/fuel quantity control	1990s

combinations of how to process specific alternative fuels will result in different requirements for the fuel delivery system.

For a period, developments in fuel delivery systems to adapt to alternative fuels have been confined to only minor modifications in conventional engines. These modifications can generally be categorized as follows:

- Material compatibility to prevent components failure;
- Proper fuel storage and supply to account for physical states;
- Adjustment of flow capacity to adapt to lower energy densities;
- Improvement of lubrication to avoid overwearing;

- Improvement of sealing to prevent leakage; and/or
- Recalibration of engine management to adjust for fuel delivery characteristics.

The above modifications were generally effective in making it possible for IC engines to use a variety of alternative fuels. However, it may be desirable to develop a dedicated fuel delivery system in order to take full advantage of alternative fuels. A new fuel delivery technology based on a dual-fuel injection concept provides profound feasibility and compatibility (Weaver and Turner 1994). Such an approach not only can be used for natural gas fuels, but also for methanol and ammonia, which are widely considered to be valuable alternatives that can reduce greenhouse gas emissions for IC engines.

4.4 Fuel Delivery System for SI Engine to Adapt Alternative Fuel

Perhaps the most popular application of alternative fuels for SI engines today is using either a pure alternative fuel or a blend of alternative fuel and gasoline. To be a proper alternative fuel for a SI engine, the candidate fuel is required to have similar physical and chemical properties as those of gasoline. Fortunately, rich sources of alternative fuels, such as LPG, natural gas and ethanol, exhibit proper properties, and aid in improving SI engine performance and emissions.

Table 4.2 provides the main combustion properties of some popular engine fuels.

Table 4.2 Combustion characteristics of various engine fuels (Mounaïm-Rousselle and Brequigny 2020)

Fuel	Formula	Storage Temp. [°C]	Storage pressure [kPa]	Lower heating value [MJ/kg]	Stoichiometric air/fuel ratio by weight	Energy content [MJ/kg-stoichiometric mixture]	Autoignition Temp [°C]
Methanol	CH_3OH	25	101.3	19.5	6.44	2.69	464
Ethanol	C_2H_5OH	25	101.3	27	8.95	2.70	423
Dimethyl Ether	CH_3OCH_3	25	1030	28.4	8.95	2.85	350
Gasoline	C_7H_{17}	25	101.3	42.5	15.29	2.58	370
Hydrogen (gas)	H_2	25	24,821	120	34.32	3.40	571
Diesel	$C_{14.4}H_{24.9}$	25	101.3	45	14.32	2.77	254
Ammonia	NH_3	25	1030	18.8	6.05	2.64	651

4.4.1 Liquefied Petroleum Gas (LPG)

As a mixture of propane and butane, LPG assumes a liquid state at pressure of 0.2–2 MPa, depending on the propane/butane ratio and the temperature (Reif 2014a). The octane number of LPG is large than 110 RON, and therefore an ideal alternative fuel for Spark ignition mode.

LPG is a byproduct from the refinery and the oil field, and it became a popular alternative fuel for IC engine since 1980s due to its low cost and rich source. In 2018, LPG was consumed by IC engine at a rate of 17 million tons annually.

Simple modifications to a conventional SI engine are required to run LPG. Figure 4.7 is schematic diagram of an LPG fuel delivery system based on MPI. As shown, an LPG fuel injection system requires, among other things, an LPG tank, a low-pressure gas shout-off valve, and an evaporation pressure regulator. The delivery process begins when fuel is injected at the intake manifold in the same way as in a conventional SI engine. In most cases, LPG is injected as a gas after evaporation in the pressure regulator. There are also systems in use which inject LPG in liquid form. In that case, an additional fuel pump is required to increase the pressure above that

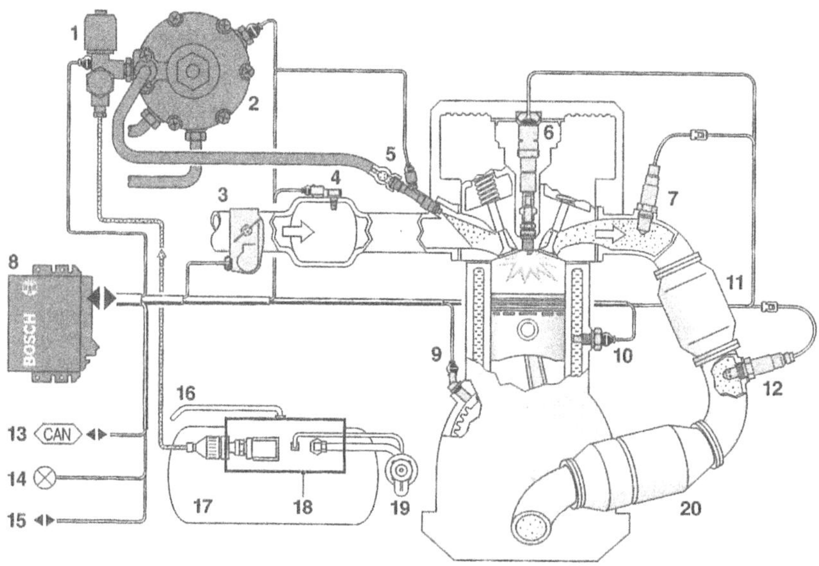

1 Gas shutoff valve (low pressure), 2 Evaporator pressure regulator, 3 Throttle device, 4 Intake manifold pressure sensor, 5 injector, 6 Ignition coil with spark plug, 7 Lambda oxygen sensor,8 ECU, 9 Speed sensor, 10 Engine temperature sensor, 11 Primary catalytic converter, 12 Lambda oxygen sensor, 13 CAN interface, 14 Diagnosis Lamp, 15 Diagnosis interface, 16 Ventilation line for tank fitting, 17 LPG tank, 18 House with tank fittings, 19 External filler valve, 20 Main catalytic converter

Fig. 4.7 Schematic diagram of an LPG fuel delivery system (Automotive Handbook 2004)

of the tank. The LPG tank, which is subject to stringent safety regulations in Europe and other regions, is made from a polymorphic composite material. The tank also requires, for the purpose of additional safety measures:

- a non-return valve;
- a fill restriction facility, designed to ensure that the fill volume is no more than 80% of maximum;
- a blow-off valve;
- a temperature safety valve; and
- a gas releasing valve with flow limit.

4.4.2 Natural Gas

Natural gas has been one of the most important alternative fuels due to its abundant supply. In recent years, major shale gas production in the United States provides an optimistic prospect.

The main component of natural gas is methane, making up approximately 80–99% of the fuel, depending on origin. Like LPG, natural gas is also an ideal fuel for spark ignition engines as its octane of can be up to 130 RON. At such levels, it is possible to raise the engine compression ratio to improve thermal efficiency. However, the engine output will still be approximately 10–15% lower due to the lower volumetric efficiency attributed to displacement caused by the injected natural gas.

A disadvantage with natural gas combustion is weak lean-burn ability, low flame speed and poor ignitibility (Alrazen and Ahmad 2018), which result in deterioration of both fuel consumption and emission at low load conditions. However, it has been found that the addition of tiny amount of hydrogen into natural gas will improve combustion process substantially. This effective measure will be discussed in later section.

Natural gas can be stored either in liquid state (LNG) below − 162C or in compressed state (CNG) at a pressure above 20 MPa. For passenger vehicles, the modification from a gasoline fueled SI engine to a natural gas fueled SI engine is quite simple, with a preference for leak-proof CNG storage. On the other hand, heavy-duty diesel engines for commercial trucks and vessels require major retrofit work. For example, the compression ratio must be decreased, and the intake and exhaust valves need different timing schemes. More importantly, the standard diesel fuel injection system must be replaced by an integrated gas injection and spark ignition system. Because of a required reduction in tank volume, LNG storage mode is preferred for heavy-duty engines.

Figure 4.8 illustrates a typical CNG fuel delivery system based on MPI for passenger vehicles. To begin, natural gas is injected into the intake manifold in the same way as in a conventional SI engine. Distinct components of this system include a high-pressure CNG tank, a high-pressure shutoff valve, and a gas pressure regulator. Natural gas stored in the gas tank at a pressure of 20 MPa flows via the high-pressure shutoff valve to the gas pressure regulator that reduces the pressure to

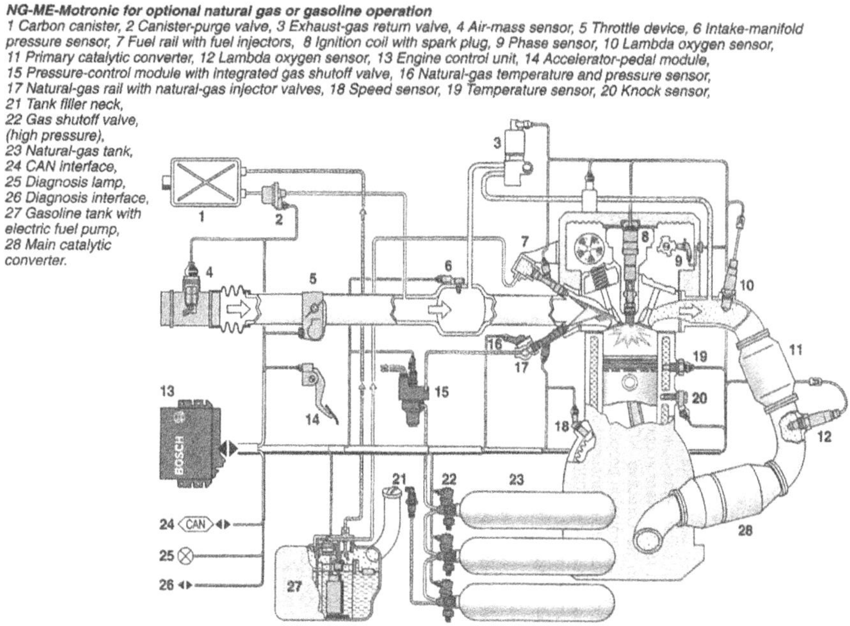

Fig. 4.8 Schematic diagram of a CNG fuel delivery system for passenger vehicle (Automotive Handbook 2004)

approximately 0.8 MPa. The natural gas then flows through a low-pressure shutoff valve to a common gas distributor rail, supplying one injector per cylinder. For LNG based systems, the fuel delivery system is slightly different as the high-pressure components are replaced by the cryogenic tank and the evaporator.

Figure 4.9 shows a natural gas delivery system on a heavy-duty engine (https://www.heinzmann.com/en/gas-engines). Here, dedicated gas admission (injection) valves are located directly on the cylinder intake ports. Integrated sensing of exhaust gas temperature for each cylinder enables accurate injection timing and quantity control, therefore ensuring improvements in emission and power output.

For marine application, the high durability of the gas admission valve is required. As shown in Fig. 4.10, the Solenoid Operated Gas Admission Valve (SOGAV) produced by Woodward is an electrically actuated, consistent opening and closing response gas valve for in-manifold (port) fuel injection. It is used in large bore marine natural gas engines, covering the output power range from 70 kW/cylinder up to 1500 kW/cylinder and all parts exposed to the gas are resistant to corrosion. The valve is a fungiform poppet with multiple concentric grooves (SOGAV^{TM} Solenoid Operated Gas Admission Valve). The number, size and distribution of the sealing ring are reasonably set according to the actual needs of the engine.

It should be mentioned that fuel delivery system for natural gas engines, either light duty or heavy duty, have been fully developed.

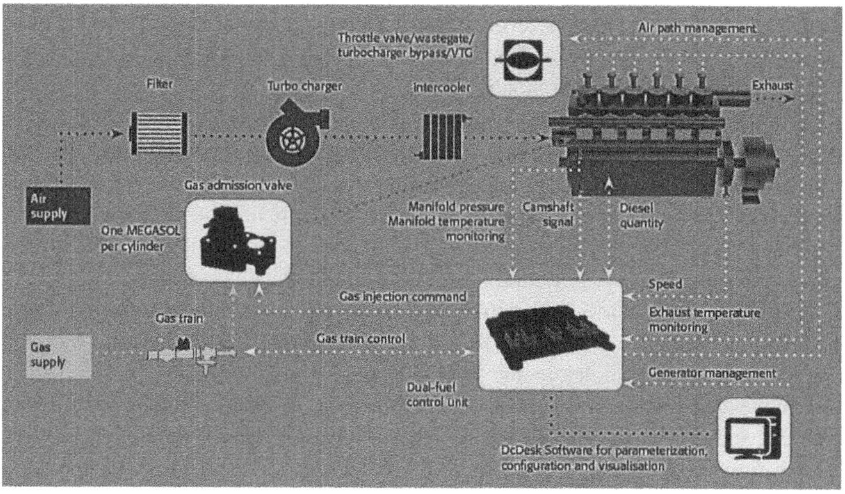

Fig. 4.9 Schematic diagram of gas injection system for heavy duty commercial engine (https://www.heinzmann.com/en/gas-engines)

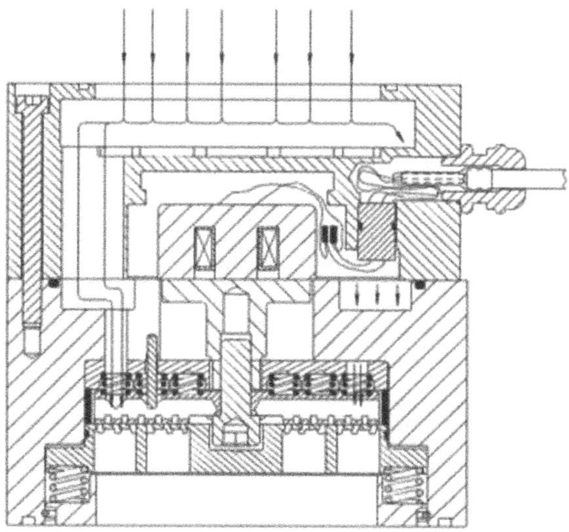

Fig. 4.10 Structural diagram of Woodward gas admission valve (SOGAV^{TM} Solenoid Operated Gas Admission Valve)

4.4.3 Alcohols

Alcohol fuels include methanol and ethanol. Although methanol can be produced from both biomass and fossil, the current production process is based on natural gas and coal resources. Ethanol, on the other hand, is mainly made in a regenerative process using biomass, and therefore could help reduce global carbon dioxide emissions. Two typical examples of ethanol production include processing sugar cane in Brazil and corn in the United States.

Alcohols are ideal fuels for SI engines. One advantage is the fuels' high octane number (> 110 RON), which allows for higher engine compression ratio. Another advantage is low stoichiometric air—fuel ratio due to high oxygen content, so more air—fuel mixture can join combustion process for same around of air. In this way, alcohols are often called as performance enhancer. Besides, alcohols are free of sulfur, and thus avoid the potential of after-treatment device poisoning.

Blending alcohol fuels with gasoline is often used to reduce emission and/or enhance performance. A blended fuel with 10% ethanol (E10) became the standard for passenger vehicles in the United States. In this case, the fuel delivering system does not require a design overhaul. However, as blended fuels become more alcohol based, particularly with high volumes of methanol, some fuel delivery system modifications become necessary. The requirements are divided into two aspects.

The first aspect relates to the lower volumetric energy content of alcohol fuels. The fuel pump, injector, and all other parts in the system require a significant redesign in order to increase flow capacity. Otherwise, the engine will not reach its rated power and peak torque.

The second aspect relates to the polarity of alcohol fuels (Verhelst et al. 2019), which presents a well-known material compatibility challenge. Both metal and elastomer (*i.e.*, the soft components used for seals and fuel lines), if not chosen properly, could be corroded by alcohols, especially methanol. And hydrous methanol corrodes materials even more seriously. For this reason, the metal components of a fuel delivery system that are in frequent contact with the fuel should preferably be made of austenitic stainless steel, and the soft components should use some methanol compatible material, such as fluorocarbon elastomer and nitrile butadiene rubbers. The industry is ripe with successful experiences which resolved these compatibility issues.

4.4.4 Hydrogen

Hydrogen has unique physical and chemical properties compared to the conventional fossil fuels widely used in the transportation sector (Yip et al. 2019). The main advantages being its CO_2-free composition and that it can be produced from renewable energy sources (Reif 2014a). And with the highest auto-ignition temperature and octane number (RON $\geq$ 130), hydrogen has found increasing significance and attention as a viable alternative fuel.

While hydrogen can be delivered to engines in a similar fashion as CNG, its low volumetric energy density, approximately 30% of CNG, presents considerable design challenges. The on-board storage pressure requires to be at 70 MPa to reach same energy density as that of CNG at 20 MPa. The thirst for high pressure places special demands on the materials used to construct the storage tank as hydrogen embrittlement is a common cause of material failure when high pressure hydrogen is used. Another possible problem relates to the high diffusivity of hydrogen that poses a high risk of leakage. These issues contribute to the need of special measures as part of the fuel delivery system.

However, a special value of hydrogen is that it, as an additive, can enhance the combustion process of some popular alternative fuels. A typical case is the addition of small amount of hydrogen into CNG. As reported (Swain et al. 1993; Mariani et al. 2012), the blends of hydrogen and CNG promotes lean-burn ability, flame speed and ignitability at low load conditions effectively. As a result, engine thermal efficiency, HC/CO emissions, and operating stability at those operating conditions are improved greatly. Another research (Morch et al. 2011) shows that the addition of hydrogen into ammonia improves ignition process and increases the combustion flame speed effectively in a SI engine uelled by ammonia, thereafter the engine achieves similar BMEP and thermal efficiency with that fuelled by gasoline. This is perhaps the most promising result with ammonia.

In the case of the blends of hydrogen and CNG, the hydrogen embrittlement, which may cause the material failure of onboard fuel storage tank, become less possible. This is because the hydrogen partial pressure in the blends is much lower than pure hydrogen. For instance, the hydrogen partial pressure will be only 7.5 in a 30% hydrogen of blends in a 250-bar storage tank. A vehicle test showed that no trace of hydrogen embrittlement was found on a commercial CNG fuel tank made of fiber-reinforced 4130 steel after one year service. Based on successful field experiences, it is possible for all the other parts in fuel delivery systems, regulators, valves, sensors, and plumbing to be identical to general used standard of CNG 250 bar (Alrazen and Ahmad 2018).

4.5 Fuel Delivery System for CI Engine to Adapt Alternative Fuel

4.5.1 Methanol

As discussed above, methanol makes an ideal alternative fuel for SI engines because of its high octane number and other physical, chemical properties. However, methanol also contains additional properties, such as higher auto ignition temperature, higher heat of vaporization, and lower energy content per unit volume than diesel fuel, that poses challenges as an alternative fuel for compression ignition engines.

In the 1980s, engine researchers have accommodated design changes to adapt to the requirements of methanol (Morch et al. 2011; Miller and Savonen 1990; Yutaka et al. 1991). Some measures that were frequently taken are summarized below:

- Raise the engine compression ratio;
- Heat a glow plug at engine start and light load conditions;

- Hot exhaust gas recirculation
- Control the scavenging ratio.

Most of these measures were obviously made to raise in-cylinder temperature to enable fuel auto-ignition at all engine operating conditions. Another approach to improve auto-ignition was to add to the methanol, up to 5% by volume, an ignition improver. However, this approach faced the problem that the methanol does not mix well with the ignition improver and thus required emulsifier agents. A better concept, known as dual fuel injection, worked by injecting the ignition improver as a pilot to ignite the methanol by in-cylinder injection or intake port injection. Dual fuel injection will be discussed in detail below.

Modifications for alternative fuel delivery systems, including those related to flow capacity and material compatibility issues discussed above with respect to SI engines, equally apply for CI engines too. Additionally, Miller (Morch et al. 2011) has reported a series of incidents related to the unit injector. The main problems were nozzle tip hole plugging, plunger scoring and seizing, and control valve wear, which were considered as the consequence of a small amount of lubricant mixing with the methanol and poor lubricity of methanol. The several solutions including the applications of fuel additive for better lubricity and modified oil formulation for decreasing the deposit on the nozzle tip hole were taken, and the improvements were found. However, today we may have better solutions for these problems, for examples, the plunger of high pressure pump can be given special surface treatment such as DLC to improve lubrication to prevent excessive wear or seizing. The special designed lip seal, see Fig. 4.11, should be applied to avoid any possible mutual penetration of fuel and lubricant through small clearance between the plunger and barrel in high pressure pump.

It would be a desired approach to develop a dedicated common rail system for methanol fuel with all modifications related to material compatibility, flow capacity increasing, and leakage prevention and lubrication enhancement.

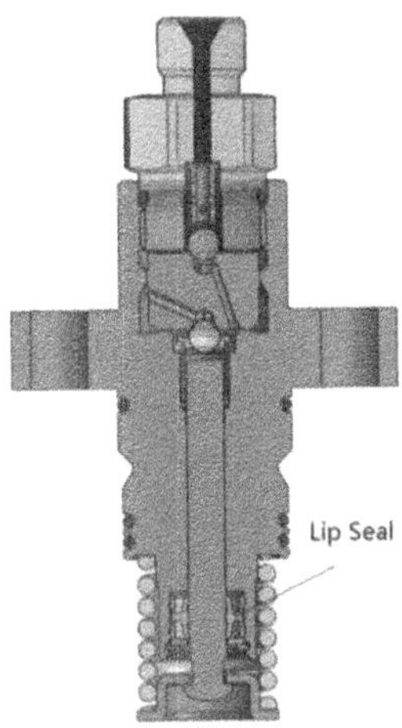

Fig. 4.11 Lip seal to prevent mutual leakage of fuel and lubricant

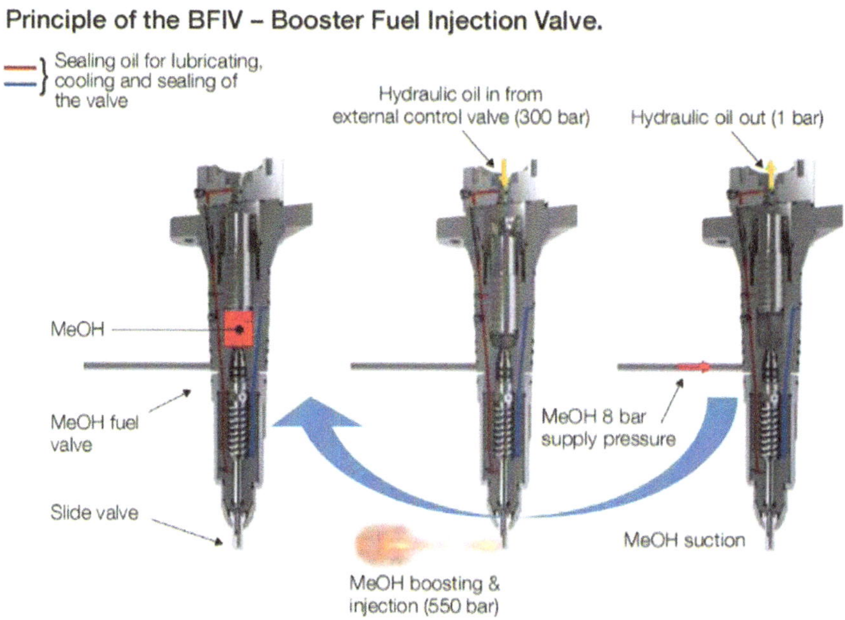

Fig. 4.12 MAN booster fuel injector for methanol direct injection (Using methanol fuel in the MAN B&M ME-LGI Series)

For large bore marine engine, the methanol direct injection was desired and more efforts were made to develop the dedicated fuel delivery system. Figure 4.12 illustrates a booster fuel injector for methanol direct injection (Baranescu et al. 1988). As shown, the methanol compatible booster injector, under the action of hydraulic oil of 300 bars, will generate a maximum injection pressure up to 550 bars. The special design features a dedicated oil circuit that plays the role of cooling, sealing and lubricating. Firstly, cooling keep methanol from vaporization; secondly, sealing prevent methanol from leak; and finally, the lubricating avoids possible excessive wear of needle valve due to poor lubricity of methanol.

4.5.2 Biodiesel

Biodiesel is produced from plant oils, animal fats, and recycled cooking oils, and has the advantages of being renewable, non-toxic, and biodegradable. Biodiesel is an ideal alternative fuel for compression ignition engines as it has similar cetane number, slightly lower energy density, and slightly higher viscosity as conventional diesel fuels.

In practice, biodiesel is most commonly used as a blend with conventional fuels. As reported (National Renewable Energy Laboratory), at concentration of up to

5% by volume (B5) in conventional diesel fuel, the mixture will meet the ASTM D975 standard and can be used in any application as if it were pure petroleum diesel. B20 is the most commonly used biodiesel blend in the United States because it provides a good balance in many criteria including material compatibility, cold weather operability, performance, emissions, and costs.

A blend with a higher percentage of biodiesel may require some physical modifications to the fuel delivery system. Non-metal material compatibility is the primary concern. It was reported that B100 will degrade, soften, or seep through some hoses, gaskets, seals, elastomers, glues, and plastics with prolonged exposure. Nitrile rubber compounds, polypropylene, polyvinyl, and Tygon are particularly vulnerable to B100. Such materials should be replaced with materials as Teflon, Viton, fluorinated plastics, and nylon. Consultation with B100 suppliers to obtain detailed information on materials compatibility is recommended for manufacturers.

Other negative effects of biodiesel fuel on fuel delivery system are the deposit formed on fuel injector tip (Using methanol fuel in the MAN B&M ME-LGI Series) and fuel filter clogging (Hoang and Le 2018). Pure biodiesel or the blends having high biodiesel content tend to form more deposit. Although using B5 blend, the test results on four test engines equipped with different fuel delivery systems showed the loss of fuel flow through injector nozzle due to nozzle hole coking, and up to 6% engine power drop were recorded. It was considered that some properties of biodiesel including oxygen presence and incomplete combustion were the main cause to affect the deposit formation. It was also found that small conicity factor can restrict the formation of deposit. Nevertheless, the blend having low content of biodiesel is recommended for better engine durability.

4.5.3 DME

Among alternative fuel for CI engine, DME has the appearance of an excellent, efficient one, with almost smoke-free combustion. This is not only because of its low auto-ignition temperature and its almost instantaneous vaporization when injected into the cylinder, but also because of its high oxygen content (around 35% by mass) and the absence of C–C bonds in molecular structure.

However, the fuel delivery system should have some special features due to other properties of DME that are different from those conventional diesel fuel, they can be summarizes as follow (Arcoumanis et al. 2008),

- Closed pressurized system—Low boiling point of DME necessitates this feature. The vapor pressure of DME, roughly the same as LPG, demands the same kind of handling and storage consideration as for LPG; moreover, the feed pressure from the storage tank to the supply pump should be kept between 1.2 to 3 MPa to prevent possible cavitation as increased temperature may cause a higher saturated vapor pressure.

- Reduced injection pressure—It is no longer necessary to adopt very high injection pressure (> 100 MPa) as in conventional diesel engine because DME gasifies immediately during injection due to low boiling point. Usually, the injection pressure of 20–30 MPa is sufficient.
- Leakage prevention—Low viscosity of DME will cause possible leakage problem that takes place at small clearance between the plunger and barrel of injection pump. This problem becomes more serious on heavy duty engine. In order to prevent the leakage, the seal made of Teflon coated O-ring and PTFE based high tension sealing should be applied at leakage location, as mentioned in methanol fuel section.
- Improving lubrication—Lower lubricity of DME than that of diesel fuel leads to wear problems. The lubricity enhancing additive can be added into DME, such as Lubrizol. Another solution might be to apply special surface coating at plunger to improve lubrication. It is expected that the experience gained in the past with the use of methanol fuel in fuel delivery system for CI engine.
- Material compatibility—DME is not compatible with most elastomers and can chemically attack some commonly used sealing materials and other plastic components, raising reliability issues. A careful selection of sealing materials is necessary to prevent deterioration after prolonged exposure to DME. Sealing of DME filled storage tank and supply lines can be achieved with PTFE.

Although conventional pump-line-nozzle fuel delivery system can be modified to adapt the requirement of DME fuel, a purpose—built common rail system is probably the best solution for a dedicated DME engine. As mentioned in previous section, the common rail system can provide more accurate and flexible injection control and can be easily adjusted to fit the requirement of DME fuel. It will become an ideal fuel delivery system for DME when combining with anti-leaking and anti-seizing measures.

4.5.4 F-T Diesel Fuel

Fischer and Tropsch described a process of producing synthetic oils from CO and H_2 in 1923. When this process is applied to renewable resources such as biomass, the synthetic oil product becomes an intriguing biofuel. Based on the specific H_2 and CO mixture percentage and the catalyst used, the reaction can produce gasoline-like or diesel-like fuels, the latter called a Fischer Tropsch (F–T) diesel fuel.

Several chemical and physical characteristics of F–T fuels, including reduced density, ultra-low sulfur levels, low aromatic content, and high cetane number (Sudeep-Kumar and Rajashekhar 2018), make it an attractive alternative fuel; F–T fuel also has reduced emissions benefits.

There was a concern that F-T fuel might have the negative impact on the elastomer components of fuel delivery system due to the lack of aromatic content in F-T fuel, a research (McMillian and Gautam 2001) including Lab scale tests and field trials was conducted to identify possible problem. However, no negative effects related to F-T fuel were found.

F–T fuel is used as a blend with biodiesel or conventional diesel fuel to improve fuel properties in most cases due to its availability and/or competitive prices. F-T fuel is not known to have special materials compatibility requirements compared to conventional fuel delivery systems for CI engines.

4.5.5 *Coal—Water—Slurry*

Coal-water-slurry (CWS) with very small diameter of pulverized coal was proven as a feasible fuel for nonroad large bore diesel engines due to its very low cost and high enough caloric value comparing with some other alternative fuel (Masko et al. 2012).

Since early 1980s, some R&D works on CWS as engine fuel had been conducted in USA and other countries.

In order to identify problems with the CWS ignition and combustion at high engine speeds and assess the effectiveness of possible design solutions, the coal combustion model was used to simulate a DDC 149 diesel engine. The modeling results did provide some guidelines to determine the parameters of both engine and CWS, such as compression ratio, injection timing, injection duration and coal droplet size. The following engine dynamometer tests verified the feasibility of CWS as engine fuel (Kakwani et al. 1990).

Reliability of fuel delivery system was found as the primary challenge of those CWS engine research projects. The problems included sticking of moving parts, packing of coal particles in tight clearances, and plugging of flow passages, most of which associated with loss of water from the slurry as it passed through the injection system components and came into contact with tight clearance.

Through these researches, a special designed fuel delivery system that has evolved as the best approach to the reliability problems involves the use of a free shuttle piston to separate a clean pumping fluid from CWS. In this approach, a standard plunger and barrel assembly is used to meter and pressurize a clean fluid such as diesel fuel. This high pressure fluid acts on one side of a free piston. The CWS is supplied through a check valve to the other side of the free piston where it is pressurized and directed through an injector nozzle. Oil is supplied to both the guide area of the needle valve in the nozzle and to the center of the free shuttle piston. The oil is supplied at pressures exceeding the peak injection pressure so that it acts as both a lubricant and a sealing fluid. As shown in Fig. 4.13, it is a significant different design from conventional system.

Another technical challenge related to fuel delivery system was the excessive wearing rate of the spray holes through which high speed CWS flows. This problem can be solved by selection of the appropriate material, such as sapphire and diamond.

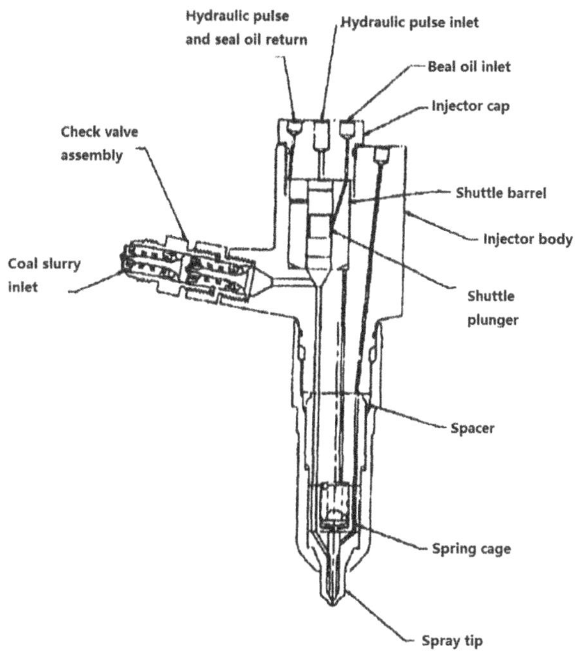

Fig. 4.13 Schematic diagram of injector for coal-water-slurry fuel (Kakwani et al. 1990)

4.6 Fuel Delivery System for Dual Fuel Engine to Adapt Alternative Fuel

Natural gas and methanol are two valuable low-carbon fuels. As having the physical and chemical properties similar to those of gasoline, they have been applied as a pure fuel or a blend agent for passenger vehicles for long time.

For heavy duty engines that usually are used as the powerplant of commercial truck, marine ship, mining machinery, and large genset, a so-called duel fuel operation mode was introduced in order to use these low carbon fuels more efficiently. Two direct advantages over throttle control spark ignition mode include improved fuel consumption and increased engine torques. In recent year, facing the great pressure to reduce greenhouse gas emission, researchers not only pay more attentions to low carbon fuels, but also explore some new carbon free fuels, such as ammonia. They find that dual fuel operation mode can be a feasible approach to use these new fuels as well.

The definition for dual fuel engine in ref. (Weaver and Turner 1994) was as follows: A dual–fuel engine is an internal combustion engine in which the primary fuel (usually natural gas) is mixed more or less homogeneously with the air in the cylinder, as in spark-ignition engine. Unlike a spark-ignition engine, however, the air/fuel mixture is ignited by injecting a small amount of diesel fuel (the pilot) as the piston approaches the top of the compression stroke. This diesel pilot fuel rapidly undergoes pre-flame reactions and ignites due to the heat of compression, just as it would in a diesel engine. The combustion of the diesel pilot then ignites the air–fuel mixture in the rest of the cylinder.

The fuel delivery system for dual fuel engines consists of two parts, one is for diesel pilot injection, and another is port injection of alternative fuel. While pilot injection only represents around 10% of the total fuel supplied at full load, the engine power output is controlled by adjusting the amount of port injection. Some research results indicate that a flexible and precise pilot injection control for timing, quantity and pressure, will improve combustion process and reduce emission. For this reason, common rail system should be required for pilot injection for dual fuel engines.

4.6.1 Natural Gas

The majority of dual fuel engines to date took natural gas as the primary fuel. The typical fuel delivery system consists of a high-pressure common rail system for diesel pilot injection and a low-pressure port injection system for gas injection. Such a configuration has the advantage of its simple adaptation, but the main disadvantage is that good diesel fuel substitution levels are only obtained a mid-load. This is because at low load, the combustion of lean pre-mixed fuel becomes instable, while at high load, the rich pre-mixed fuel tends to knock.

It has been found that high pilot injection pressure is required to raise combustion rate at light load. At some large bore marine engines that may operate in either dual fuel mode or normal diesel mode, the special injector designs were adopted to meet the requirements of both pilot diesel injection mode and normal diesel injection mode. Figure 4.14 shows an injector design from Wärtsilä (Kakwani et al. 1990), in which two fuel injection needle valves were integrated into an injector body. Among two

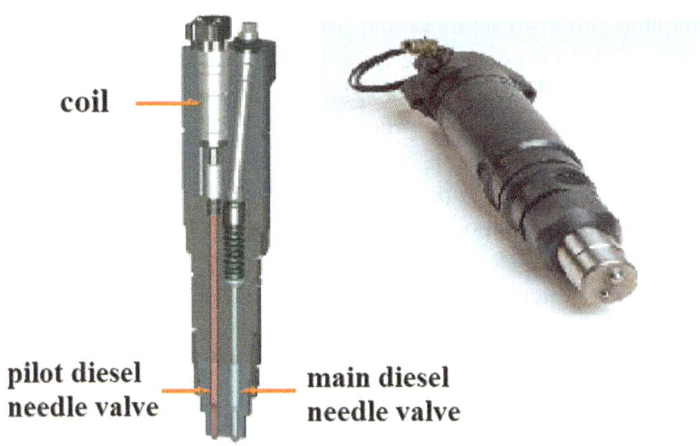

Fig. 4.14 Structural diagram of Wärtsilä integrated injector (Dvornik and Dvornik 2014)

fuel injection needle valves, one controls diesel pilot injection and another controls normal diesel injection. In this way, high pilot injection pressure can be achieved and the engine performance at lean pre-mixture condition is improved.

MAN adopted another fuel delivery system configuration to meet the requirements of both pilot diesel injection mode and normal diesel injection mode. As shown in Fig. 4.14 (Mitianiec 2018), two pilot injectors with built-in pre-combustion and one main injector were located on engine cylinder head. With such a design, the fuel injected by the pilot injector is ignited in the built-in prechamber, and then the flame jet flows into the main combustion chamber and promote the combustion process at light load condition.

A more valuable concept to improve dual fuel operation mode is high pressure direct injection (HPDI) (Kakwani et al. 1990). The natural gas is directly injected into cylinder shortly before the end of the compression stroke. This technique provides better fuel economy and more efficient of an equivalently-sized conventional diesel engine, however, it requires the development of special high-pressure gaseous injector (Fig. 4.16).

A special state-of-the-art HPDI multi-fuel injector was invented by Westport (Using methanol fuel in the MAN B&M ME-LGI Series). As shown in Fig. 4.15, this injector, called concentric needle multi-fuel injector, was used to replace the conventional diesel injector. It directly injects both the diesel pilot and the natural gas into the engine cylinder. This injector features two solenoid-controlled valves to precisely control the injection timing and the metering of each fuel to the combustion

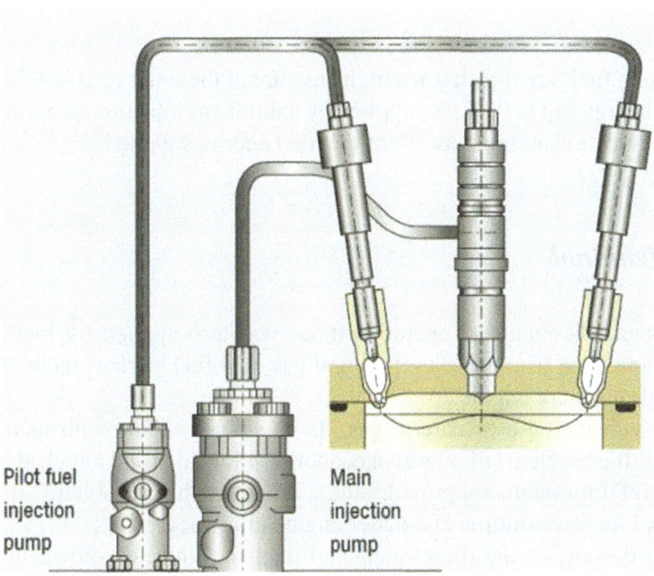

Fig. 4.15 MAN DF fuel injection system (L+V32/40DF-Environmentally Friendly and economical)

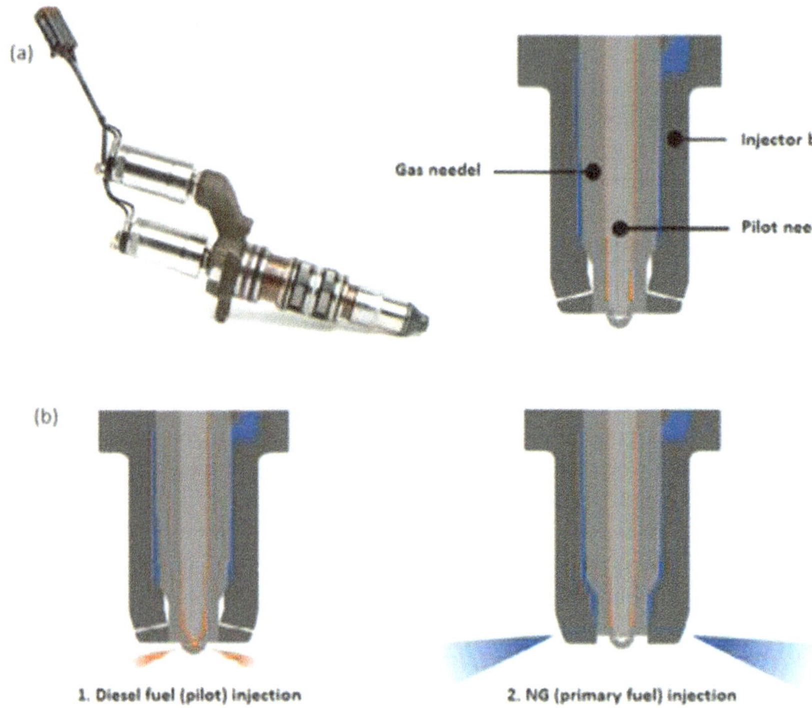

Fig. 4.16 Concentric—needle multi-fuel injector by Westport (Westport HPDI 2.0)

chamber. Both fuels are injected at a high pressure of the order of 20–30 MPa, where the bulk of energy (up to 95%) is supplied by natural gas injection, preceded by pilot diesel pilot injection as small as 5% of the total energy supplied.

4.6.2 *Methanol*

In recent years, the dual fuel operation mode was also applied for methanol. The experience obtained from the diesel-natural gas dual fuel engine can be transferred to the diesel-methanol engine.

Similar with the case of natural gas, the dual fuel engine with methanol port injection has the problems of poor fuel economy at low load and knock at high load.. However, HPDI for methanol provides the solution for these problems, allowing for higher diesel fuel substitution and better engine efficiency.

The two designs of the diesel-methanol dual-fuel injector shown in Fig. 4.17 (Dvornik and Dvornik 2014) are inspired by the diesel-natural gas dual-fuel injector investigated by Westport Research Institute and Columbia University, respectively. The first design in Fig. 4.17a is a concentric needle multi-fuel injector, which has

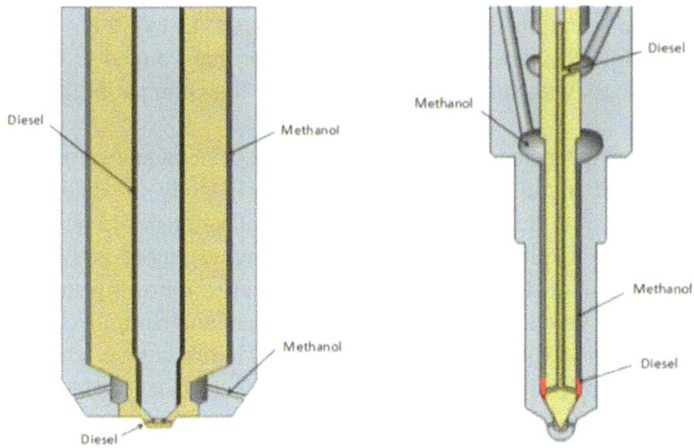

(a) Dual-channel (b) Single – channel

Fig. 4.17 Two designs of injector nozzle for diesel-methanol dual fuel engine (https://wfsinc.com/our-solutions/hpdi-2.0; Li and Wang (2021))

two fuel channels with the inner channel for diesel fuel and the outer channel for methanol fuel. The second design in Fig. 4.17b is a modification on the conventional diesel injector, in which the original fuel channel is for methanol fuel and a new passage located inside the needle is for diesel pilot fuel; the diesel pilot injection and the methanol main injection are controlled by two different solenoid valves and a small amount of pilot diesel is first stored in the methanol channel to achieve the separate injection of methanol and diesel.

4.6.3 Ammonia

The ability of ammonia to act as a hydrogen carrier, without the drawbacks of hydrogen gas—storage costs and low stability—renders it a potential solution to the decarburization of transport.

However, there are some challenges for engines to use ammonia as fuel, because ammonia has very high auto-ignition temperature (651 °C), lower flame temperature, and narrow flammability limits (16–25% by volume in air). Ammonia is also corrosive to copper, copper alloys, nickel and plastics so that these materials must be avoided in an ammonia fuel engine.

Some studies found, it is not feasible to use the pure ammonia for SI engine, but blending ammonia with gasoline could assist in the combustion quality. On the other hand, it is almost impossible to realize a controlled auto-ignition process with pure ammonia, only dual fuel operation mode can make ammonia be combusted successfully.

It should be mentioned that dual fuel engines tested with ammonia and (bio)diesel/DME are at research level, which is primary aimed to obtain a stable combustion process with such fuel systems rather than compatibility with existing engine designs.

Kong (https://wfsinc.com/our-solutions/hpdi-2.0) conducted several diesel-ammonia dual fuel engine investigations. At an early study, the dual fuel operation mode was tried by introducing ammonia into the intake manifold on a conventional diesel engine. The results indicated the stable engine output and high ammonia combustion efficiency when using 50% ammonia. At a following study, a different technical approach was tested, that was, ammonia and another high cetane fuel were mixed in separating facility and then were provided to test engine, a highly modified common rail system with a GDI injector, were used to inject the blend into engine cylinder (Fig. 4.18).

It should be noted that ammonia due to its low flame speed is more suitable for lower engine speeds. MAN regards ammonia as a potential long-term fuel for two-stroke marine engines and adopts high-pressure liquid ammonia injection (Veltman and Kong 2009). The liquid ammonia low-flashpoint fuel supply system (LFSS) follows the MAN B&W ME LGIP fuel supply system and the key components have been replaced to ensure the system's corrosion resistance. WinGD's Dual Fuel ammonia engine uses the low pressure gas injection system (Fig. 4.19).

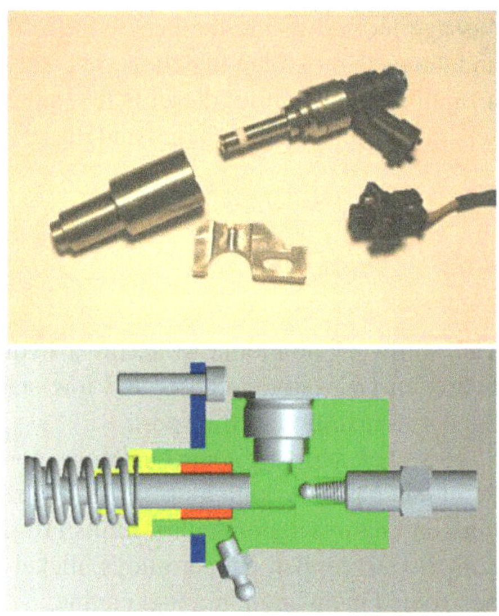

Fig. 4.18 The main components of fuel delivery system from Iowa State University (Veltman and Kong 2009)

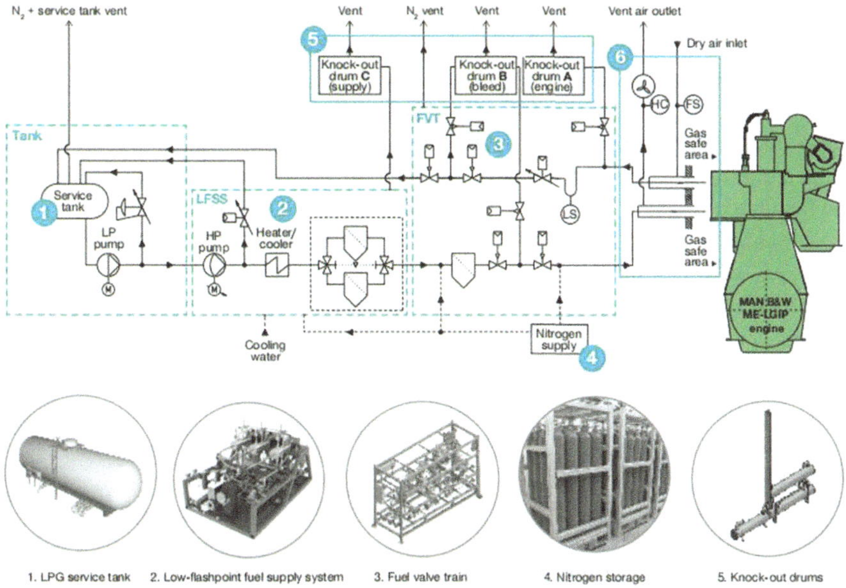

Fig. 4.19 MAN B&W ME-LGIP engine modified for ammonia supply system (MAN B&W two-stroke engine operating on ammonia)

4.7 Summary

Fuel delivery system plays a key role for alternative fuel engines just as it does for conventional engines. A minor modification or a major modification or a brand new design might be required comparing with conventional fuel delivery system depending on the physical and chemical properties of alternative fuels, and the combustion mode in which alternative fuel is used. To adapt common E10 bending fuel available in most gas stations in North America, fuel delivery system in SI engines does not need any change. On the other hand, fuel delivery system should be modified in term of material compatibility when using pure methanol on SI engine. SI engines fueled by natural gas, including CNG and LNG, have become popular powerplants for long time, and the specially designed fuel injector and other major components of fuel delivery system has been well developed. Hydrogen can be delivered to SI engines in a similar fashion as CNG, but major challenge is ultra-high pressure required for hydrogen on-board storage, which may result in a special material failure called as hydrogen embrittlement. In CI engine case, bio-diesel fuel requires some minor modifications on fuel delivery system due to material compatibility issue. However, methanol fuel necessitates a series of major modifications on fuel delivery system to prevents possible oil leaking into fuel, parts wearing, and plunger scuffing besides material compatibility. Similarly, fuel delivery system for DME needs most modifications adopted for methanol, moreover, a closed pressurized condition should be established to allow DME to assume liquid state. As dual fuel combustion mode provides more promising future for some major low carbon fuel, such as natural gas, methanol and ammonia, the great efforts have been made to develop new generation fuel delivery system. A typical design is the concentric needle multi-fuel injector, for natural gas + diesel fuel application, it is reported that improved emission and higher diesel substitution rate. Another new design concept is established on one conventional common rail injector for dual fuel injection. That is, the diesel pilot injection and the methanol main injection are controlled by two different solenoid valves and a small amount of pilot diesel is first stored in the methanol channel to achieve the separate injection of methanol and diesel. It is expected that more innovative design of fuel delivery system will promote application of low carbon fuel.

References

Alrazen HA, Ahmad KA (2018) HCNG fueled spark-ignition(SI) engine with its effectson performance and emissions. Renew Sustain Energy Rev, Feb 2018. https://www.researchgate.net/publication/322863066

Arcoumanis C, Bae C, Crookes R, Kinoshita E (2008) The potential of di-methyl ether (DME) as an alternative fuel for compression—ignition engines: a review. Fuel 87(7):1014–1030

Automotive Handbook (2004) Bosch

Baranescu R et al (1988) Conversion of a Navistar DT466 diesel engine to methanol operation. The 8th ISAF

Darade P, Dalu R (2012) Performance and emission of internal combustion engine fueled with CNG—a review. Int J Eng Innov Res 1(5)

Dual-fuel upgrade. MAN Diesel & Turbo

Dvornik J, Dvornik S (2014) Dual-fuel-electric propulsion machinery concept on LNG carriers. Trans Marit Sci. https://doi.org/10.7225/toms.v03.n02.005

Engineering the future two-stroke green-ammonia engine. MAN energy solutions

Fan LY, Long WQ, Zhu YX, Xue YY (2008) A characteristic study of electronic in-line pump system for diesel engines. SAE Techn Papers. https://doi.org/10.4271/2008-01-0943

Fan LY, Zhu YX, Ma XZ, Tian BQ, Song EZ, Li WH (2010) Quantitative analysis on cycle fuel injection quantity fluctuation of diesel engine electronic in-line pump system. SAE Techn Papers. https://doi.org/10.4271/2010-01-0875

Gas engine management, HEINZMANN Product Catalogue. https://www.heinzmann.com/en/gas-engines

Heywood JB (1988) Internal combustion engine fundamentals, TJ755.H45

Hoang AT, Le AT (2018) A review on deposit formation in the injector of diesel engines running on biodiesel. Energy Sour Part A: Recovery, Utilization Environ Eff 41:5, 584–599. https://doi.org/10.1080/15567036.2018.1520342

Kakwani M, Winsor R, Ryan T, Schwalb J, Wahiduzzaman S, Wilson R (1990) Coal—fueled high speed diesel engine development. Final report, Sept 28 1990–Nov 30 1993, DOE/MC/27222-3624

Li ZY, Wang Y, Yin ZB, Gao ZB, Wang YJ, Zhen XD (2021) Parametric study of a single-channel diesel/methanol dual-fuel injector on a diesel engine fueled with directly injected methanol and pilot diesel. Fuel 302:15. https://doi.org/10.1016/j.fuel.2021.121156

Mariani A, Morrone B, Unich A (2012) A review of hydrogen-natural gas blend fuels in internal combustion engines. Fossil Fuel Environ. https://www.researchgate.net/pulication/224624319

Masko M, Sakamoto YS, Showa NO, Uchida N, Kitano K, Sakata I (2012) Effect of Fischer-Tropsch diesel on fuel supply system. SAE 2011-01-1950

McMillian M, Gautam M (2001) Combustion and emission characteristics of Fischer–Tropsch and standard diesel fuel in a single-cylinder diesel engine. SAE Int Fall Fuels Lubricants Meet Exhib. https://doi.org/10.4271/2001-01-3517

Miller S, Savonen C (1990) Development status of the Detroit diesel corporation methanol engine. SAE Techn Paper 901564. https://doi.org/10.4271/901564

Mitianiec W (2018) Combustion process of coal-water mixture in heavy duty diesel engines for power engineering. In: 5th international conference on contemporary problems of thermal engineering. https://doi.org/10.4271/2001-01-3517

Morch CC, Bjerre A, Gottrup MP, Sorenson SC, Schramm J (2011) Ammonia/hydrogen mixtures in an SI engine: engine performance and analysis of a proposed fuel system, 2011. Fuel 90(2):854–864. https://doi.org/10.1016/j.fuel.2010.09.042

Mounaïm-Rousselle C, Brequigny P (2020) Ammonia as fuel for low-carbon spark-ignition engines of tomorrow's passenger cars. Univ. Orléans, INSA-CVL, PRISME, EA 4229, Orléans, France. https://doi.org/10.3389/fmech.2020.00070

National Renewable Energy Laboratory. Biodiesel handling and use guide, 4th edn. https://www.nrel.gov/

Reif K (2014a) Gasoline engine management system and component

Reif K (2014b) Diesel engine management system and component. https://doi.org/10.1007/978-3-658-03981-3

SOGAV™ Solenoid Operated Gas Admission Valve. Product Manual 26114 (Revision J) Original Instructions

Sudeep-Kumar KS, Rajashekhar CR (2018) Impact of bio-diesel fuel on durability of CI engines—a review. IOP Conf Ser Mater Sci Eng 376:012011. https://doi.org/10.1088/1757-899X/376/1/0120

Swain MR, Yusuf MJ, Dulger Z, Swain MN (1993) The effects of hydrogen addition on natural gas engine operation. SAE paper 932775

Torres FA, Doustdar O, Herreros JM, Li RZ., Poku R, Tsolakis A, Martins J, Vieira de Melo S (2021) A comparative study of biofuels and Fischer—Tropsch diesel blend on the engine combustion performance for reducing exhaust gaseous and particulate emissions. Energies 14:1538. https://doi.org/10.3390/en14061538

Using methanol fuel in the MAN B&M ME-LGI Series. MAN Diesel & Turbo. https://www.man-es.com/

Varde KS, Frame GM (1984) A study of combustion and engine performance using electronic hydrogen fuel injection. Int J Hydrogen Energy 9(4):327–332. https://doi.org/10.1016/0360-3199(84)90085-5

Veltman M, Kong S (2009) Developing fuel injection strategies for using ammonia in direct injection diesel engines. Lowa State University

Verhelst S, Turner J, Sileghem L, Vancoillie J (2019) Methanol as fuel for internal combustion engines. Prog Energy Combust Sci 70:43–88. https://doi.org/10.1016/j.pecs.2018.10.001

Weaver C, Turner S (1994) Dual fuel natural gas/diesel engines: technology, performance and emissions. SAE Techn Paper 940548. https://doi.org/10.1007/978-3-658-03981-3

Westport updates HPDI 2.0 dual fuel system with new Delphi injectors, upgraded LNG storage and supply. https://www.greencarcongress.com/2014/10/20141001-hpdi.html

Westport HPDI 2.0. https://wfsinc.com/our-solutions/hpdi-2.0

Yip H, Srna A, Yuen A, Kook S, Taylor R, Yeoh G, Medwell P, Chan Q (2019) A review of hydrogen direct injection for internal combustion engine: towards carbon–free combustion. Appl Sci 9:4842. https://doi.org/10.3390/app9224842

Yutaka Y et al (1991) Improved performance and durability of a heavy-duty methanol DI diesel engine. The 9th ISAF

Part III
Alcohol as a Fuel/Additive

Chapter 5
Alcohols as Alternative Fuels for Transport

Byunghchul Choi

Abstract Owing to the increasing global demand for energy, the consumption of fossil fuels has increased; however, the utilization of such fuels negatively affects the environment. This has also led to major issues, including global warming, climate change, rising sea levels, and the depletion of carbon-based fuels. Petroleum-based energy supplements are mainly responsible for these problems. More specifically, environmental concerns have peaked after the critical CO_2 level was exceeded. Thus, now more than ever, it is imperative to increase the percentage of clean and renewable energy sources used to address energy demands. In this regard, this study intends to reveal the performance, combustion, and emission characteristics of various means of transportation using alcohols, such as methanol, ethanol, propanol, and butanol, as alternative fuels. The primary means of transportation include passenger cars, commercial vehicles, ships, locomotives, non-road vehicles, and alternative powering systems employing fuel cells supplied with alcohols. The effects of using alcohols as alternative fuels and carbon neutral characteristics, combustion characteristics, and emissions such as NO_x, CO, HC, particulate matter, and CO_2 were also investigated.

Keywords Alcohol fuels · Transport · Methanol · Ethanol · Propanol · Butanol

5.1 Introduction

Carbon neutrality (or net-zero emissions) refers to all greenhouse gas (GHG) emissions released by fuel combustion being counterbalanced by removing GHGs from the atmosphere via a process known as carbon removal. Since the Paris Agreement in 2016, many countries have joined the 'Carbon Neutrality by 2050 for Climate Alliance Emission Targets'; this has become a global area of concern (The Government of the Republic of Korea 2020). In addition, as awareness of the seriousness of climate change is increasing due to the COVID-19 outbreak in 2020, the

B. Choi (✉)
School of Mechanical Engineering, Chonnam National University, Gwangju 61186, Republic of Korea
e-mail: bcchoi@jnu.ac.kr

A. V. Agarwal and H. Valera (eds.), *Potential and Challenges of Low Carbon Fuels for Sustainable Transport*, Energy, Environment, and Sustainability,
https://doi.org/10.1007/978-981-16-8414-2_5

declaration toward achieving carbon neutrality in major countries is becoming more common. There exists a wide consensus that the best-positioned fuels to achieve carbon neutrality in the transportation sector are methanol, ethanol, biomethanol, biomethane, hydrogen, natural gas, and ammonia.

After the first and second oil shocks, the use of alcohol fuels for automobiles became widespread in the general automobile industry. Owing to the recent concerns regarding the emission of GHGs such as CO_2 in the automobile as well as off-road industry, the use of bio-alcohol fuels is gradually being considered. Many countries are presenting the 2050 carbon neutral strategy, including the transportation sector. Expanding the utilization of alcohol fuels in the transportation sector will be an important strategy to adhere to the carbon neutral policy. Alcohol fuels have also certain advantages as alternative fuels in terms of their performance, combustion characteristics, and emissions (such as NO_x, CO, hydrocarbons (HC), particulate matter (PM), and particularly CO_2).

About the upcoming regulations from the International Maritime Organization (IMO) on CO_2 reductions (Van et al. 2019), it is incumbent on ship-owners to identify and employ pathways for a clean energy future. Given the major investment required for purchasing a new merchant ship with a working life of 20–25 years, it is imperative for the shipping industry to implement long-term, innovative, cost-effective, future-proof, and environmentally friendly solutions. Methanol offers many advantages when used as a marine fuel; these include its clean burning properties, cost-effectiveness, and simpler storage and handling, as compared to other alternative fuels. Methanol is a widely available and future-proof marine fuel that can be adapted to existing ships and engine technologies at a lower adaptation cost.

Alcohols have high knock resistance in the automobile engine, which can be attributed to the absence of negative temperature coefficient behaviors. Methanol was first used in the aviation industry in the 1940s for military aircraft, in order to realize the additional thrust that was required for full payloads (Verhelst et al. 2019). In the aviation field, methanol was used owing to its benefits in terms of the octane number and latent heat; however, it was used only during take-off and to a certain extent when maximum power was required. The widespread use of pure methanol in the United States Auto Club (USAC) Indy car competition commenced in 1965. Methanol was used by the Championship Auto Racing Team circuit throughout its campaign (1979–2007) (Verhelst et al. 2019). The performance-enhancing attributes of both ethanol and methanol were realized early in motor racing. Blends containing methanol and benzene were often used in Grand Prix cars, especially after supercharging was developed for the maximum performance. Meanwhile, a change from employing internal combustion engine power systems to using fuel cells as the main powertrain or as auxiliary power systems is also being considered. Most fuel cells use hydrogen; however, some alcohol fuels are also used.

As shown in Fig. 5.1, although the net gravimetric energy densities of ethanol and particularly methanol are significantly lower than those of diesel and gasoline, the values are considerably higher than those of electricity and hydrogen, even when higher tank-to-wheel efficiencies of battery electric vehicles and proton exchange

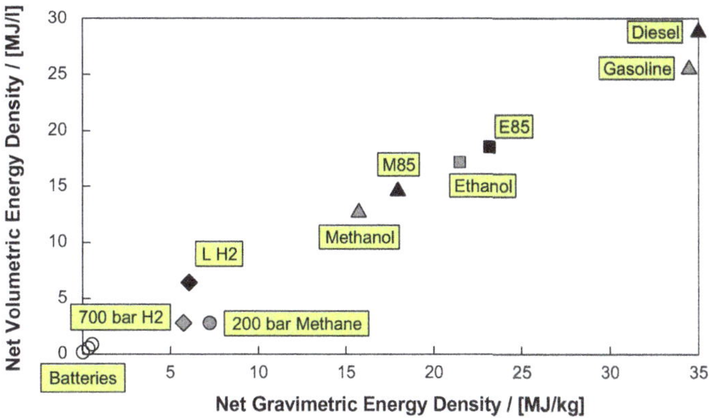

Fig. 5.1 Net system volumetric and gravimetric energy densities for various on-board energy carriers (based on lower heating values) (Turner et al. 2013)

membrane fuel cell vehicles are included (Turner et al. 2013). The energy density of alcohol fuels lies between that of conventional petroleum fuels and hydrogen energy.

This review aims to disseminate knowledge regarding the key role of various alcohol fuels in enabling a secure, eco-friendly, and affordable energy future in the transportation sector. First, the physical, chemical properties, and economic status of alcohol fuels are reviewed. Thereafter, the history of alcohol use in various means of transportation, the characteristics and advantages of this use, and the current status and related problems are reviewed. It should be noted that the respective methods of producing alcohol fuels have been excluded, because these are beyond the scope of the current study.

5.2 Characteristics of Alcohol Fuels

5.2.1 Physical and Chemical Properties

Table 5.1 lists the physical properties, while Table 5.2 lists the chemical properties of alcohol fuels. Gasoline, iso-octane, and diesel are also included as primary reference fuels. Methanol and ethanol have a higher heat of vaporization; thus, they provide longer ignition delay during combustion compared to other fuels. Methanol and ethanol have a lower heating value (LHV) because of a low carbon number and high ratio of OH radical. Propanol and butanol have a higher flash point compared to methanol and ethanol; therefore, they are safe for use in the transportation sector. With regard to the elemental composition, the hydrogen to carbon ratio of methanol is higher than that of gasoline and equal to that of methane. Thus, when calculating the CO_2 emissions on an energy-specific basis (CO_2 g/MJ), the specific CO_2 emission of methanol is 7% lower than that of gasoline. However, this is assuming a similar

Table 5.1 Comparison of physical properties of methanol, ethanol, propanol, 1-butanol, isooctane, gasoline, and diesel fuel (Verhelst et al. 2019; Choi 2021)

Property	Methanol	Ethanol	Propanol	1-butanol	i-octane	Gasoline	Diesel
Chemical formula	CH_3OH	C_2H_5OH	C_3H_7OH	C_4H_9OH	C_8H_{18}	C_4–C_{12}	C_8–C_{25}
Density (STP) (kg/m^3)	790	790	803	810	692	740	850
Flash point (°C)	11–12	12	15	35	−12	−43	60–80
Boiling point at 1 bar (°C)	65	79	97.1	118	99	25–215	210–235
Heat of vaporization (kJ/kg)	1100	838	727.8	585	270	180–350	180–375
Surface tension (20 °C) (mN/m)	22.1	22.3	23.7	24.57	18.6	21.6	25.8
Dynamic viscosity (20 °C) (mPa s)	0.57	1.2	1.959	2.8	0.5	0.6	1.3

conversion efficiency (brake thermal efficiency), although methanol affords increased efficiencies. The density of methanol is higher than that of gasoline, despite its low molecular mass. This is because the methanol molecule is polar owing to its OH group, resulting in hydrogen bonding (Verhelst et al. 2019). These hydrogen bonds explain the characteristics of methanol, such as high latent heat, infinite miscibility with water, and low vapor pressure. A significant disadvantage of the OH group is that it separates the hydrocarbon phases in methanol. When methanol is mixed with hydrocarbons, its low vapor pressure is replaced by significantly high vapor pressures. This results from the intramolecular forces between the pure hydrocarbon molecules and methanol, which break the hydrogen bonds of the cyclic tetramers, and the small molecules rising through the mixture and easily leaving the surface of the mixture. The oxygen content (50% of methanol's molecular mass) leads to a low (mass-based) stoichiometric air/fuel ratio (A/F; Table 5.2). This also indicates a high fraction of methanol in the stoichiometric ratio. It is important to understand the volumetric efficiency used, considering the substantial differences based on the inclusion of the fuel vapor fraction. Despite the high methanol fraction, fuel–air mixture properties such as viscosity and thermal diffusivity are similar to those of gasoline-air mixtures (Verhelst et al. 2019).

Finally, the average global warming potential (GWP) of ethanol and petroleum fuels at well to tank (WtT) and well to wheel (WtW) perimeters were compared in Table 5.3 (Carneiro et al. 2017). The GWP at the WtW perimeter of commercial ethanol is 61.6–63% lower as compared to gasoline and diesel fuel.

The polarity of alcohol fuels results in challenges concerning material compatibility, which necessitate modifications to engine fuel systems. Both metals and elastomers can be attacked by alcohols. This is true for all alcohols except methanol, which is the most aggressive agent. Light alcohols are more corrosive to both ferrous and non-ferrous metals, as compared to gasoline. The polarity of methanol and

Table 5.2 Comparison of chemical properties of methanol, ethanol, propanol, 1-butanol, isooctane, gasoline, and diesel fuel (Choi 2021; Yates et al. 2010)

Property	Methanol	Ethanol	Propanol	1-butanol	i-Octane	Gasoline	Diesel
Chemical formula	CH_3OH	C_2H_5OH	C_3H_7OH	C_4H_9OH	C_8H_{18}	C_4–C_{12}	C_8–C_{25}
Molecular weight (kg/kmol)	32.04	46.07	60.10	74.12	114.23	100–105	~200
Oxygen content by mass (%)	49.94	34.73	26.23	21.58	0	0	0
Hydrogen content by mass (%)	12.58	13.13	14.75	13.60	15.88	~14	13
Carbon content by mass (%)	37.48	52.14	59.02	33.08	84.12	83.3–82.5	87
Lower heating value (MJ/kg)	20.09	26.95	30.68	64.82	44.30	42.90	42.8
Higher heating value (MJ/kg)	22.88	29.85	33.60	36.07	47.80	48.00	45.50
Volumetric energy content (MJ/m^3)	15,871	21,291	26,902	26,795	30,656	31,746	38,600
Stoichiometric A/F (kg/kg)	6.50	6.00	10.35	11.10	15.10	14.70	12.63
Stoichiometric A/F (kmol/kmol)	7.14	9.52	21.42	28.56	59.50	54.49	71.4
Specific CO_2 emissions (g/MJ)	68.44	70.99	71.71	71.90	69.57	73.95	~75.2[a]
Specific CO_2 emissions relative to gasoline (−)	0.93	0.96	0.97	0.97	0.94	1.00	1.01[a]
Autoignition temperature (K)	738	698	623	616	690	465–743	~573
Adiabatic flame temperature (K)	2143	2193	2328	2262	2276	~2275	2397

[a]Stoichiometric condition (at $C_{12}H_{20}$)

Table 5.3 Average global warming potential (GWP) of ethanol and petroleum fuels at well to tank (WtT) and well to wheel (WtW) perimeters (Carneiro et al. 2017)

	GWP (g CO_2/MJ)	
	WtT	WtW
Ethanol (commercial)	13	34
Diesel	87.1	91.9
Gasoline	81.0	88.6

ethanol results in dry corrosion; however, this corrosion is often reinforced by ionic impurities such as the chloride ions in the fuel (Yuen et al. 2010). Owing to the hygroscopic nature of alcohols, dissolved or separated water molecules can trigger wet corrosion phenomena. When hydrous methanol is used as a fuel, a combination of three contaminants (chloride ion, acetic acid, and ethyl acetate) produces a synergistic effect in hydrous ethanol; the resulting corrosion is many times greater than that due to any single contaminant (Walker and Chance 1983).

In addition to the physical and chemical properties summarized above, other important parameters include the volume versus and mole concentration (Sileghem 2015), vapor pressure (Andersen et al. 2010), research octane number (Anderson et al. 2010), density (Turner et al. 2012; Atkins 1994), water tolerance of alcohol–gasoline blends (Sayin and Balki 2015; Awad et al. 2017), burning velocity (Sileghem et al. 2012; Hirasawa et al. 2002), autoignition delay time (Warnatz et al. 1996; Sileghem et al. 2015), and ternary blend property (Waqas et al. 2017).

5.2.2 Economic Status

One important factor in selecting alternative fuels for carbon neutrality is the price. Figure 5.2 shows global methanol pricing in key regional markets (the United States, Gulf Coast, Rotterdam, and coastal China) on a spot and contract basis. The data were obtained from Methanol Market Services Asia (https://www.methanolmsa.com/additional-mmsa-services/price-forecasts/).

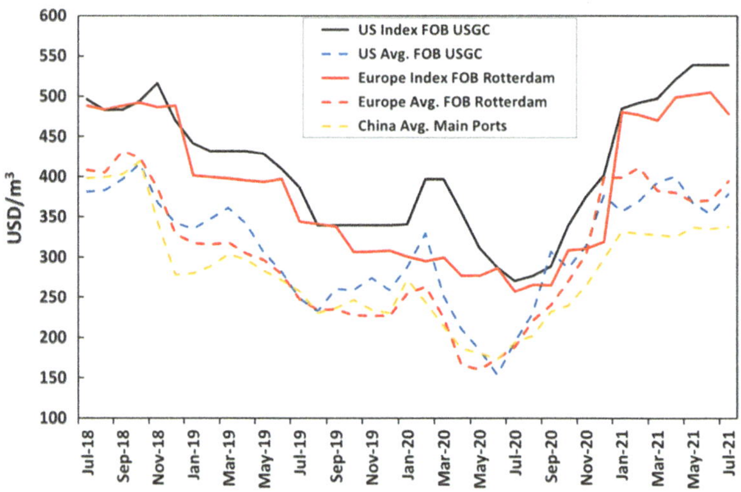

Fig. 5.2 Global methanol pricing in key regional markets (https://www.methanolmsa.com/additional-mmsa-services/price-forecasts/)

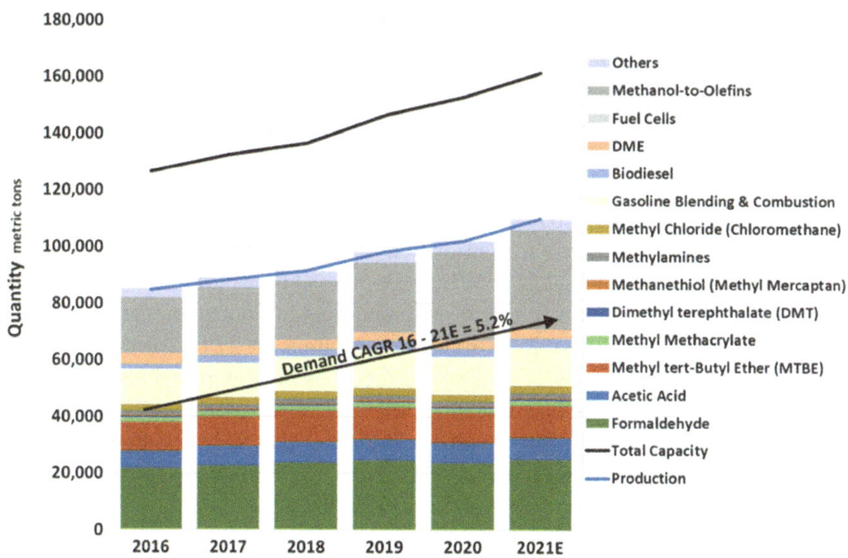

Fig. 5.3 Global methanol supply and demand balance from 2016 to 2021 (https://www.methanolmsa.com/additional-mmsa-services/price-forecasts/)

Figure 5.3 showcases the historic supply and demand of methanol from 2016 to the present (https://www.methanolmsa.com/additional-mmsa-services/price-forecasts/). The supply of methanol, a representative alcohol fuel with diverse uses, has increased by an average of 5.2% over the past five years. In particular, the use of methanol-to-olefins processing, gasoline blending, and combustion methods are increasing; novel applications also include the use of fuel cells.

5.3 Engine with Alcohol Fuels for Transports

This section presents primary examples of various alcohol fuels (methanol, ethanol, propanol, butanol) used in transportation (vehicles, ships, aviation, locomotives, non-road vehicles, and fuel cells).

5.3.1 Vehicles

Notably, many countries have established fuel economy and CO_2 regulations for light-duty vehicles and medium and heavy-duty vehicles; this is because over 20% of the global CO_2 emissions are generated by the transportation sector, especially by on-road vehicles (Oh et al. 2016). Therefore, in the field of transportation, spark ignition (SI) and compression ignition (CI) engines are required to decrease GHG emissions, which have detrimental effects on the environment and on human health. The utilization of biofuels from the alcohol family is vital for reducing these GHG emissions. Therefore, over the last couple of decades, many researchers have attempted to investigate the effects of alcohol on engine performance and exhaust emissions in SI and CI engines (Yusri et al. 2017; Zacharof et al. 2021). The reduction of CO_2 emissions from road transport is a critical priority worldwide. Emissions from heavy-duty vehicles constitute approximately 25% of the total road transport emissions, and this trend is expected to increase if appropriate measures are not implemented (Regulation (EU) 2017/2400 2018). Road transport accounts for 22.8% of the total CO_2 emissions in the United States, according to a report by the Environmental Protection Agency (EPA) (Inventory of U.S. Greenhouse Gas Emissions and Sinks: 1990–2013 2015). Therefore, a vehicle energy consumption tool that could calculate the CO_2 emissions from heavy-duty vehicles was developed. The EPA also developed a GHG emissions model (Regulation (EU) 2017/2400 2018; Inventory of U.S. Greenhouse Gas Emissions and Sinks: 1990–2013 2015). GHG emissions must be reduced by 50–80% worldwide by 2050 to avoid severe environmental problems (Leighty et al. 2012). Under these circumstances, the Korean Ministry of Environment reported a new administrative notice of CO_2 emission regulation for light-duty vehicles in 2030 (The Government of the Republic of Korea 2020; Ministry of Environment 2020). For example, it is predicted that the greenhouse warming gas (CO_2) standard in 2030 for passenger cars is 70 g/km. Yusri et al. summarized the application examples and technical characteristics of alcohol fuels with respect to automobile engines prior to 2015 (Yusri et al. 2017).

(1) Methanol (CH_3OH)

Methanol (CH_3OH) is one of the alternative fuels for transportation; this was demonstrated through fleet tests in the 1980s and 1990s. It is currently being introduced again in various locations and for different applications. It can be produced from fossil fuels (such as coal and natural gas); however, it has recently been produced from biomass and renewable energy sources through the carbon capture process. It can be used as a pure fuel or as a blend component in internal combustion engines or in direct methanol fuel cells (DMFCs).

Since the early days of the SI engine, means to extend the octane rating of fuels have been researched. This led to the early development of octane enhancers such as aniline compounds and ultimately tetraethyl lead. For the autoignition resistance of fuels (Wagner and Russum 1973; Kovarik and Charles 1994), isomerization is an important parameter. The isomerization of long-chain alkanes can hold octane numbers. Simple alcohols, including methanol and ethanol, also have high knock resistances, which can be attributed to the absence of negative temperature coefficient behaviors (Verhelst et al. 2019).

Considering the inherent characteristics of the fuel used in SI engines, which has high autoignition resistance and high heat of vaporization, methanol can be used as a pure fuel or as a blend component. In this case, the necessary modifications for operating an SI engine are not significant. Operational parameters would need to be altered considering that methanol condenses at low temperatures. A significant number of gasoline/alcohol flex-fuel engines have been reported.

Under cold starting conditions, using gasoline with high blend rates of alcohols has remained challenging. To successfully start a cold engine, a combustible fuel–air mixture is required at the spark timing, and the combusted mixture needs to generate sufficient work to keep the engine running (Pearson and Turner 2012). When 15% gasoline is blended with methanol, its ignitability is lowered from being close to that of diesel to approximately half that of gasoline (Machiele 1987).

Balki et al. reported the effect of methanol and ethanol blended fuels on the performance, emission, and combustion characteristics (Balki et al. 2014; Balki and Sayin 2014). Figure 5.4 shows that the use of alcohol fuels increased the engine torque, brake specific fuel consumption (BSFC), thermal efficiency, and combustion efficiency; furthermore, NO_x emission decreased, whereas CO_2 emission increased because of improving combustion (Balki et al. 2014).

Blaisdell et al. investigated the potential of light-duty methanol vehicles (Blaisdell et al. 1989). Engines optimized to operate on methanol offer significant fuel efficiency and performance benefits over typical gasoline engines. Using methanol instead of gasoline is estimated to potentially improve the fuel efficiency by 15–20%. The effects of ethanol–gasoline (E5 and E10) and methanol–gasoline (M5 and M10) fuel blends on the performance characteristics of an SI engine were investigated by Eyidogan et al. (2010). A vehicle with a four-cylinder, four-stroke, multi-point injection SI engine was tested at two different speeds (80 and 100 km/h) and four

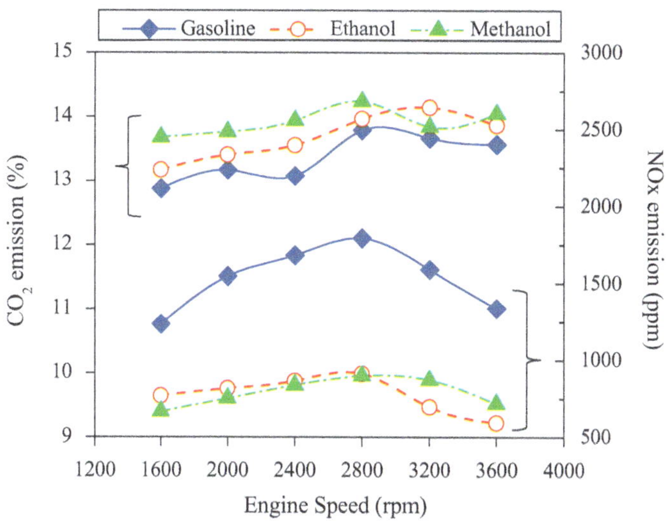

Fig. 5.4 Comparison of CO_2 and NO_x emissions for gasoline and alcohol engines (Balki et al. 2014)

different wheel powers (5–20 kW). The results obtained when using the alcohol–gasoline blends were compared with those obtained when using gasoline. The BSFC of the methanol/ethanol–gasoline blends was found to be higher.

Zhao et al. examined the effects of different mixing ratios on the emissions from passenger cars supplied with methanol/gasoline blends (Zhao et al. 2011). The effects of methanol blends, specifically M15, M20, M30, M50, M85, and M100, were evaluated over the New European Driving Cycle (NEDC). Compared to the baseline (pure gasoline), the use of M85 reduced BTEX (benzene, toluene, ethylbenzene, and p-, m-, o-xylene) emissions by 97.4%, whereas the use of M15 reduced them by 19.7%. The M15, M20, M30, and M50 fuels induced a slight increase in the level of formaldehyde emissions; by contrast, when using the fuels with high mixing ratios (M85 and M100), the emissions were three times higher than those of the baseline fuel (pure gasoline). When the vehicles were retrofitted with new three-way catalytic converters, the emissions of CO, THC, and NO_x were decreased by 24–50%, 10–35%, and 24–58%, respectively. Thus, employing new three-way catalytic converters with conventional vehicles could decrease the emissions of both formaldehyde and BTEX.

Elfasakhany (2015) investigated the engine performance and exhaust emissions of an SI engine fueled with blended fuels. The CO and HC levels were significantly decreased when using with the ethanol–methanol–gasoline blended fuels compared to the pure gasoline. Methanol–gasoline blends afforded the lowest levels of CO and HC among all the tested fuels. CO and HC emissions decreased, however CO_2 emission increased with the blending ratios of ethanol and methanol because combustion was improved. Finally, the methanol–gasoline blends afforded the highest engine performances and low HC, CO emissions.

Balki and Sayin (2014) reported the effect of the compression ratio (CR) on the performance of an SI engine and the exhaust emissions when using pure ethanol, methanol, and gasoline. Experiments were conducted under four different CRs of 8.0–9.5 with a wide-open throttle and the original ignition timing. Under all the CRs, the brake mean effective pressure, brake thermal efficiency (BTE), and BSFC obtained using ethanol and methanol were generally higher than those when using pure gasoline. Three kinds blended fuels—M10, M20, and M85—were used (Yanju et al. 2008). It was shown that the brake torque decreases when using M10 and M20, as compared to that when using pure gasoline; however, it increases when using M85. The BTEs for M10, M20, and M85 are 6.7%, 8.9%, and 10.9% higher than that for pure gasoline, respectively. Moreover, on increasing the methanol ratio, CO decreased by 25% for M85. Due to the increase in flame propagation speed, the combustion temperature increased, resulting in an increase of NO_x. Although increasing the methanol ratio induces higher HC emissions, a noticeable reduction in HC was observed when using M85.

Bilgin et al. researched the effects of methanol blends and pure gasoline on the performance of an SI engine (Bilgin and Sezer 2008). The fuel blends were used by mixing 5–20 vol% of methanol with a pure gasoline. The results shown that the M5 blend yielded the best the brake mean effective pressure, while the M20 suggested the best performance in terms of brake thermal efficiency.

Hong et al. investigated emissions of particulate matters (PM) and polycyclic aromatic hydrocarbons (PAHs) from a gasoline direct injection (GDI) engine using gasoline and gasoline mixed with methanol in 15% volume (M15) (Hong et al. 2018a). PM mass emissions decreased while the coolant's temperature increased. Compared with M0 fuel, the PM mass with M15 fuel reduces by 80% at 20 °C of the coolant. The application of M15 fuel reduced the concentrations of most PAH species compared with M0 fuel, except those with smaller aromatic rings. In addition, benzo(a)pyrene equivalent toxicity decreased when the GDI engine starts with higher coolant's temperature or with M15 fuel.

Prasad et al. investigated the effect of the wide-open throttle condition in an SI engine at different speeds ranging from 1200 to 1800 rpm, using a single-cylinder four-stroke variable compression ratio with three different CRs (8–10) (Prasad et al. 2020). Increasing the CR from 8 to 10 for the methanol/gasoline blend improved the combustion efficiency owing to the peak pressure and net heat release increasing by 27.5% and 30%, respectively. The vehicle's performance indicates an improvement of 25% in the BTE and a reduction of 19% in the BSFC at a CR of 10:1. At a CR exceeding 10:1, there was a significant decrease of 30–40% in the CO and HC emissions; however, the NO_x emissions increased with the CR.

Alcohols have also been used for CI engines, but these engines require more significant modifications depending on the application; this is owing to the high autoignition resistance of alcohols. In such applications, methanol is introduced separately to the diesel engine and is then used as a source of ignition. This, however, requires significant modifications to the fueling system (Yusri et al. 2017).

To solve the problem of cold start, engine operation and simplification of complex systems have been the primary focus areas. Advanced mechanical components and electro/mechanical controls such as for the engine block (Balki and Sayin 2014; Gu et al. 2012), injector (Kabasin et al. 2009; Gong et al. 2011; Colpin et al. 2009), valves (Colpin et al. 2009), cranking rpm (Brusstar et al. 2002; Aikawa et al. 2009), ignition timing control (Markel and Bailey 1998), and direct injection (Agarwal et al. 2019; Marriott et al. 2009) can help improve the cold-start ability.

There are two applications for the methanol using in CI engines, fumigation and methanol–diesel blended fuels. Qi et al. researched the effects of using methanol as additive to biodiesel–diesel blends on the engine performance, emissions and combustion characteristics (Qi et al. 2010). Methanol was added to BD50 (50% biodiesel and 50% diesel in vol.) as an additive by volume percent of 5% and 10% (denoted as BDM5 and BDM10). The results indicated that the power and torque outputs of BDM5 and BDM10 are slightly lower than those of BD50. BDM5 and BDM10 show dramatic reduction of smoke emissions. CO emission was slightly lower, and NO_x and HC emissions were almost similar to those of BD50 at speed characteristic of full engine load.

Zhang et al. investigated the effects of fumigation methanol on the combustion and particulate emissions of a diesel engine (Zhang et al. 2013). Fumigation methanol increased the peak heat release rate and ignition delay but does not significantly change the combustion duration. The fumigation method resulted in a significant decrease in particulate mass and number concentrations from medium to high engine loads, due to the increase of fuel burned in the premixed mode and a reduction of diesel fuel involved. Fumigation methanol also slightly decreased the fraction of accumulation mode particles and thus the particulate geometric mean diameter.

Sayin et al. tested the influence of operating parameters on the performance and emission characteristics of a direct injection diesel engine with various methanol ratios from 5 to 15% (Sayin et al. 2010). Phase separation was avoided by using 1 vol% of dodecanol (Sayin 2010). The results indicated that the BSFC and NO_X emissions increased with the BTE, smoke opacity, and CO. The total HC decreased with an increase in the amount of methanol in the fuel mixture. The optimal results for the BSFC, and BTE were obtained at the original injection pressure and timing. For all the fuels tested, increasing the injection pressure and timing resulted in a decrease in the smoke opacity and CO and THC emissions; however, this led to an increase in the NO_X emissions.

Fan et al. reported that the chemical structure of soot particles has a significant influence over the concentration and accessibility of potential radical sites for the reactions governing soot growth and oxidation behaviors during in-cylinder combustion and thus affects the particle emissions of engines (Fan et al. 2021). Their results showed that epoxy, ether, and C vacancies as well as ester groups were the dominant surface species, whereas the carboxyl and carbonyl groups contributed less, among the carbon–oxygen groups on soot surfaces for both pure diesel and methanol–diesel blends. At low engine load conditions, an increase in the methanol blending content led to the unstable behavior of the C–O epoxy concentration. At a high engine load, an increase in the methanol content of fuel blends led to greater increments in the hydroxyl species with respect to the epoxy groups on the soot surfaces, especially when 5 and 10% of methanol were added. Relative to the aromatic C=C, the concentration of aliphatic C–H reduced with an increase in the methanol content of the fuel blends.

Masimalai et al. evaluated the effects of methanol induction on the performance, emission, and combustion behavior of a dual-fuel engine (Masimalai 2014). The BTE during dual-fuel operation was better than that during normal diesel operation with methanol induction, mainly at high power outputs. It increased from 30.3% with pure diesel to a maximum of 32.7% when methanol contributed approximately 44% of the energy share. During dual-fuel operation, smoke was significantly reduced under all the methanol induction rates and power outputs considered, while diesel was used as the pilot fuel. NO emissions were reduced during the dual-fuel operation under all the loads and methanol admission rates. However, there was an increase in the HC and CO emissions with the methanol induction during dual-fuel operation.

Guo et al. evaluated the effects of methanol in a single-cylinder, four-stroke, water-cooled direct injection diesel engine (Guo et al. 2011). There were minor differences in the brake torques of the diesel and blended fuels. Compared to those when using diesel, the BSFC increased and the emissions of NO_X, CO, HC, and PM decreased when using the diesel fuel mixed with 10–30% of methanol, as illustrated in Fig. 5.5. Wu et al. investigated the effect of cetane number improver on performance and emissions, including particulate number concentration and size distribution, of a turbocharged, common-rail diesel engine fueled with biodiesel-methanol (Wu et al.

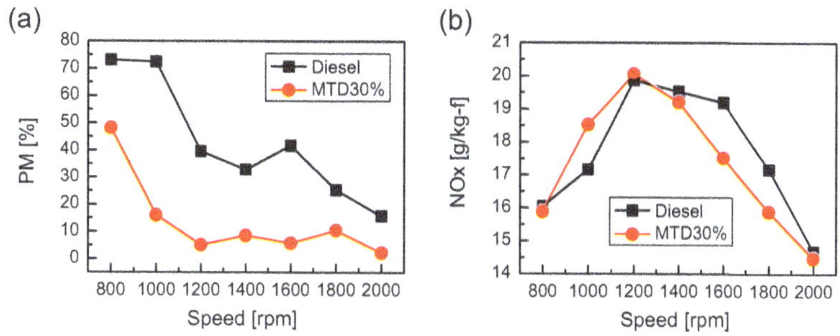

Fig. 5.5 PM, NO_x emissions with 30% methanol to diesel fuel at full load condition (Guo et al. 2011)

2011). Compared with those of biodiesel-methanol blend, the fuel economy and thermal efficiency are improved when CN improver is added to biodiesel-methanol blend. Besides, CO and HC emissions decreased, NO_x emission increased little and smoke emissions increased slightly. PM number concentration decreased and peak of size distribution profile shifted toward large size direction.

Ma et al. conducted a parameter study on the energy balance of a diesel methanol dual-fuel (DMDF) engine, including load characteristic tests for both the pure diesel mode and the DMDF mode and a comparison of the energy flow conversion under different load ratios (Ma et al. 2019). It was found that increasing the intake air temperature had the most significant effect in terms of reducing the DMDF engine's replacement ratio. Moreover, higher cooling water and methanol temperatures contributed to an improvement in the DMDF engine's thermodynamic performance. It can be explained based on the energy flow transformation when switching to DMDF combustion.

A summary of the advantages and shortcomings of using methanol as a vehicle fuel is presented in Fig. 5.6 (Verhelst et al. 2019). The challenges pertaining to a general schematic linking methanol's properties to engine performance are (Verhelst et al. 2019): To compare methanol-fueled vehicles with internal combustion engines and other alternatives, it is necessary to assess the efficiency of dedicated methanol engines at peak and part loads via state-of-the-art engine technologies, develop strategies to enable the use of gasoline in these engines, evaluate the potential of onboard methanol reforming using engine waste heat, and determine optimal strategies for using methanol in CI engines considering that another method of introducing methanol is to blend it with gasoline. It is also necessary to acquire additional data pertaining to factors such as the water tolerance and phase separation of these blends; these data would help clarify the PM and aldehyde emissions during methanol-fueled operation and improve simulation tools, enabling them to better reflect the effects of using methanol fuels. Moreover, data from practical scenarios would help identify methanol's potential and investigate the cost-effectiveness of different octane numbers.

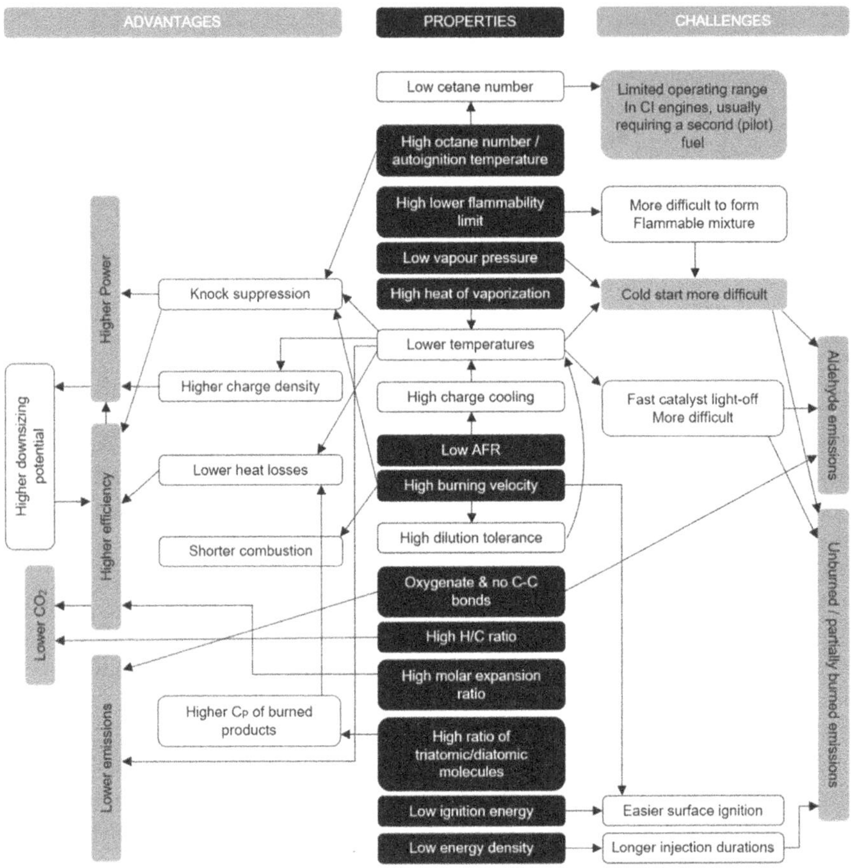

Fig. 5.6 Effect of methanol properties on engine performance (Verhelst et al. 2019)

(2) Ethanol (C_2H_5OH)

Ethanol flex-fueling has had a long history in Brazil, and sophisticated vehicle technologies have been developed to cater to the wide variety of blends available (Policarpo et al. 2018). In the Brazilian market, OEMs have developed novel technologies, such as systems starting with gasoline and then switching to ethanol once the engine is sufficiently warm. Ethanol is a renewable fuel because, if higher hydrocarbon electrofuels can be synthesized from renewable carbon and hydrogen, their hydrogen mass storage capability would be similar to that of methane (Verhelst et al. 2019).

Herein, the research and application results pertaining to the use of ethanol in automobile SI engines are presented (Yusri et al. 2017). Baek et al. investigated the impact of engine control parameters on combustion behaviors and particle number emissions was investigated with a spark ignition direct injection engine used various gasoline-ethanol blended fuels (Baek et al. 2021). The peak pressure in the cylinder was similar regardless of the type of fuel and the particle emissions were reduced by

92% compared to gasoline at the engine out position in the E85 fuel. It was found that gasoline-ethanol blends have potential as alternative fuels and are effective in reducing the particle emissions in the SI engine.

Yang et al. studied the impacts of the chemical structure and physical properties of gasoline on the particulate emissions from a fleet of Tier 3 compliant gasoline direct injection (GDI) vehicles (Yang et al. 2019). Fuels with higher aromatic contents result in an increase in the PM mass and black carbon emissions. The total and solid particle number emissions also showed an increasing trend when using fuels with higher aromatic contents, as compared to those for fuels with lower aromatic contents. The high molecular weight and low volatility of the hydrocarbon species (especially aromatics) also strongly influenced the formation of particulate emissions in GDI vehicles.

Yusuf et al. reported the effects of low content rates of Mbwazirume bioethanol blends (5, 10, and 15%) with gasoline in a modernized electronic fuel injection engine (Yusuf and Inambao 2021). The results indicated that the E15 fuel ratio induced an increase in combustion duration with minimum premixing combustion duration, hence exhibited low emissions, high indicated power, and efficiency with low fuel consumption (at 12°–18° ATDC). Because the low content of carbon in bioethanol, which eliminates the occurrence of soot formation and requires less air to burn fuel blends with low luminosity and radiation. The rate of NO_x formation rose higher with E5 and E10; this was due to the combustion advanced which led to higher temperature and in-cylinder pressure than that of E0 and E15. In addition, the HC decreased at all injection timings for E5, E10, and E15. With regard to emissions, Dhande et al. reported that E10 to E25 perform from gasoline engine (Dhande et al. 2021). It was found that the ethanol enrichment increased the fuel consumption and power for braking while the thermal efficiency decreased. CO and HC have decreased, but ethanol concentrations have increased the NO_x and CO_2.

Choi reported the vehicle fleet test involving gasoline blending with 3% (E3) and 5% (E5) ethanol. The Korea Petroleum Quality and Distribution Authority (K-Petro) reported the production costs of bioethanol using very high gravity and back-set methods; they conducted 20,000 km durability tests on four vehicles and developed an anti-corrosion fuel pump (Choi 2018). It was found that E3 and E5 vehicles provide similar driving environments that are as stable as those afforded by gasoline vehicles. The CO, HC, NO_x, and CO_2 emissions were also lower than those of conventional gasoline vehicles.

Policarpo et al. analyzed the CO, CO_2, NO_x, and HC emission from a flex-fuel vehicle over an urban route in Brazil (Policarpo et al. 2018). The analyses were performed with different ethanol/gasoline blends (27, 85, and 100% ethanol). It was found that the overall profile of the CO, CO_2, and NO_x emissions during deceleration, idling, and acceleration were quite similar, even though the E85 blend resulted in emissions 100 times lower than those afforded by the other two blends, as summarized in Table 5.4.

Goktas et al. reviewed the HC, CO, CO_2, and NO_x emissions and combustion characteristics when using alcohol fuel in an SI engine (Goktas et al. 2021). The use of alcohol fuels in SI engines induced an increase in the BTE, whereas the HC

Table 5.4 Mean and standards deviation related to emissions (Policarpo et al. 2018)

Fuel blend	CO (g/s)		CO_2 (g/s)		NO_x (g/s)	
	Mean	SD	Mean	SD	Mean	SD
E27	4.6×10^{-3}	1.2×10^{-3}	7.5×10^{-1}	1.1×10^{-1}	1.9×10^{-4}	1.3×10^{-4}
E85	1.8×10^{-4}	2.1×10^{-6}	1.1×10^{-2}	2.3×10^{-4}	3.95×10^{-5}	1.3×10^{-6}
E100	4.1×10^{-3}	1.0×10^{-3}	1.0	0.2	1.5×10^{-4}	1.1×10^{-4}

and CO emissions were reduced, as compared to those for pure gasoline. When the carbon content in the fuel–air mixture was relatively lower than that of gasoline, the CO_2 emissions were also reduced. Moreover, pure methanol and ethanol had a positive effect on the NO_x emissions, i.e., the NO_x emissions were lower than those when using pure gasoline.

With regard to CI engines (Yusri et al. 2017), ethanol is soluble in diesel fuel at contents of approximately 0–30% ethanol and 70–100% diesel (Caro et al. 2001). An increase in the ethanol blend percentages further accelerates the separation of the ethanol–diesel fuel blends. Therefore, when ethanol is considered for a diesel engine, many aspects need to be considered, namely the ethanol fuel properties, emissions, and engine durability (Hansen et al. 2005). Most applications of ethanol in a CI engine pertain to the fumigation of ethanol (Zhang et al. 2011) and emulsification (using an emulsifier to mix the fuels in order to prevent separation) (Asfar and Hamed 1996).

Vali et al. focused on improving the performance and emission characteristics of a diesel engine through the addition of mixed nano-additives to the diesel–ethanol blend (Vali and Wani 2021). Mixed nano-additives of 50 and 100 ppm (Al_2O_3 and TiO_2) were added to the diesel–ethanol blend in the presence of a surfactant through the aid of ultrasonic agitation. Significant improvements in performance parameters, such as the BTE and BFSC, were observed after the addition of the nano-additives. Reductions in the NO_x, CO, and HC emissions were also observed on adding the nano-additives; however, the CO_2 emissions were increased.

Han et al. investigated ethanol/diesel dual-fuel combustion in a heavy-duty diesel engine (Hana et al. 2020). Their results show that ethanol mass ratios of up to 80% can be achieved at low-to-medium loads without misfiring. The addition of ethanol could help reduce the emission of soot, with no consistent effect on the NO_x emissions. As the ethanol mass ratio increased, the dual-fuel operation progressively suffered from incomplete combustion. Moreover, the HC and CO emissions increased. However, it is believed that this can be managed using a diesel oxidation catalyst at higher loads. When ethanol is introduced, both the combustion and thermal efficiencies decreased at lower loads. However, the thermal efficiency at medium loads increased from 49.1 to 50%. For medium to high loads, the thermal efficiency first increased to 50.7% and then decreased to 49.7%. Subsequently, it continuously decreased owing to the sub-optimal combustion phasing at high ethanol mass ratios. It is noteworthy that the pressure rise rate, intensity range, and peak pressure appear to limit the ethanol ratio to below 40% under medium to high loads.

Gomasta et al. studied the effects of the lower heating value (LHV) of ethanol on engine performance and emissions (Gomasta and Mahla 2012). The results shown that a higher ethanol content yields a decreased LHV, reflecting a lower BSFC and BTE, as compared to that of diesel. In emissions, CO and PM were significantly reduced with an increase in the ethanol ratio. These phenomena were attributed to the same physical and chemical characteristics. HC was reduced with an increase in the ratio of ethanol–diesel fuel blends, as compared to the pure diesel.

Shanmugam et al. (2021) researched a CI engine to assess the engine characteristics fueled with the blend of diesel and high-oxygenated additives such as ethanol. It was found that there was an improvement in the performance characteristics, such as BTE and SFC, with the enrichment of diethyl ether in ethanol–diesel blend. It is also noticed that the blend without diethyl ether (DEE) exhibited lower magnitude because of higher energy content and cetane number of DEE. Emission characteristics, HC and CO were drastically increased with the increase in the ethanol concentration in the diesel blend. This is attributed to higher latent heat of vaporization of ethanol present in the blend. Combustion pressure and heat release rate of the DEE-enriched ethanol blends were higher by 2.2% and 2.4%, respectively, when compared with their corresponding blends without DEE.

A mixture of biodiesel–diesel–ethanol was also investigated by many researchers (Yilmaz et al. 2013; Paul et al. 2017; Hebbar and Bhat 2013; Luiz et al. 2018; Kim and Choi 2008). Ethanol and diesel can be blended together but only up to a certain limit, as both fuels tend to separate owing to their immiscible behaviors. Kim and Choi tested three types of ethanol–diesel blends used in a common rail direct injection diesel engine (Kim and Choi 2008). They reported on the engine performance and the HC, CO, NO_x, smoke, and PM emissions. When using the ethanol–diesel blends, the HC and CO emissions increased slightly; a mean conversion of approximately 50–80% for the HC and CO on catalysts was achieved during the ECE 13-mode cycle. Smoke emission was decreased by more than 42% throughout this mode cycle. In the particle size range of 10–385 nm, the total number and total mass of the PM for the ethanol–diesel blends were decreased by approximately 11.7–15% and 19.2–26.9%, respectively. As biodiesel is miscible in alcohols and diesel, it is suitable for use as an emulsifier to blend alcohols and diesels; hence, it is directly used as biodiesel–ethanol–diesel blends in diesel engines. Using biodiesel is more cost-effective than using an emulsifier (Kim and Choi 2008).

Park et al. reported that biodiesels are capable of improving the low CN of diesel–ethanol blended fuels owing to their higher CN (Park et al. 2009). Luiz et al. reported the study of oxidized biodiesel as a cetane improver for diesel–biodiesel–ethanol mixtures in a vehicle engine (Luiz et al. 2018). The volumetric percentage of ethanol in the mixtures of diesel, biodiesel and ethanol was 0, 2, 4 and 6%, while the biodiesel volumetric content was 0, 5 and 35%. The results have revealed no significant changes in the wheel power for small percentages of ethanol and biodiesel. Emission tests using a mixture containing 90% of conventional diesel, 4% ethanol, 5% biodiesel and 1% oxidized biodiesel have shown reduction in NO_x and PM emissions in comparison to conventional diesel.

Yilmaz et al. investigated the effects of varying ethanol percentages (from 3 to 25%) in biodiesel–diesel–ethanol (BDE), while the biodiesel and diesel percentages were maintained (BDE3–BDE25) (Yilmaz et al. 2013). Results indicated that emissions strongly depended on not only engine operating conditions, but also fuel blends ratio. Alcohol blended fuels increased CO emission as compared to diesel fuel for all operating conditions. While ethanol blended fuels reduced NO emission for all concentrations, unburned HC emission depended not only on ethanol concentrations, but also operating conditions.

Paul et al. reported that the D35E15B50 blend with 15% ethanol afforded the best engine performance, leading to an increase of approximately 21.2% in the BTE and a decrease of approximately 4.6% in the BSFC at full load (Paul et al. 2017). Combustion analyses also revealed an increase in the cylinder pressure and heat release rate, indicating an improvement in the combustion conditions for the above-mentioned blend. The D35E15B50 blend could substantially suppress the unburned hydrocarbon and CO emissions; however, it resulted in a marginal increase in the NO_x emissions.

Shrivastava et al. optimized the diesel engine performance and emission parameters of biodiesel–ethanol–diesel blends at optimal operating conditions (Shrivastava et al. 2021). Compared to diesel at full load, the experiments revealed marginal reductions of approximately 2% in the BTE and 3% in the BFC. The exhaust gas temperature also increased by 3%. Moreover, the biodiesel–ethanol–diesel blend increased CO_2 emissions by 0.86%, whereas it decreased HC emissions by 12 ppm; it also decreased the CO and NO_x emissions by 0.029% and 8%, respectively. Tse et al. investigated the combustion characteristics and particulate emissions of a diesel engine fueled with DBE (diesel–biodiesel–ethanol) blended fuels. It was found that the DBE blends offer the added advantage of reducing NO_x emissions. It is worth noting that the use of DBE blends can weaken the PM-PN-NO_x trade-off relationship, as shown in Fig. 5.7 (Tse et al. 2015).

(3) Propanol (C_3H_7OH)

The solutions for cold start when using alcohol fuels include the use of additives such as highly volatile gasoline, butane, or pentane; the dissociation of methanol to hydrogen and CO; or the conversion of methanol to dimethyl ether (Yusri et al. 2017).

For SI engines, Wallner et al. analyzed the effect of gasoline-higher alcohols (C2–C8) blends (Wallner et al. 2012). It was found that higher alcohols have certain physical properties that might be desirable for blending with gasoline. Due to their oxygen content all alcohols have an inherent disadvantage in terms of energy content compared to non-oxygenated fuels. While this disadvantage becomes less pronounced with increasing carbon count, and reduced knock resistance become more dominant with longer chain length alcohols.

Mourad et al. conducted the performance of a gasoline engine using different fuel blends of propanol/gasoline (Mourad and Mahmoud 2018). The results indicated that the use of propanol/gasoline fuel blend can improve the fuel economy by 2.8% for the

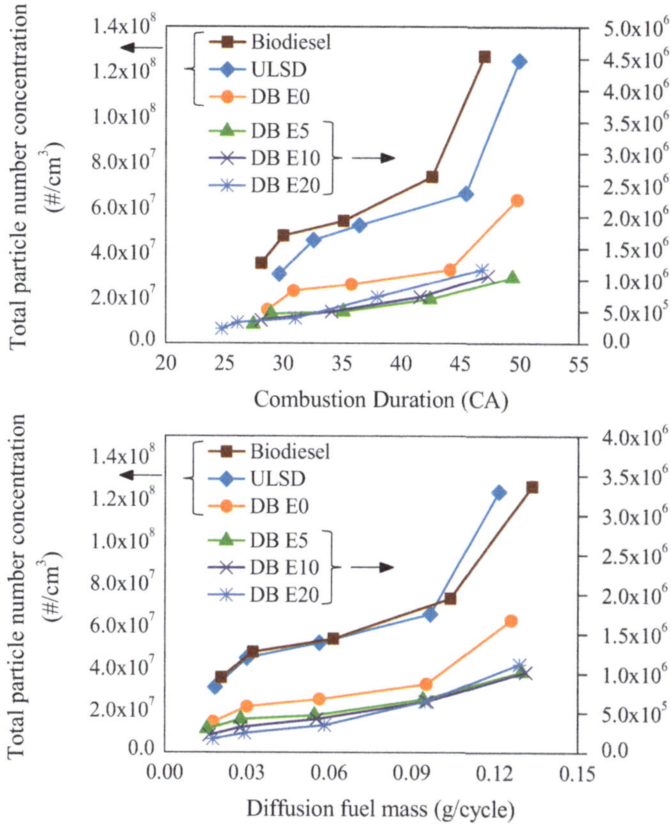

Fig. 5.7 Total PM number concentration against combustion duration and diffusion fuel mass (Tse et al. 2015)

blend ratio (propanol/gasoline 15%). It decreased the pollutants emission of vehicle engine, especially HC, CO by 14.2% and 10.9% respectively. Also, the vehicle SFC was improved and the pollutants emissions of for HC and CO were decreased.

Gravalos et al. researched the emission characteristics of lower to higher molecular mass alcohol-blended gasoline, as shown in Fig. 5.8 (Gravalos et al. 2013). The CO emission was decreased by higher ethanol blended fuels, while lower molecular mass blended fuels induce slightly higher CO emission than that of lower and higher alcohol molecular blend fuels. They revealed that the CO and HC levels in the engine exhaust were reduced when using alcohol–gasoline blends. NO_x emissions with the alcohol–gasoline blends were higher than those with pure gasoline. Qian et al. examined the combustion and emissions of n-propanol/gasoline surrogates were studied (Qian et al. 2018). The proportions of n-propanol in blends are 10, 30 and 50% by volume. Results shown that the blends with higher proportions of n-propanol lead to higher maximum in-cylinder pressure, in-cylinder averaged temperature, and shorter rapid combustion duration. The CO, THC, alkane, acetylene and aromatic emissions

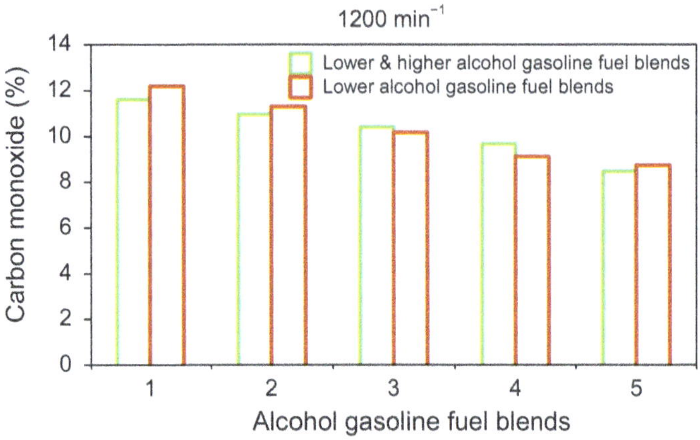

Fig. 5.8 Comparison of CO exhaust emission for alcohol and gasoline blend fuels (Gravalos et al. 2013)

decrease as the proportion of n-propanol increases. However, 30% propanol fuel had the highest alkene and aldehyde emissions. Both of the geometric mean diameter of PM and the accumulation-mode particle ratio in PM decrease as the proportion of n-propanol increases. The ETH of the engine increases with the increase of the proportion of n-propanol.

In CI engines, propanol affords better fuel properties than methanol and ethanol; however, studies evaluating the effects of propanol on the performance and exhaust emissions of CI engines remain limited. Most previous studies have focused on the use of propanol as a solvent in diesel blends in order to form ternary alternative fuels (Haigh et al. 2014). Isopropanol was also used as an additive to improve ethanol–diesel blends by enhancing the lower cetane number of ethanol blends; this enabled better ignition properties (Hu et al. 2020). Additives could help improve viscosity to ensure adequate lubrication in the injection pump.

Moreover, they stabilize the mixture in the presence of high water contents, resulting in fuel homogeneity under all conditions. Lavi et al. studied the effect of including ethylhexyl nitrate in a diesel-waste plastic oil–propanol mixture on the performance and exhaust emissions of a CI engine (Ravi and Karthikeyan 2021). It was concluded that the inclusion of propanol and ethylhexyl nitrate decreases the NO_x and CO emissions and also the brake power, while increasing the CO_2 emissions and the BSFC.

Hu et al. analyzed i-propanol-n-butanol-ethanol (IBE) as an assuring alternative biofuel of internal combustion engines (Hu et al. 2020). However, only a few studies have focused on this topic, most of which focused on the use of propanol in comparison to ignition engines. A reduced mechanism was included 151 species and 775 reactions to track the combustion and emissions behavior of engine fueled with IBE/diesel blends. The mechanism was validated against ignition delay, laminarflame speed, and species mole fractions over a wide range of engine relevant conditions. Results show that the mechanism can reliably predict the combustion characteristics and emissions characteristics of IBE/diesel blends in the CI engine.

Deep et al. investigated the impact of adding orange peel oil methyl ester and propanol to diesel at different ratios (Deep et al. 2014). On using the propanol blends, the BSFC was improved, as compared to that when using pure diesel. Zhao et al. researched the results of indicated specific fuel consumption (ISFC) as a function of intake pressure, blend ratio, and engine load (Zhao et al. 2022). The ISFC of all the tested fuels increased when lowering the intake pressure because the thermal efficiency became lower under the low intake pressure. The ISFC values of diesel/alcohol blends were higher than that of diesel under all the tested conditions. The smaller lower heating value of propanol and pentanol would be the reason for that. Both propanol and pentanol had smaller lower heating values. A higher blend ratio led to a smaller lower heating value, so more fuel was consumed under the same operating condition, resulting in a larger ISFC value. Pentanol/diesel blends showed lower ISFC compared to propanol/diesel blends because of the higher lower heating value of pentanol (Fig. 5.9).

(4) Butanol (C_4H_9OH)

Recently, bio-butanol(n-butanol) fuel was produced and refined from lignocelluloses. It plays deterministic roles in bio-butanol production. Yusri et al. reviewed the engine performance and emissions for butanol-blended fuels in 2014 (Yusri et al. 2017). Alasfour also reported the emission characteristics of an SI engine using 30%

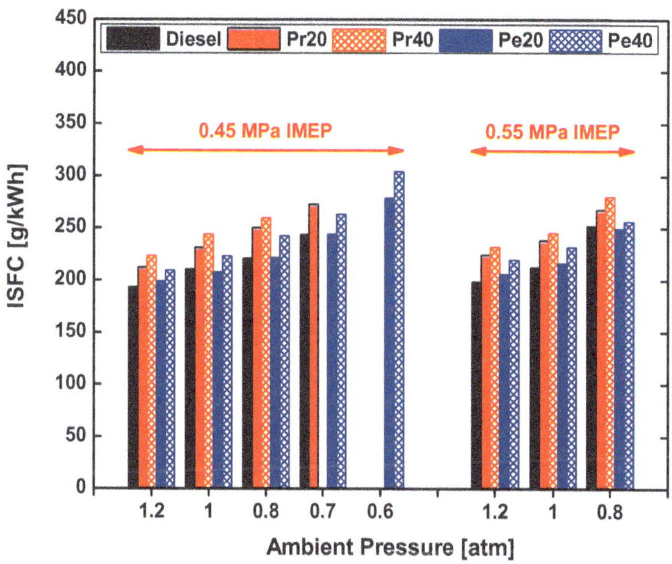

Fig. 5.9 Indicated specific fuel consumption (ISFC) of tested fuels under different loads and intake pressures (Zhao et al. 2022)

i-butanol–gasoline blend fuel (Alasfour 1998). From the experimental results, when using the 30% i-butanol–gasoline blend, NO_x and HC emissions were suppressed by 9 and 17%, respectively; furthermore, the brake torque was decreased, whereas the BSFC was increased. Zaharin et al. investigated the effects of i-butanol additives in ethanol–gasoline blends on the fuel properties, performance, and emission characteristics of an SI engine (Zaharin et al. 2018). An improvement in the BTE was observed when using the blended fuel, with a maximum increment of 18.91% achieved when using E10B15. The CO and HC emissions were reduced with the addition of i-butanol, as compared to those when suing pure gasoline.

Recent studies have led to significant improvements in the prediction of PM emissions from GDI vehicles based on differences in fuel compositions (Barrientos et al. 2016). The effects of fuel volatility are considered, including the characteristics derived from molecular weights, distillation, vapor pressures, and heat of vaporization. n-butanol is more effective in suppressing soot formation than i-butanol, whereas the PM index predicts similar sooting tendencies. Smoke point-based approaches revealed highly similar trends of the PM index for gasoline and blends with ethanol and butanol.

Mourad et al. investigated the incorporation of ethanol/butanol additives in gasoline to reduce pollutant emissions and enhance fuel economy (Mourad and Mahmoud 2019). The blending ratios of ethanol and butanol to gasoline were 2, 5, 10, 15, and 20%. The experimental results showed a clear reduction in the pollutants emitted from the engine; CO and HC emissions were reduced by 13.7% and 25.2%, respectively, while fuel consumption was reduced by 8.22%. Figure 5.10 shows the NO_x emissions with different engine speed, which is increased with both the blend ratios and engine speeds.

Sayin and Balki analyzed the consequences of different CRs (9, 10, and 11) for different iso-butanol–gasoline blends (10, 30, and 50%) (Sayin and Balki 2015). Throughout the experiment, it was reported that a higher CR of 11 was capable of

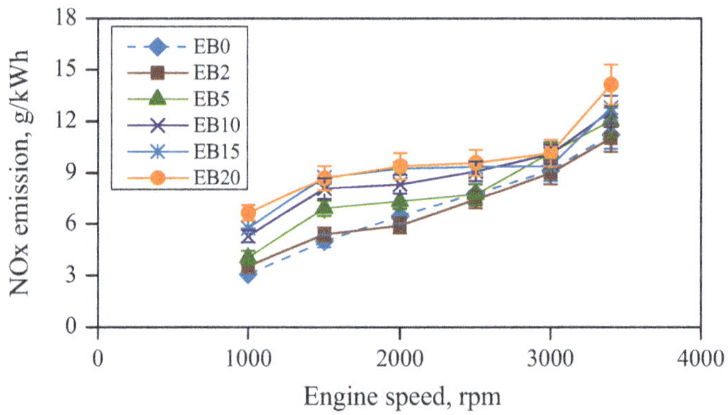

Fig. 5.10 NO_x emission of various ethanol and butanol ratios according to different engine speeds (Mourad and Mahmoud 2019)

maintaining low HC and CO emissions; however, this was accompanied by higher CO_2 emissions. It was noted that, when fusel oil (ethanol 3.22%, iso-butylalcohol 17.5%, amylalcohol 60.4%, and n-propylalcohol 0.53%) was used as a blend with gasoline, the engine torque was increased slightly, and the volumetric efficiency and specific fuel consumption also increased. On an average, the HC and CO emissions increased. Furthermore, knocking and nitrogen oxides (NO_x) were observed to decrease when using fusel oil (Calam et al. 2015).

Tang et al. investigated the effects of the spark timing and lambda on a high-speed SI engine fueled with n-butanol/gasoline blends (Tang et al. 2021). The output power, BTE, and BSFC increased with the n-butanol ratio. The addition of n-butanol to gasoline could significantly decrease the NO emissions; however, it slightly increased the CO emissions.

Short et al. conducted an experiment on the particle speciation of emissions when using i-butanol and ethanol-blended gasoline in light-duty GDI vehicles (Short et al. 2015). The results showed that the port fuel injection (PFI) vehicles, fuel composition, and testing conditions impacted the PM emissions. The GDI vehicles emitted more PM and twice as much black carbon, as compared to the PFI vehicles. Furthermore, the vehicle's operating conditions (steady state or transient) could significantly impact the average composition of the particles, irrespective of the fuel composition. Gu et al. reported that, when the EGR rate was increased by 5–20%, the NO_x emissions decreased and the CO and HC emissions increased for blends of gasoline and n-butanol (Gu et al. 2012). With regard to the performance, increasing the n-butanol ratio in the gasoline resulted in a deterioration of the engine brake torque and BSFC. Singh et al. measured the HC, CO, and NO_x emissions at higher engine speeds for the fraction of n-butanol blends with 5, 10, 20, 50, and 75% in the gasoline in a four-cylinder SI engine (Singh et al. 2015). Especially for Bu50 and Bu75, the BSFC was lower than that when using pure gasoline, particularly at higher engine speeds and loads.

Zhen et al. summarized bio-butanol as a new generation of clean alternative fuels for SI and CI engines (Zhen et al. 2020). Lapuerta et al. investigated cold- and warm-temperature emissions when using n-butanol blends in a Euro 6 vehicle (Lapuerta et al. 2018). As compared to ethanol, the higher heating value of n-butanol, its superior miscibility with diesel fuel, and its lower hydrophilicity indicate that butanol is a better option for use as a blending component in diesel fuels. Butanol blends of up to 16% afforded benefits in terms of the particle number and PM emissions upstream of the diesel particulate filter under all ambient conditions, thus implying a reduction in the frequency of regeneration. N-butanol blends led to better engine efficiency at low ambient temperature, and even to slightly lower SFC with respect to diesel fuel, with no significant benefits at warm ambient temperature. An increase in the NO_x emissions was observed only at a low ambient temperature (−7 °C), whereas an increase in the CO and hydrocarbon emissions was noted at all temperatures. Blends with n-butanol contents exceeding 13% were associated with start-ability issues under cold ambient conditions.

For CI engines, butanol is an excellent alternative fuel. The mixtures of n-butanol/i-butanol and diesel have been proven to reach a steady state. The employment of butanol-diesel fuel blends in diesel engines and its effects on the engine performance and exhaust emissions have been investigated by number of studies.

Han et al. investigated reactive controlled CI by using n-butanol/n-heptane in a heavy-duty diesel engine (Han et al. 2021). In their study, n-butanol was injected at the port as a low-reactivity fuel, while n-heptane was directly injected into the cylinder as the high-reactivity fuel in order to achieve high thermal efficiency and suppress soot/NO_x emissions. The results showed that a single direct injection caused either early combustion phasing or excessive HC/CO emissions.

Choi et al. investigated the effect of diesel fuel blended with n-butanol (5–20 vol%) on the emission of a turbocharged common rail direct injection diesel engine (Choi et al. 2015). It was found that, for the n-butanol blend, NO_x emissions increased as compared with those for the pure diesel fuel case at the European Stationary Cycle. In the case of 20% butanol, both THC and CO emissions increased significantly, and both HCHO and HCOOH increased moderately under low load conditions, as shown in Fig. 5.11 (Choi et al. 2015). The BU5 (5 vol% butanol) blend could be a better option for reducing the PM mass and emissions of nanosized PM under 50 nm. In another study (Choi and Jiang 2015), the influence of a diesel fuel blend with n-butanol on individual hydrocarbons and the PM of a turbocharged common rail direct injection engine was studied. It was reported that more ethylene and benzene are emitted under low load conditions compare to high load conditions, and their amounts increase with the n-butanol blend ratio. The fuels blended with over 10% butanol produced higher non-regulated HCHO than the pure diesel fuel under low engine load conditions. The PM numbers for particles smaller than 50 nm were

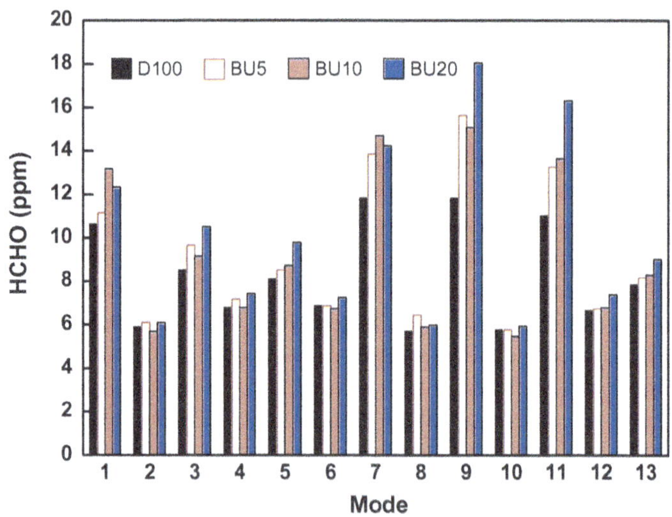

Fig. 5.11 HCHO emission according to the test cycles (Choi et al. 2015)

higher when using 20 vol% butanol in diesel, as compared with those when using pure diesel. Further, the emission characteristics of a single-cylinder diesel engine using a blended fuel of diesel with hydrated butanol were investigated. It was found that the blended butanol fuel slightly increased the THC and CO with an increase in the water content. The number of PM decreased by 81% and 67% when blended fuels DB10W2 and DB10W5 were used, respectively. The carbonyl compounds increased with lower exhaust temperatures, shorter mixing times for the air and fuel, and an increase in the water content of the blended fuel (Kim et al. 2019).

Xiao et al. investigated the combustion performance and emission characteristics of a diesel engine fueled by biodiesel blended with n-butanol (Xiao et al. 2019). The ignition delay period was prolonged in the case of n-butanol. However, the combustion duration was significantly shortened with an increase in the butanol ratios. When n-butanol was added, more HC, NO_x, and acetaldehyde and less CO, 1,3-butadiene, and benzene were emitted. About soot emission, the addition of n-butanol could reduce the formation of soot. The addition of n-butanol improved the combustion performance.

Liang investigated the emissions and power tradeoffs of biodiesel and n-butanol in diesel blends for fuel sustainability (Liang 2021). It was reported that the binary diesel–biodiesel blend D80B20 (80% D100 and 20% B100 by volume) offers reduced PM and black carbon emissions but higher NO_x emissions in the engine exhaust. Both biodiesel and n-butanol fuel blends result in an increase of 3.0–5.6% in the BSFC. The use of biodiesel and n-butanol additives in petroleum diesel can decrease PM emissions; the resulting undesired NO_x can be managed by optimizing the tertiary composition of the petroleum diesel, biodiesel, and fuel additives. Rakopoulos et al. performed experimental tests on a single-cylinder, CI engine, direct injection, naturally aspirated diesel engine (Rakopoulos et al. 2014).

Gainey et al. investigated the autoignition characteristics of methanol, ethanol, propanol, and butanol over a wide range of operating conditions in low-temperature combustion (LTC)/homogeneous charge CI(HCCI) engines (Gainey et al. 2021). The reactivity of the seven fuels investigated was in the following order: isopropanol, sec-butanol, ethanol $\approx$ n-propanol $\approx$ i-butanol, methanol, and n-butanol.

Chen et al. tested butanol/diesel blends with ratios of 20–40% by volume of butanol with a turbocharged, inter-cooled, high-speed, direct injection diesel engine for a passenger car equipped with common rail and a variable geometry turbocharger (VGT) (Chen et al. 2013). A higher butanol ratio resulted in an increase in the BSFC, HC, and BTE. Under high loads, the NO_x emissions increased but the CO emissions decreased as the butanol blending ratio was increased. Conversely, under low load conditions, the CO increased but the NO_x decreased. The soot produced decreased as the butanol content was increased.

5.3.2 Ships

Regarding the fuel consumption of the shipping industry worldwide in 2012, 84% was consumed for international shipping, and over 70% of this was consumed by main engines. In addition, bulk carriers, container ships, and oil tankers accounted for more than half of the total fuel consumption (IMO 2015). The resulting GHG emissions such as CO_2, N_2O, and methane and the emissions of air pollutants such as NO_x, SO_x, PM, and CO have attracted public attention (Corbett et al. 2007).

In particular, SO_x, NO_x, and CO_2 emissions from ships have attracted more attention owing to their large amounts. It is estimated that global shipping contributed an average of 5.6×10^{12} g of sulfur, 6.3×10^{12} g of nitrogen, and 277.1×10^{12} g of carbon per year from 2007 to 2012; this accounts for approximately 13.0%, 15.0%, and 3.1% of the global SO_x, NO_x, and CO_2 produced from anthropogenic sources (IMO 2015). The IMO estimates that shipping emits approximately 2.8% of the annual global emissions or 1036 Mt of CO_2; this is predominantly from container ships, bulk carriers, and oil tankers that travel internationally (The Government of the Republic of Korea 2020; Olmer et al. 2017).

Over the next decades, a significant reduction in the CO_2 emissions from the shipping sector is necessary. To this end, over the last two decades, the scientific community and the shipping sector have devoted significant efforts toward technologies for reducing the CO_2 emissions from ships.

Figure 5.12 shows alternative solutions to conventional marine fuels for reducing the CO_2 emissions from shipping (IMO 2011, 2015; Corbett et al. 2007). Among these solutions, biofuels and/or alcohol fuels are included in the category of eco-friendly fuels. The pathways to decarburization should not involve an increase in NO_x, SO_x, and PM emissions, and the trade-off between CO_2 emissions and other air pollutants should not be neglected (Xing et al. 2020).

The IMO is a specialized branch of the UN; it issues regulations on the safety, security, and environmental performance of global shipping. In particular, Annex VI of the International Convention for the Prevention of Pollution from Ships–the Marine Pollution Convention was adopted in the 1973 convention and then modified according to the 1978 protocol with regard to limiting the harmful impacts of ship emissions on air quality (IMO 1997). These regulations were effective as of May 19th, 2005, and aimed to reduce the NO_x, SO_x, and PM from marine engines. These regulations expected to limit the marine fuel sulfur content to 0.5% globally by 2020 (IMO 2016a).

To control the emissions of SO_x and PM, the fuel sulfur content must be reduced because it significantly increases the emissions. SO_x emitted from marine diesel engines will form sulfate aerosols or sulfur-containing particles, which are a major component of PM. SO_x emissions can condense to form small-size particles in the nucleation mode, which may then adhere to coarse particles, thus resulting in the growth of particle composites. These secondary sulfate particles constitute an important contribution to land-based air pollution and also play a role in the marine

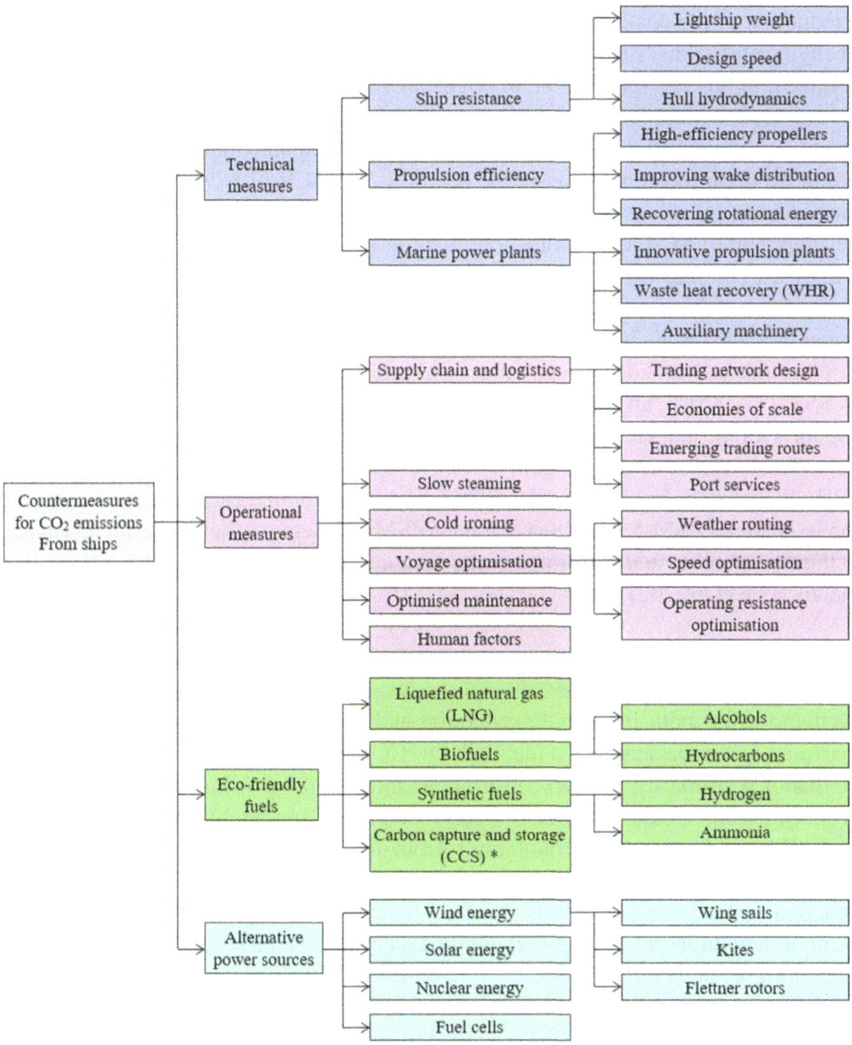

Fig. 5.12 Alternative solution to conventional marine fuels for shipping CO_2 emissions reduction (Xing et al. 2020)

aerosol budget. Therefore, the IMO does not particularly limit PM emissions directly; however, it does regulate the sulfur-related portion of PM formation.

NO_x emissions are mainly formed via the reaction of atmospheric nitrogen with oxygen through the Zeldovich mechanism, which is strongly influenced by the combustion temperature (Choi 2021). EGR arrangements and selective catalytic reduction (SCR) systems seem to be feasible abatement technologies to achieve NO_x reduction. By recirculating a part of the exhaust gas to the scavenging air intake, EGR decreases oxygen availability in the engine combustion chamber, resulting in a reduction of the peak combustion temperature. This system can cause an increase in engine emissions, such as black carbon, owing to the lack of oxygen in the engine combustion process (Nielsen et al. 2018). SCR is a method of converting NO_x into N_2 and H_2O with the aid of a catalyst. Ammonia or urea solution is added to a fuel stream or the exhaust gas as a reductant and is adsorbed onto the catalyst. CO_2 is then produced upon the conclusion of the reaction. The SCR of NO_x using ammonia as the reducing agent was patented in the United States by Engelhard Co. in 1957. The development of SCR technology continued in Japan and the US in the early 1960s, with research focusing on less expensive and more durable catalyst agents. The SCR method is suitable for heavy-duty diesel vehicles owing to its higher NO_x reduction potential (Choi et al. 2020).

Over the last few decades, copper (Cu) and iron (Fe)-exchanged zeolites (such as SSZ-13, ZSM-5, and BETA) (Lee et al. 2020) have emerged as popular catalytic supports for the abatement of NO_x emissions from vehicles and ships. These catalysts have demonstrated excellent activity for selective catalytic reduction by ammonia (NH_3-SCR) or hydrocarbons (HC-SCR). The most studied catalysts are Cu and Fe ion-exchanged ZSM-5 (Cu/ZSM-5 and Fe/ZSM-5), owing to their superior resistance to thermal stability and sulfur poisoning, as compared to those of the other transition metals (Lee et al. 2021).

(1) Methanol (CH_3OH)

Methanol from both industrial production and biomass is being considered in the maritime sector, because it is a fuel with low CO_2 and pollutant emissions. Currently, methanol is primarily produced from coal and natural gas; however, biomethanol produced from biomass has also received considerable attention as a sustainable fuel. The production of alcohol fuels from renewable energy sources has been increasing globally. However, methanol has a lower energy density than fossil fuels; hence, a larger storage space is required for maritime applications. The development of methanol–diesel dual-fuel engines is currently underway (Imran et al. 2013; Balamurugan and Nalini 2014).

Methanol has recently attracted research attention as a low-carbon alternative to conventional fuels (Dolan 2021). The total life cycle emissions of methanol use were compared with those of marine gas oil and heavy fuel oil to identify the environmental benefits of using methanol (IMO 2016b). Generally, emissions of SO_x, NO_x, PM, and GHGs (CO_2, CH_4, and N_2O) are used in the total life cycle model. Although methanol is slightly cheaper than marine gas oil, it is not the preferred fuel for shipping in the long term. This is because its use does not afford a favorable payback period for the capital investment and also fails to meet the price of production. Methanol use is expected to reduce SO_x, NO_x, CO_2, and PM emissions by up to 99%, 60%, 25%, and 95%, respectively (MKC 2018).

Ammar proposed the use of methanol as an alternative fuel to comply with the IMO emission regulations. The results indicate environmental benefits, as the NO_x, SO_x, CO, CO_2, and PM emissions were reduced by 76.78%, 89%, 55%, 18.13%, and 82.56%, respectively, as shown in Fig. 5.13 (Ammar 2019). Currently operated

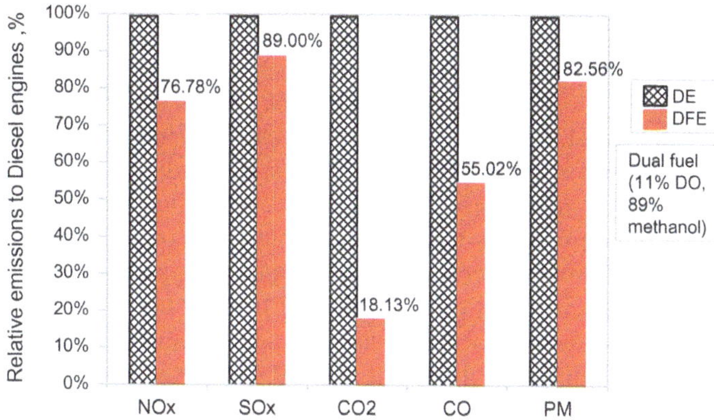

Fig. 5.13 Emission reduction ratio for dual methanol-diesel engine for ship (Ammar 2019)

diesel engines use SCR catalysts to comply with the IMO emission regulations. When the benefits of ship slow steaming and the saved SCR costs are combined, the cost-effectiveness of dual-fuel engines for reducing NO_x, CO, and CO_2 emissions will be 385.2 \$/ton, 6548 \$/ton, and 39.9 \$/ton, respectively (Ammar 2019). Thus, existing economic barriers could be overcome by strengthening the environmental targets for shipping or by offsetting higher fuel oil prices (Alonso et al. 2010).

(2) Ethanol (C_2H_5OH)

Bioethanol, which is the most abundantly produced biofuel, can be produced from a variety of feedstocks such as agricultural feedstocks, energy crops, and even waste products, including waste wood, waste paper, sugarcane waste, rice straw, and corn stalks (Alonso et al. 2010). The available and indigenous sources, easy handling, low emissions, and high energy density make it a favorable alternative fuel for the future, especially when compared with the hydrogen generation required for fuel cells. In addition to ethanol, methanol is another common alcohol fuel (Imran et al. 2013). The benefits and potential barriers for using methanol and ethanol as alternative fuels in shipping, in terms of the technical and operational factors, availability, environmental impacts, safety regulations, and economic considerations, have been evaluated by the European Maritime Safety Agency (Ellis and Tanneberger 2015). Both methanol and ethanol are significantly attractive fuels from an environmental perspective. However, the safety regulations, operational experience, and infrastructure for bunkering need to be strengthened. Biomethanol and bioethanol are technically viable options for reducing emissions.

5.3.3 Locomotives

Rail transportation is considered the best form of land transportation owing to its many economic, social, energetic, and environmental advantages. Rail transport is characterized by congestion-free traffic and high safety, reduced emissions and fuel consumption per passenger transported, reduced use of land resources (the land requirement is three times less than that for motorways), and high transportation speed. A significant proportion of rail transport operations pertains to freight transport, which generally involves cargo carried by light or heavy rails. In addition, the goods are transported via trams, funiculars, and monorails. For example, the GHG emissions resulting from diesel–electric rail locomotive operations amount to approximately 120 kt CO_2 equivalent per year in Canada, with a projected increase rate of 2.8 kt/year over the next 15 years (Dincer and Zamfirescu 2016). The clean rail initiatives in various jurisdictions focus on using electric locomotives as opposed to their diesel counterparts. However, because most freight and passenger railway fleets worldwide run on diesel fuel, many jurisdictions have implemented immediate measures to lower the emissions from current diesel locomotives, without applying any major changes. Current policy measures have led to the enforcement of new regulations to reduce the amount of sulfur, soot, and other pollutants. Most project initiatives emphasize reducing emissions such as NO_x by applying an NH_3/urea-SCR system and PM by using diesel particulate filters.

In the 1980s, alcoholic fuels received significant attention, given that research at this time was motivated by various fuel crises (Cataldi 1988; Onion and Bodo 1983). Methanol has since emerged as a strong alternative that can meet future fuel requirements for locomotive traction. Methanol is a promising replacement for conventional diesel because it can be prepared from renewable and waste resources; its usage in IC engines can lead to significantly lower emissions. Kumar et al. investigated a high-pressure, coaxial, direct injection system with methanol fueling by using a 1D simulation approach (Kumar et al. 2021). Methanol can be used in locomotive engines as a traction fuel, and the coaxial injector design offers new avenues for the introduction of higher quantities of methanol, thus enabling a larger displacement of mineral diesel. Simulation results showed that locomotives using methanol as a fuel exhibited superior performance, emissions, and combustion characteristics, as compared to diesel-based locomotive engines.

Seyam et al. studied alternative fuels (hydrogen, methanol, ethanol, and dimethyl ether) for a cleaner locomotive powering system, as shown in Fig. 5.14 (Seyam et al. 2021). The abbreviations in the figure are as follows. ARS: absorption refrigeration system, GT: gas turbine, ICE: Internal combustion engine, MCFC: molten carbonate fuel cell. Using these fuels, a powering system can produce 4200 kW, which is twice that produced by internal combustion engines, accompanied by thermal and energy efficiencies of 43% and 55%, respectively. The CO_2 emissions reduced by over 60% when using the alternative fuels. In addition, parametric studies have focused on the operating pressure of the MCFC and GT. The best performances for the MCFC and

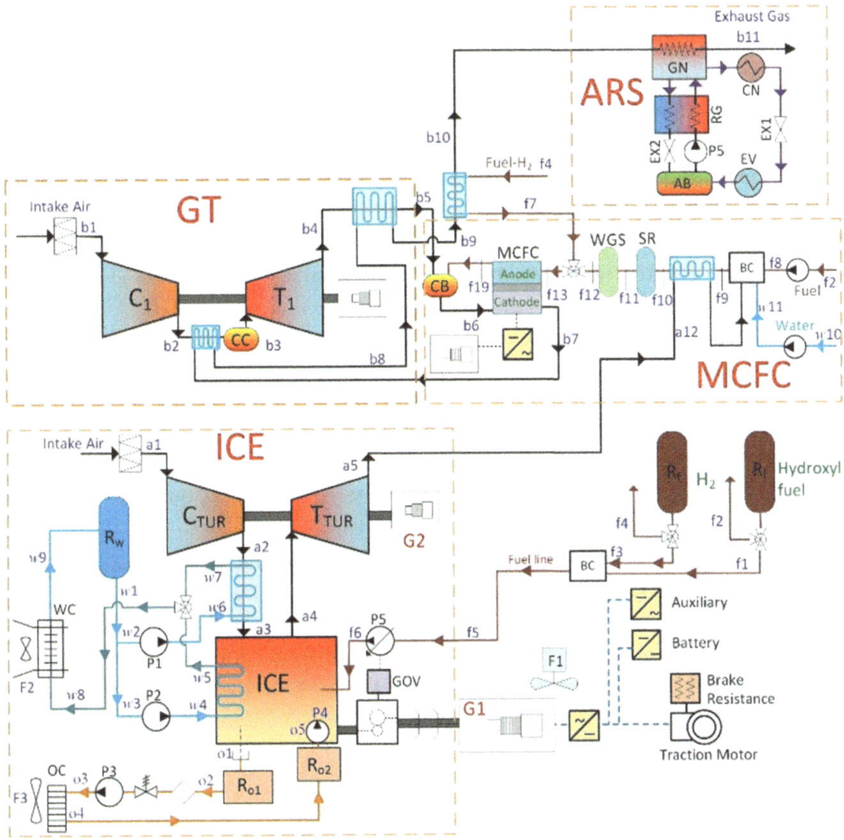

Fig. 5.14 The configuration of the hybrid combined engine system (Seyam et al. 2021)

gas turbine can be achieved at 200 kPa and 900 kPa, respectively. Therefore, a hybrid combined engine can provide high power and performance, with less CO_2 emissions.

5.3.4 *Aviation*

Aviation is one of the fastest-growing sources of energy-related emissions and also the most climate-intensive means of transportation; the global aircraft fleet is expected to increase from approximately 27,000 in 2019 to 39,000 by 2030 (Gray et al. 2021). Aviation is responsible for approximately 2% of the global emissions or 900 Mt CO_2 (International Renewable Energy Agency 2020). The aviation sector releases significant amounts of CO_2 from combustion, and it is also responsible for considerable amounts of non-CO_2 warming effects (Stratton et al. 2011).

Further, in the field of aviation, methanol was used for its benefits in terms of the octane number and latent heat, although this was solely during take-off and to a certain extent when maximum power was required. Energy density is an overriding consideration for aircraft propulsion; thus, straight hydrocarbon fuels were mostly used in highly boosted aircraft engines. However, during take-off, maximum performance is required at sea level; thus, supplementary mixtures of methanol and water were used as a knock suppressant, which is injected with the main fuel into the compressor entry.

Figure 5.15 summarizes the direct life cycle analysis (LCA) emissions from a selection of alcohol to jet (ATJ) pathways (Pavlenko and Searle 2021). The various ATJ pathways result in direct LCA emissions ranging from 23.8 to 65.7 g CO_2/MJ. Sugar cane-derived ATJ pathways, which typically include either lignocellulosic residues or energy crops, could have low emissions. This is because of the significantly lower feedstock production emissions, as compared to most purpose-grown food crops, and because combusting the lignin in these crops, which cannot be hydrolyzed into sugars, produces renewable electricity as a by-product (Pavlenko and Searle 2021).

Another potential source of feedstocks for ATJ are the flue gases from steel mills. Although the typical emissions for ATJ produced from flue gases have not been characterized by the International Civil Aviation Organization thus far, a recent LCA for the Lanzatech process estimated that the ethanol production emissions from blast furnace gases amount to 31.4 g CO_2/MJ (Handler et al. 2016).

With regard to the advantages and technical and/or economic features of using ethanol as fuel in agricultural aviation, the effect of ethanol may be even greater if the aircraft engine is specially designed (Hausen et al. 2010). The use of ethanol shows considerable technical and economic feasibility when compared to aviation gas, even with the low efficiency of current applications. Operational improvements and advantages in cost reduction can already be observed in existing applications. Moreover, with a specific engine design, these advantages can be further improved

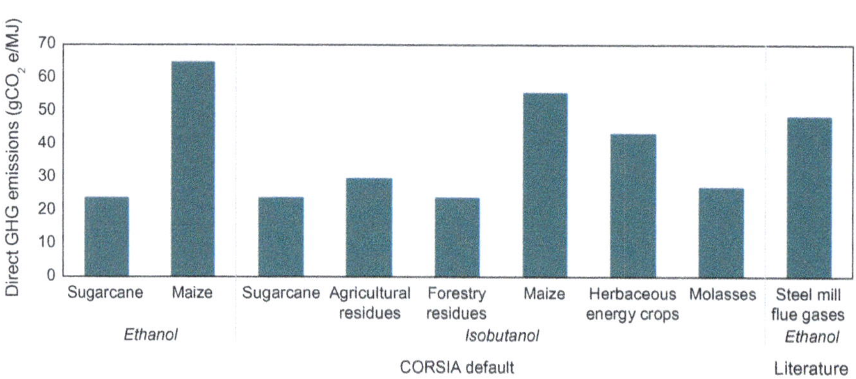

Fig. 5.15 Summary of direct LCA GHG emissions for ATJ production pathways (Pavlenko and Searle 2021)

owing to the increased fuel efficiency and power. The emission rates of CO, HC, and NO_x are also lower than those when using aviation gas. CO_2 emissions are not a concern because ethanol can be obtained from a renewable source.

Some analyses indicate that the use of an alternative jet fuel (AJF) could reduce the lifecycle of GHG emissions from aviation by a maximum of 68.1% by 2050 (Staples et al. 2018). However, these scenarios require prices or policies to emphasize AJF production over other potential uses for the primary bioenergy resources. In addition to the limitations on feedstock availability, the existence of sufficient bio-refining infrastructure for AJF production could act as a binding constraint on emission reduction.

5.3.5 Non-road Vehicles

The production of green energy on farms could accelerate the application of electric vehicles (EVs), which would lead to a decrease in GHG emissions. Although more novel technologies are being employed in farm machinery and equipment, certain drawbacks hinder the development of EVs, such as the high cost of EVs and the easy accessibility of fossil fuels. In addition to supplying power for EVs, on-site generation of renewable energy could compensate for the farms' electricity demands, especially at locations far from the network. Thus, it can be concluded that solar energy, biomass, and hydrogen show good potentials for on-site energy production in the agricultural sector (Ghobadpour et al. 2019).

Fan et al. investigated the carbonyl emissions from a non-road diesel engine of machinery running on diesel or a methanol–diesel blend fuel under a series of steady-state operating conditions (Fan et al. 2018). The results indicated that formaldehyde and acetaldehyde are the most abundant carbonyls. The methanol/diesel produces significantly higher amounts of formaldehyde, acetaldehyde, acrolein, acetone, and crotonaldehyde than diesel at low loads, whereas there are marginal differences in the emission levels of these carbonyls between the methanol/diesel and diesel at medium and high loads. Furthermore, using the methanol/diesel decreases propionaldehyde emissions under all test conditions. Compared with the results obtained using diesel, the total CBC emissions increase by 45.3%, and the ozone formation potential increases by 57% when burning the methanol/diesel at low engine loads. In general, the methanol/diesel increases both the ozone formation potential and the total carbonyl compound emissions to a greater extent than biodiesel-blended fuels but to a lesser degree than ethanol- or butanol-containing blends.

Saxena et al. studied the effect of the fuel premixing ratio, direct fuel injection timings, and engine CR on soot particle emissions in the nano-size range from a non-road CI engine (Saxena and Maurya 2017). Experiments were conducted to run the engine in the dual-fuel mode, and a PFI system was installed by modifying the intake manifold of the engine and developing a PFI controller. The experiments involved various fuel premixing ratios of gasoline/methanol–diesel at different engine loads, diesel fuel injection timings, and CRs. The results revealed that unburned HC and

CO emissions increase with the premixing of both gasoline and methanol, whereas the NO emission decreases. It was also observed that the using fuel premixing ratio was limited by the HC emission and the combustion stability; the total PM number concentration was higher at full engine loads and increased with the fuel premixing ratio, especially in the case of methanol–diesel dual-fuel operation.

Ning et al. investigated the effects of adding methanol, ethanol, and n-butanol on the combustion characteristics and performance of a common rail dual-fuel engine with diesel direct injection and alcohol fuel port injection (Ning et al. 2020). The experimental results demonstrated that slower flame development and faster flame propagation can be obtained by mixing any of the three alcohol fuels with diesel, as compared with those when using pure diesel. With an increase in the alcohol percentage, the coefficient of variation of the indicated mean effective pressure and the BTE decreased. The addition of primary alcohol fuels in the dual-fuel mode also increased the total THC and NO_x emissions; however, the CO and soot emissions were decreased. The addition of methanol afforded the lowest indicated mean effective pressure and the highest BTE, among the three alcohol fuels; it also results in the lowest CO, NO_x, and soot emissions and the highest THC emissions, as shown in Fig. 5.16 (Ning et al. 2020).

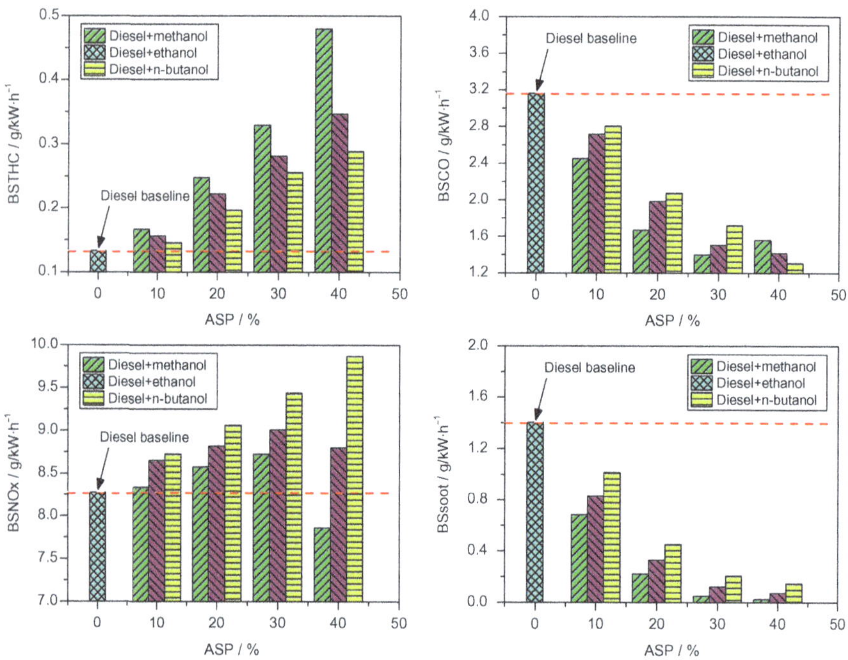

Fig. 5.16 Effect of alcohol fuels additions on HC, CO, and NO emission for non-road common rail dual fuel engine (Ning et al. 2020)

5.4 Fuel Cell Powertrain with Alcohols for Transports

A fuel cell (FC) converts fuels and oxygen into electricity through an electrochemical reaction; it offers the advantages of higher electrical efficiency (approximately 40–65%), higher system reliability, and lower maintenance costs (Koscher and Kordesch 2003; Akikur et al. 2013). Moreover, FCs have a lower noise level because the basic fuel cell stack has no moving components; hence, FCs facilitate comfortable and silent environments during operation (Hemmati and Saboori 2016; Biert et al. 2016). Based on their working temperature ranges, FCs can be divided into three categories: high-temperature, intermediate temperature, and low-temperature FCs (Sharaf and Orhan 2014). Solid oxide fuel cells (SOFCs) and MCFCs are high-temperature FCs that can operate in a temperature range of 650–1000 °C (Koscher and Kordesch 2003; Sharaf and Orhan 2014). Phosphoric acid fuel cells, whose working temperature is approximately 200 °C (Sharaf and Orhan 2014) are considered as intermediate temperature fuel cells. Alkaline fuel cells, DMFCs, and proton exchange membrane fuel cells (PEMFCs) are low-temperature fuel cells that operate below temperatures of 120 °C (Sharaf and Orhan 2014). Furthermore, PEMFCs can be classified into low-temperature PEMFCs (LT-PEMFCs) and high-temperature PEMFCs (HT-PEMFCs); the working temperature ranges of these cells are 60–80 °C and 110–180 °C, respectively.

FCs can operate continuously with the supply of suitable fuels such as hydrogen, LNG, methanol, or biogas. Compared conventional fuels, FCs offer notable benefits to shipping such as high electrical efficiency (Koscher and Kordesch 2003), no pollutant emissions, lower noise and vibration (Koscher and Kordesch 2003; Pachauri and Chauhan 2015), lower maintenance costs, and higher system reliability (Singh et al. 2017; Pan et al. 2021). MCFCs, SOFCs, and PEMFCs combined with a reformer can be employed depending on traditional fuels such as diesel and natural gas, which are more appropriate for use in large-scale ships (Han et al. 2014a). However, most existing FC-powered ships are equipped with low-temperature FCs, owing to their zero emissions and quicker start-up capability (Singh et al. 2017).

5.4.1 Vehicles

FCs offer several advantages over combustion engines. For example, they are smaller, more efficient, silent, have no or low environmental impacts, and can cover a wide range of applications ranging from a few watts to several hundred megawatts (Wilberforce et al. 2019). Thus, in the transportation sector, FCs are considered as an ideal alternative to overcome the aforementioned problems associated with existing internal combustion engines. By 2040, it is expected that the EV market will reach 350 million units of passenger cars. Consequently, EVs would account for one-fourth of all passenger vehicles (Olabi et al. 2021).

The first mass-produced hydrogen FC cars were introduced by Hyundai and Toyota. Additionally, trucks and heavy-duty vehicles are now being produced to aid in decarbonizing the transport sector (Hardman and Tal 2018). Most of these FC cars use hydrogen fuel; thus, this work only covers cases involving alcohol FC cars.

Two types of FC systems using methanol exist: the methanol-reform type and DMFCs. The methanol-reform type system is large and complicated. DMFCs are a viable option for portable energy sources (Alias et al. 2020). In regard to DMFCs, the liquid state of methanol is easier to handle than hydrogen; it possesses a high energy density and is environmentally friendly (Schulze Lohoff et al. 2016; Yuan et al. 2013). Methanol is also easy to store, easy to refuel, and safe to handle at standard temperatures and pressures (Calabriso et al. 2015; Atacan et al. 2017). A DMFC is a simple system that is suitable for use as a power source in portable and low-power electrical devices (Lee et al. 2013). Therefore, DMFCs are not suitable as the main powering device for the propulsion of a ship; they can instead be used as a power source for auxiliary devices.

Moreover, the DMFC system can be divided into active and passive systems (Kamarudin et al. 2009), which differ in terms of their power outputs and components. However, certain drawbacks of passive systems arise owing to their low energy density, power, and output voltage caused by the limited supply of the reactant to the cell (Wang et al. 2015; Lee et al. 2017; Masdar et al. 2017). This limited power emanating from passive DMFCs can be improved by using an active DMFC system; several studies have focused on this device modification. A DMFC requires sufficient fuel to produce the optimal energy output; thus, fuel management is a key parameter. Fuel management in active DMFC systems enhances the design of flow fields that enable the transport of fuels, such as methanol, air, and oxygen, to the anodes and cathodes (Yuan et al. 2017; Maslan et al. 2015; Wong et al. 2006; Jung et al. 2007; Li et al. 2009). An active system employs a design with a serpentine (Maslan et al. 2015; Wong et al. 2006; Jung et al. 2007; Li et al. 2009; Yang and Zhao 2005) or parallel (Maslan et al. 2015; Wong et al. 2006; Jung et al. 2007; Li et al. 2009; Yang and Zhao 2005) shape and a grid pattern (Jung et al. 2007). Kunimatsu et al. developed a method for measuring the methanol concentration to analyze the characteristics of the power output and thermal efficiency of a DMFC using a nondispersive infrared sensor (Kunimatsu et al. 2001).

For FCs, the utilization of biofuels such as methanol and ethanol instead of hydrogen alleviates problems pertaining to fuel transportation and storage and also provides a CO_2-neutral power generation technology, thus leading to a reduction in CO_2 and other pollutants. Bioethanol is particularly suitable as it is nontoxic, inexpensive, renewable, and readily available. These factors make ethanol viable from the economic and environmental perspectives, especially when used with a low emission technology such as FCs (Kamarudin et al. 2009).

Most studies on liquid electrolyte-based alcohol FCs have focused on methanol or ethanol mixed with alkaline electrolyte media. In 1965, an alkaline/methanol FC system (methanol/KOH (4.5 M/10 N)) with Pt (2–5 mg/cm^2) as the anode catalyst and electrochemically active carbon (surface area 250 cm^2/g) as the cathode catalyst was investigated (Kamarudin et al. 2009). Using Pt-Ru (9:1 ratio by weight) as the anode catalyst instead of Pt, the current density was found to increase by some parameters. Prabhuram et al. studied the methanol oxidation on unsupported Pt electrodes under various ratios of methanol and KOH solutions (Wang et al. 2015). Methanol oxidation rates were also studied by employing cyclic voltammetry and via steady-state polarization measurements. It was found that the methanol oxidation activity increased as the solutions changed from acidic to basic. The highest oxidation rate was noted for the 6 M methanol/6 M KOH solution, and it was concluded that, through the appropriate selection of the hydroxyl ion to methanol concentration, the formation of intermediate reaction products can be ceased completely.

Another direct-ethanol FC is based on a proton-conducting polymer electrolyte membrane fuel cell (PEMFC), where ethanol is fed directly to the cell instead of hydrogen. The membrane electrode assembly comprises a polymer electrolyte membrane electrolyte sandwiched between the anode and cathode. The polymer electrolyte membrane may be based on Nafion or a different polymer, but it still transports protons. At the cathode, oxygen is reduced by protons migrating through the electrolyte membrane to water. However, compared to those for the direct hydrogen-based PEMFC, the ethanol conversion and utilization rates are significantly low. Problems pertaining to slow or considerably sluggish electrode kinetics at the anode, ethanol crossover, and the formation of many intermediate products rather than the desired complete conversion of ethanol to CO_2 were noted. The C–C bond breakage is an arduous process during the electrochemical oxidation of ethanol. In the material of construction (cathode, anode, electrolyte); lifetime operational issues including cell/stack degradation and their causes, the performance achieved, and the current status of the technology have been discussed. Other FC components (such as bipolar plates or interconnects and seals), the cell and stack designs, and the membrane electrode assembly processes are somewhat similar to those for conventional hydrogen-based PEMFCs (Badwal et al. 2015).

SOFCs operate in the intermediate to high-temperature range (400–1000 °C). SOFCs are based on solid oxide electrolytes, which may be either an oxygen ion (O^{2-}) or a proton-conducting (H^+) ceramic material. The oxygen-ion conductors are the most popular electrolytes and are used in many commercialized SOFC systems. Figure 5.17 shows a schematic of an SOFC with various possible fuels. In a typical, commercial planar SOFC, ceramic electrolytes are prepared using ceramic fabrication methods such as tape casting, followed by the screen printing of the anode and

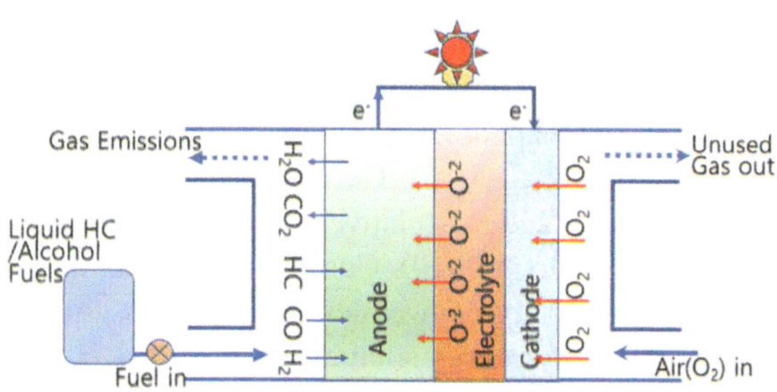

Fig. 5.17 Principle of typical SOFC

cathode materials (Lee et al. 2012). The stacking is completed using suitable metallic, ceramic, or cermet interconnects and compressive seals. Other SOFC designs include anode- or cathode-supported planar or tubular SOFCs. Details regarding the SOFC geometry, fabrication, and design can be found in literature (Singhal and Kendal 2003; Stambouli and Traversa 2002). Most SOFC systems are operated using either pure hydrogen or hydrogen/CO fuels; however, number of studies have used various other hydrocarbons such as methane to fuel SOFCs (Lee et al. 2012). Owing to the relatively higher operating temperature of most advanced SOFCs (>800 °C) (Badwal et al. 2014), the direct internal utilization of ethanol is possible, without requiring external reforming. The fuel electrode (anode) of an SOFC can comprise materials such as Ni and CeO_2; these materials are known to be highly catalytic toward hydrocarbon oxidation and may cause ethanol break-up. Several possible reactions and reaction mechanisms via which ethanol can be consumed in these high-temperature FCs exist. The overall reaction mechanism in an ethanol-fueled high-temperature FC can be a highly complex combination of thermochemical and electrochemical reactions occurring at various locations in the FC. These locations include the inlet fuel flow channels, anode chamber, anode surface, or triple-phase boundary between the fuel, anode, and electrolyte. However, this technology is still in its infancy owing to the issues discussed above, and its commercialization will require significant efforts (Badwal et al. 2015). Thus, very few systems are employing the prototype fuel cells exist, either directly fed with ethanol solutions (DEFC).

5.4.2 Ships

With the development of FCs, various FC power systems with different power levels have been used in ships. Therefore, selecting the appropriate FC power system and fuel is expected to have significant impacts on the suitability of ship power systems (Pan et al. 2021). For choosing a suitable FC power system for ship use, the overall efficiency, system complexity, power density, fuel selection, and fuel processing devices should be considered (Biert et al. 2016; Pachauri and Chauhan 2015). Most FCs can operate on pure hydrogen (Kee et al. 2015). However, the cost of converting diesel to hydrogen onboard would be rather high, and this would result in a low energy density (Liu et al. 2021; Zhu et al. 2021). Furthermore, incorporating a diesel fuel processor would increase the complexity and size of the entire FC system. Alternative fuels such as biofuels or even solar fuels can serve as a solution to alleviate the aforementioned issues (Gray 2009; Danl and Chorkendorff 2012). Biert et al. reported that alternative logistics fuels could be used in the FC power systems of ships (Biert et al. 2016). The main factors restricting the widespread use of FCs are related to fuels and the relatively low output power range of a single fuel cell stack (Vogler and Würsig 2010). Currently, FCs can only supply a maximum of 350 kW. Considering the high power demand of ocean-going ships, existing FCs can only be used as auxiliary energy sources.

Many countries have already or will soon develop commercial products or plans for FC-powered ships. Pan et al. reported on the FC-powered ships in the world as an example of Fig. 5.18 (Pan et al. 2021). In the future, the ship power system will be integrated with fuel cells, solar energy, wind energy, wave energy, batteries and diesel engines. PEMFCs running on hydrogen are mainly used in the existing FC-powered ships. High-temperature fuel cells running on diesel, LNG, or methanol are more appropriate for ships with greater power demands (Han et al. 2014b).

FCs can meet the power demands of loads under steady-state working conditions; however, they are unable to dynamically respond to transient power demands (Vogler and Würsig 2010). Under a varying load, the generation capacity of an FC power system cannot meet the constantly changing electrical power needs (Kim et al. 2004). Dynamic changes in the power requirements exert severe stress on the FC membrane, which reduces the service life (Kickulies 2005). Consequently, the introduction of FCs in ships in hindered by this lack of a dynamic response.

Different FCs and logistics fuels can constitute potential choices, and the specific combination selected has a significant influence on the system characteristics. Biert et al. reviewed power systems integrated with different FCs for maritime applications and indicated that the electrical efficiency, power, energy density, load transients, system start-up, environmental impact, safety, reliability, and economics should be considered when selecting a suitable FC power system (Biert et al. 2016).

Straza et al. evaluated the use of methanol in SOFCs employed as auxiliary power systems for commercial ships by using LCA (Strazza et al. 2010). The LCA methodology enables an assessment of the potential environmental impacts throughout the life cycle of the process. The unit considered was a 20 kW FC system. The different

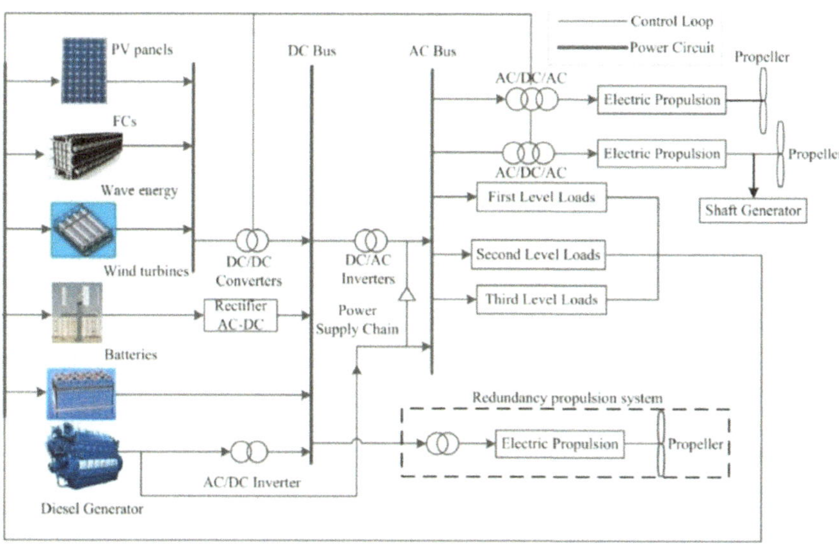

Fig. 5.18 Structure of the ship power system integrated with new energy sources (Pan et al. 2021)

fuel options were compared (methanol; biomethanol; natural gas; and hydrogen from cracking, electrolysis, and reforming); furthermore, the operation of a cell fueled by methanol was compared with that of a traditional auxiliary power system, i.e., a diesel engine. The environmental benefits of using FCs were assessed considering different impact categories. Based on the results, the fuel production phase has a strong influence on the life cycle; the results also indicated that feeding with biomethanol is a highly attractive solution from a life cycle perspective (Strazza et al. 2010).

5.4.3 *Locomotive*

Scott et al. proposed deploying FC locomotives three decades ago (Scott et al. 1993). They reported a matrix of technological pathways, including onboard storage of H_2 sources, the range, duty cycles, and novel train technologies such as smart trains. They considered the prospect of legislation requiring the incorporation of anthropogenic CO_2 emissions in life cycle costing.

Al-Hamed et al. investigated a powering system for clean locomotives with CH_4 SOFCs (Al-Hamed and Dincer 2020). Some FC-powered locomotives employing hydrogen PEMFCs (Hong et al. 2018b) and SOFCs (Al-Hamed and Dincer 2020) exist. A few studies presented alcohol-fueled FC for locomotives; for example, Elleuch et al. reported a Nickel-SDC anode for the direct utilization of biomethanol in SOFCs (Elleuch et al. 2016). Cell performance when using methanol was slightly better than that when using hydrogen. The better performance afforded by methanol could also be a consequence of the improved electronic conductivity in the presence of carbon deposits. The results indicated that the catalytic decomposition of methanol in the anodic three-phase boundaries was dependent on the temperature and governed by complex anodic electrochemical dynamics.

5.4.4 *Aviation*

Based on statistics, CO_2 emissions from the aviation sector account for over 2% of the total global CO_2 emission (Masiol and Harrison 2014), and this value is expected to grow at an annual rate of 5% (Fernandes et al. 2018a). The development of a commercial all-electric aircraft is, therefore, of increasing global interest to reduce both emissions and the cost per passenger-mile. Battery-powered, short-range aircraft that may be in service as early as 2021 are being introduced by certain companies (Collins and McLarty 2020). Current state-of-the-art batteries store only 2.5% of the energy stored by standard jet fuel on a mass basis. This low energy storage density of batteries limits the payload and range of planned electric aircraft prototypes to less than 20 passengers and 600 nautical miles, respectively. NASA has also invested heavily in electric aircraft research, and larger prototypes are expected to be test-ready by 2025 (Jansen et al. 2017).

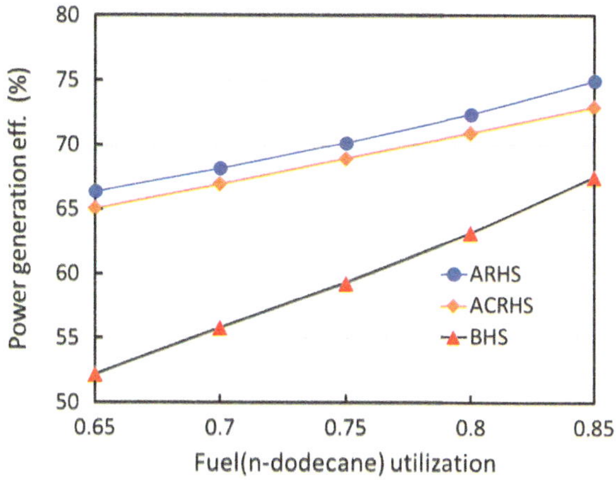

Fig. 5.19 Various power generation efficiency with fuel use (Liu et al. 2021)

SOFC systems have been considered for supplemental power generation in aviation owing to their high potential fuel-to-electricity conversion efficiency (Fernandes et al. 2018b). Previous studies have established that existing SOFC technology cannot match the power density of turbine engines. However, SOFC technology offers the benefit of increased efficiency, in conjunction with airframe concept designs (Bradley and Droney 2012). Recently, the use of hybrid power generation system, like as FC–gas turbine, is increasing in the aviation. Liu et al. reported an advanced SOFC–gas turbine hybrid power generation system based on aviation hydrogen or kerosene (Collins and McLarty 2020; Liu et al. 2021).

The fuel (n-dodecane) utilization of solid oxide fuel cells (SOFCs) requires a particular ratio of fuel with electro-chemical reaction to total inlet fuel, as shown in Fig. 5.19. Chemical energy in SOFCs can be directly converted to high-grade electrical energy, whereas certain fuels entering the combustor must undergo three energy conversion processes to yield electric energy. Figure 5.19 shows the power generation efficiency of the three configurations of hybrid system with fuel use (Liu et al. 2021). In the figure, BHS refers to basic fuel cell gas turbine hybrid power generation system, ARHS refers to the anode exhaust recirculation hybrid power generation system, and ACRHS refers to the anode and cathode exhaust recirculation hybrid power generation system. More fuel reacts in SOFC with the increase of fuel use. The power generation efficiency of the three types of hybrid systems increased with the increase in fuel utilization, with that of ARHS being greater than the others.

Based on the study of Zhu et al., hydrogen-powered PEMFCs appear promising for the main propulsion power in future aviation applications (Zhu et al. 2021). However, few studies have reported on alcohol-fueled FCs for aviation.

5.5 Conclusions

This work aims to disseminate information regarding the key role of various alcohol fuels in promoting a secure, eco-friendly, and affordable energy future for transportation facilities, including vehicles, ships, locomotives, aviation, and non-road vehicles. The physical and chemical properties of major alcohol fuels, which are methanol, ethanol, propanol, and butanol, were reviewed. Thereafter, the history of alcohol fuels used in transportation, characteristics and advantages, as well as current status and problems associated with their use were extensively reviewed. The impacts of using alcohols as alternative fuels in terms of their performance, combustion characteristics, and emissions were also investigated.

(1) In vehicles, spark ignition (SI) and compression ignition (CI) engines are required to decrease the CO_2 and GHG emissions, wherein the utilization of alcohol fuels can lower GHG emissions. It is necessary to assess the efficiency of methanol engines from a broad perspective and acquire additional data pertaining to the influence of water tolerance and phase separation of methanol-fueled vehicles. Ethanol can afford clear decrements in NO_x, CO, and HC emissions in SI engines, while ethanol requires appropriate blending for use in CI engines to avoid undesirable phase separation of ethanol–diesel blending along with an increase in the ethanol ratio. It is worth noting that the use of biodiesel ethanol diesel (DBE) could weaken the PM-PN-NO_x trade-off relationship. Propanol emits the highest concentration of NO_x at full load, but the introduction of exhaust gas recirculation (EGR) could reduce NO_x emissions. In CI engines, propanol can be used as an additive to improve the alcohol-diesel mixture by enhancing the lower cetane number of ethanol blends; it also stabilizes the mixture in the presence of high content of water, resulting in fuel homogeneity under all conditions. In high-speed SI engines, an increase in butanol ratio improved the output power, BTE, and BSFC, whereas CO emissions were slightly increased and NO emissions were significantly decreased. For CI engines, butanol is an excellent alternative fuel, which can significantly shorten the combustion duration via an increase in butanol ratio. Most butanol-diesel blended fuels can reduce PM and CO emissions but lead to greater NO_x emissions as compared to pure diesel fuel.

(2) In ships, SO_x, NO_x, and CO_2 emissions have attracted more attention due to their large amounts. To comply with International Maritime Organization (IMO) emission regulations, the use of methanol could be an alternative fuel to reduce NO_x, SO_x, CO, CO_2, and PM emissions. From an environmental perspective, both methanol and ethanol are suitable fuels. However, some measures, such as safety regulations, operational experience, and infrastructure for bunkering, need to be strengthened. In locomotives, alcohol fuels have received considerable attention owing to the fuel crises over the past year. The use of methanol in locomotives presents superior performance, low emissions, and improvement of combustion characteristics, as compared to pure diesel-based locomotive engines.

(3) In the field of aviation, methanol can be used for its numerous benefits, such as a high octane number and latent heat, although this is solely applicable during take-off. The supplementary mixtures of methanol and water can also be used as a knock suppressant, which is injected with the primary fuel. Despite the low efficiency of current applications, the use of ethanol has considerable technical and economic feasibility as compared to current aviation gas. In addition, methanol can reduce NO_x, CO, and HC emissions; further, CO_2 is not a concern as ethanol can be produced from renewable sources.

(4) In non-road vehicles, methanol-diesel blended fuel increases both the ozone formation potential and the total carbonyl compounds, as compared to biodiesel-blended fuels. There are numerous factors, including fuel premixing ratios, engine loads, and fuel injection timings, which can be optimized to reduce HC, PM, and NO_x emissions. Fuel cells are considered as an alternative to overcome the current problems associated with existing internal combustion engines. For vehicle applications, liquid electrolyte-based alcohol FCs have focused on methanol or ethanol mixed with alkaline electrolyte. Owing to the high operating temperature of advanced SOFCs (>800 °C), the internal direct utilization of ethanol is possible without any external reforming device. The direct-ethanol proton-exchange membrane fuel cells (PEMFCs) exhibit low ethanol conversion and low utilization rates, resulting in difficulties in practical applications. For ship applications, many FC power systems with different power levels have been developed. Considering the high-power demand of ships, current FCs can only be used as auxiliary energy sources owing to their limited maximum power of 350 kW. Methanol-based SOFCs are employed in auxiliary power systems of commercial ships; the power of the FC system is 20 kW using LCA methodology. For a suitable FC power system for commercial maritime applications, the electrical efficiency, power, energy density, load transients, system start-up, environmental impact, safety, reliability, and economics should be considered.

(5) For locomotive applications, some FC-powered locomotives employing H_2 PEMFCs and SOFCs were investigated, and a few studies were conducted for alcohol-fueled FCs over the locomotives. Using methanol-fueled SOFCs presents a slightly improved cell performance than H_2-based SOFCs, which is attributed to the improved electronic conductivity in the presence of carbon deposits. In the field of aviation, SOFCs system have been considered for supplemental power generation owing to their high fuel-to-electricity conversion. Recently, H_2-powered PEMFCs have shown promising potential for main propulsion power in future aviation applications, while few studies have been reported on the use of alcohol-fueled FCs in this field. The appropriate utilization of alcohol fuels with existing fossil fuels, including diesel, gasoline, and biofuels, can provide promising alternative fuels for compliance with the stringent emission standards for transportation facilities, owing to the significant reduction in exhaust gases and GHGs. However, it is imperative to develop suitable modifications for control, material compatibility, and fuel injection systems, to help realize the practical utilization of alcohol fuels. Thus, the use

of alcohols as alternative fuels for transportation could help reduce emissions, establish an eco-friendly society, and facilitate the design of a pollution-free and sustainable society in the future.

References

Administrative notice of automobile GHG (CO_2) emission, fuel efficiency standard. Ministry of Environment, Republic of Korea, 23 Aug 2020.

Agarwal AK, Anirudh G, Sharma N, Singh AP (2019) Methanol and the alternate fuel economy. Springer

Aikawa K, Sakurai T, Hayashi A (2009) Study of ethanol-blended fuel (E85) effects under cold-start conditions. SAE Int J Fuels Lubr 2(1):196–203

Akikur RK, Saidur R, Ping HW, Ullah KR (2013) Comparative study of stand-alone and hybrid solar energy systems suitable for off-grid rural electrification: a review. Renew Sust Energ Rev 27:738–752

Alasfour F (1998) NO_x emission from a spark ignition engine using 30% iso-butanol-gasoline blend: part 1-preheating inlet air. Appl Therm Eng 18:245–256

Al-Hamed KHM, Dincer I (2020) Investigation of an integrated powering system for clean locomotives with solid-oxide fuel cell with heat recovery organic rankine cycle. Energy Conv Manag 219:112857

Alias MS, Kamarudin SK, Zainoodin AM, Masdar MS (2020) Active direct methanol fuel cell: an overview. Int J Hydrogen Energy 45(38):19620–19641

Alonso DM, Bond JQ, Dumesic JA (2010) Catalytic conversion of biomass to biofuels. Green Chem 12:1493–1513

Ammar NR (2019) An environmental and economic analysis of methanol fuel for a cellular container ship. Transp Res Part D Transp Environ 69:66–76

Andersen VF, Anderson JE, Wallington TJ (2010) Vapor pressures of alcohol–gasoline blends. Energy Fuels 24(6):3647–3654

Anderson JE, Kramer U, Mueller SA (2010) Octane numbers of ethanol and methanol gasoline blends estimated from molar concentrations. Energy Fuels 24(12):6576–6585

Asfar KR, Hamed H (1996) Combustion of fuel blends. J Energy Convers Manag 39:1081–1093

Atacan OF, Ouellette D, Colpan CO (2017) Two-dimensional multiphase non-isothermal modeling of a flowing electrolyte e direct methanol fuel cell. Int J Hydrogen Energy 42(4):2669–2679

Atkins PW (1994) Physical chemistry, 5th edn. Oxford University Press. ISBN 0198557302

Awad OI, Ali OM, Rizalman M, Abdullah AA, Najafi G, Kamarulzaman MK, Yuri IM, Noor MM (2017) Using fusel oil as a blend in gasoline to improve SI engine efficiencies: a comprehensive review. Renew Sust Energ Rev 69:1232–1242

Badwal SPS, Giddey S, Munnings C, Kulkarni A (2014) Review of progress in high temperature solid oxide fuel cells. J Aust Ceram Soc 50:23–36

Badwal SPS, Giddey S, Kulkarni A, Goel J, Basu S (2015) Direct ethanol fuel cells for transport and stationary applications—a comprehensive review. Appl Energy 145:80–103

Baek S, Cho J, Kim K, Myung CL, Park S (2021) Effect of engine control parameters on combustion and particle number emission characteristics from a SIDI engine fueled with gasoline-ethanol blends. J Mech Sci Tech 35:1289–1300

Balamurugan T, Nalini R (2014) Experimental investigation on performance, combustion and emission characteristics of four stroke diesel engine using diesel blended with alcohol as fuel. Energy 78:356–363

Balki MK, Sayin C (2014) The effect of compression ratio on the performance, emissions and combustion of an SI (spark ignition) engine fueled with pure ethanol, methanol and unleaded gasoline. Energy 71:194–201

Balki MK, Sayin C, Canakci M (2014) The effect of different alcohol fuels on the performance, emission and combustion characteristics of a gasoline engine. Fuel 115:901–906
Barrientos EJ, Anderson JE, Maricq MM, Boehman AL (2016) Particulate matter indices using fuel smoke point for vehicle emissions with gasoline, ethanol blends, and butanol blends. Combust Flame 167:308–319
Biert VL, Godjevac M, Visser K, Aravind PV (2016) A review of fuel cell systems for maritime applications. J Power Sources 327:345–364
Bilgin A, Sezer İ (2008) Effects of methanol addition to gasoline on the performance and fuel cost of a spark ignition engine. Energy Fuels 22:2782–2788
Blaisdell TB, Jackson MD, Smith KD (1989) Potential of light-duty methanol vehicles. SAE technical paper 891667
Bradley MK, Droney CK (2012) Subsonic ultra green aircraft research phase II: N+4 advanced concept development. NASA/CR-2012-217556. National Aeronautics and Space Administration
Brusstar M, Stuhldreher M, Swain D, Pidgeon W (2002) High efficiency and low emissions from a port-injected engine with neat alcohol fuels. SAE technical paper 2002-01-2743
Calabriso A, Cedola L, Del Zotto L, Rispoli F, Santori SG (2015) Performance investigation of passive direct methanol fuel cell in different structural configurations. J Clean Prod 88:23–28
Calam A, İçingür Y, Solmaz H, Yamık H (2015) A comparison of engine performance and emissions of fusel oil and gasoline mixtures at different ignition timings. Int J Green Energy 12(8):767–772
Carneiro NMML, Pradelle F, Braga SL, Gomes MSP, Martins ARFA (2017) Potential of biofuels from algae: comparison with fossil fuels, ethanol and biodiesel in Europe and Brazil through life cycle assessment (LCA). Renew Sust Energ Rev 73:632–653
Cataldi R (1988) Broadened diesel fuel specifications reduce railroad costs. Association of American Railroads TR News Magazine, pp 18–19
Chen Z, Liu J, Han Z, Du B, Liu Y, Lee C (2013) Study on performance and emissions of a passenger-car diesel engine fueled with butanol–diesel blends. Energy 55:638–646
Choi BC (2018) Status of alcohol fuels as renewable energy in Korea. In: Proceedings of the international symposium on alcohol fuels (ISAF)
Choi BC (2021) Combustion engineering. Munundang, Seoul, Korea. ISBN 9791156925453
Choi BC, Jiang X (2015) Individual hydrocarbons and particulate matter emission from a turbocharged CRDI diesel engine fueled with n-butanol/diesel blends. Fuel 154:188–195
Choi BC, Jiang X, Kim YK, Jung GS, Lee CH, Choi IC, Song CS (2015) Effect of diesel fuel blend with n-butanol on the emission of a turbocharged common rail direct injection diesel engine. Appl Energy 146:20–28
Choi BC, Lee KS, Son GS (2020) Review of recent after-treatment technologies for de-NO_x process in diesel engines. Int J Automot Tech 21(6):1597–1618
Collins JM, McLarty D (2020) All-electric commercial aviation with solid oxide fuel cell-gas turbine battery hybrids. Appl Energy 265:114787
Colpin C, Leone T, Lhuillery M, Marchal A (2009) Key parameters for startability improvement applied to ethanol engines. SAE Int J Fuels Lubr 2(1):180–188
Corbett JJ, Winebrake JJ, Green EH, Kasibhatla P, Eyring V, Lauer A (2007) Mortality from ship emissions: a global assessment. Environ Sci Technol 41(24):8512–8518
Danl S, Chorkendorff I (2012) Solar-fuel generation: towards practical implementation. Nat Mater 11:100–101
De Caro PS, Mouloungui Z, Vaitilingom G, Berge JC (2001) Interest of combining an additive with diesel–ethanol blends for use in diesel engines. Fuel 80:565–574
Deep A, Kumar N, Gupta D, Sharma A, Patel JS, Karnwal A (2014) Potential utilization of the blend of orange peel oil methyl ester and isopropyl alcohol in CI engine. SAE technical paper 2014-01-2778

Dhande DY, Sinaga N, Dahe KB (2021) The study of performance and emission characteristics of a spark ignition (SI) engine fueled with different blends of pomegranate ethanol. Inter J Energy Environ Eng 12:295–306
Dincer I, Zamfirescu C (2016) A review of novel energy options for clean rail applications. J Nat Gas Sci Eng 28:461–478
Dolan G (2021) Methanol: a future proof marine fuel. Methanol Institute, Sea Commerce, 28 Apr 2021
Elfasakhany A (2015) Investigations on the effects of ethanol-methanol-gasoline blends in a spark-ignition engine: performance and emissions analysis. Eng Sci Technol Int J 18(4):713–719
Elleuch A, Halouani K, Li Y (2016) Bio-methanol fueled intermediate temperature solid oxide fuel cell: a future solution as component in auxiliary power unit for eco-transportation. Mater Des 97:331–340
Ellis J, Tanneberger K (2015) Study on the use of ethyl and methyl alcohol as alternative fuels in shipping, Report prepared for the European Maritime Safety Agency (EMSA)
Eyidogan M, Ozsezen AN, Canakci M, Turkcan A (2010) Impact of alcohol–gasoline fuel blends on the performance and combustion characteristics of an SI engine. Fuel 89:2713–2720
Fan C, Song C, Lv G, Wang G, Zhou H, Jing X (2018) Evaluation of carbonyl compound emissions from a non-road machinery diesel engine fueled with a methanol/diesel blend. Appl Therm Eng 129:1382–1391
Fan C, Wei J, Huang H, Pan M, Fu Z (2021) Chemical feature of the soot emissions from a diesel engine fueled with methanol-diesel blends. Fuel 297:120739
Fernandes MD, Andrade STP, Bistritzki VN (2018a) SOFC-APU systems for aircraft: a review. Int J Hydrogen Energy 43(33):16311–16333
Fernandes MD, Andrade STDP, Bistritzki VN, Fonseca RM, Zacarias LG, Gonçalves HNC, de Castro AF, Domingues RZ, Matencio T (2018b) SOFC-APU systems for aircraft: a review. Int J Hydrogen Energy 43(33):16311–16333
Gainey B, Yan Z, Lawler B (2021) Autoignition characterization of methanol, ethanol, propanol, and butanol over a wide range of operating conditions in LTC/HCCI. Fuel 287:119495
Ghobadpour A, Boulon L, Mousazadeh H, Malvajerdi AS, Rafiee S (2019) State of the art of autonomous agricultural off-road vehicles driven by renewable energy systems. Energy Proc 162:4–13
Goktas M, Balki MK, Sayin C, Canakci M (2021) An evaluation of the use of alcohol fuels in SI engines in terms of performance, emission and combustion characteristics: a review. Fuel 286:119425
Gomasta S, Mahla SK (2012) An experimental investigation of ethanol blended diesel fuel on engine performance and emissions. Int J Emerg Technol 3:74–79
Gong CM, Li J, Li JK (2011) Effects of ambient temperature on firing behavior and unregulated emissions of spark-ignition methanol and liquefied petroleum gas/methanol engines during cold start. Fuel 90(1):19–25
Gravalos I, Moshou D, Gialamas T, Xyradakis P, Kateris D, Tsiropoulos Z (2013) Emissions characteristics of spark ignition engine operating on lower–higher molecular mass alcohol blended gasoline fuels. Renew Energy 50:27–32
Gray HB (2009) Powering the planet with solar fuel. Nat Chem 1:7–7
Gray N, McDonagh S, O'Shea R, Smyth B, Murphy JD (2021) Decarbonizing ships, planes and trucks: an analysis of suitable low-carbon fuels for the maritime, aviation and haulage sectors. Adv Appl Energy 1:100008
Gu X, Huang Z, Cai J, Gong J, Wu X, Lee CF (2012) Emission characteristics of a spark-ignition engine fuelled with gasoline-n-butanol blends in combination with EGR. Fuel 93:611–617
Guo Z, Li T, Dong J, Chen R, Xue P, Wei X (2011) Combustion and emission characteristics of blends of diesel fuel and methanol-to-diesel. Fuel 90:1305–1308
Haigh KF, Vladisavljević GT, Reynolds JC, Nagy Z, Saha B (2014) Kinetics of the pretreatment of used cooking oil using Novozyme 435 for biodiesel production. Chem Eng Res Des 92:713–719

Han JG, Charpentier JF, Tang TH (2014a) An energy management system of a fuel cell/battery hybrid boat. Energy 7:2799–2820
Han JG, Charpentier JF, Tang TH (2014b) An energy management system of a fuel cell/battery hybrid boat. Energy 7:2799–2820
Han J, Bao H, Somers LMT (2021) Experimental investigation of reactivity controlled compression ignition with n-butanol/n-heptane in a heavy-duty diesel engine. Appl Energy 282:116164
Hana J, Somersa LMT, Cracknell R, Joedicke A, Wardle R, Mohan VRR (2020) Experimental investigation of ethanol/diesel dual-fuel combustion in a heavy-duty diesel engine. Fuel 275:117867
Handler RM, Shonnard DR, Griffing EM, Lai A, Palou-Rivera I (2016) Life cycle assessments of ethanol production via gas fermentation: anticipated greenhouse gas emissions for cellulosic and waste gas feedstocks. Ind Eng Chem Res 55(12):3253–3261
Hansen AC, Zhang Q, Lyne PW (2005) Ethanol-diesel fuel blends—a review. Bioresour Technol 96:277–285
Hardman S, Tal G (2018) Who are the early adopters of fuel cell vehicles? Int J Hydrogen Energy 43(37):17857–17866
Hausen RB, Romano LN, Schlosser JF, Martins M, Trindade A (2010) Efficient ethanol engines in agricultural aviation. SAE technical paper 2010-36-0398
Hebbar GS, Bhat AK (2013) Control of NO_x from a DI diesel engine with hot EGR and ethanol fumigation: an experimental investigation. Int J Automot Technol 14(3):333–341
Hemmati R, Saboori H (2016) Emergence of hybrid energy storage systems in renewable energy and transport applications: a review. Renew Sust Energ Rev 65:11–23
Hirasawa T, Sung CJ, Joshi A, Yang Z, Law CK (2002) Determination of laminar flame speeds using digital particle image velocimetry: binary fuel blends of ethylene, n-butane, and toluene. Proc Combust Inst 29(2):1427–1434
Hong W, Yuan C, Xie F, Su Y, Chen J (2018a) Particulate matter and particle-bound PAHs emissions from gasoline direct injection (GDI) engine with methanol-gasoline blended fuel during start. Int J Automot Technol 19:717–726
Hong Z, Li Q, Han Y, Shang W, Zhu Y, Chen W (2018b) An energy management strategy based on dynamic power factor for fuel cell/battery hybrid locomotive. Int J Hydrogen Energy 43(6):3261–3272
Hu J, Abubakar S, Li Y (2020) A novel reduced i-propanol-n-butanol-ethanol (IBE)/diesel mechanism for engine combustion and emissions prediction. Fuel 278:118291
IMO (1997) International Convention for the Prevention of Pollution from Ships MARPOL 73/78: the regulations for the prevention of air pollution from ships (Annex VI). IMO, London
IMO (2011) Marginal abatement costs and cost effectiveness of energy-efficiency measures, MEPC 62/INF.7. Submitted by the Institute of Marine Engineering, Science and Technology
IMO (2015) Third IMO GHG study 2014. Technical report. International Maritime Organization
IMO (2016a) Marine environment protection committee (MEPC), 70th session, 24–28 Oct 2016. IMO, London
IMO (2016b) Methanol as marine fuel: environmental benefits, technology readiness, and economic feasibility. IMO, London
Imran A, Varman M, Masjuki HH, Kalam MA (2013) Review on alcohol fumigation on diesel engine: a viable alternative dual fuel technology for satisfactory engine performance and reduction of environment concerning emission. Renew Sust Energ Rev 26:739–751
International Renewable Energy Agency (2020) Global renewables outlook: energy transformation 2050
Inventory of U.S. Greenhouse Gas Emissions and Sinks: 1990–2013 (2015) (Reports and assessments No. EPA 430-R-15-004). EPA Washington DC
Jansen R, Bowman C, Jankovsky A, Dyson R, Felder J (2017) Overview of NASA electirified aircraft propulstion (EAP) research for large subsonic transports. In: 53rd AIAA SAE/ASEE joint propulsion conference. AIAA 2017-4701, 10–12 July 2017

Jung GB, Su A, Tu CH, Lin YT, Weng FB, Chan SH (2007) Effects of cathode flow fields on direct methanol fuel cell-simulation study. J Power Sources 171:212–217

Kabasin D, Hoyer K, Kazour J, Lamers R, Hurter T (2009) Heated injectors for ethanol cold starts. SAE Int J Fuels Lubr 2(1):172–179

Kamarudin SK, Achmad F, Daud WRW (2009) Overview on the application of direct methanol fuel cell (DMFC) for portable electronic devices. Int J Hydrogen Energy 34(16):6902–6916

Kee RJ, Zhu H, Goodwin DG (2015) Solid-oxide fuel cells with hydrocarbon fuels. Proc Combust Inst 30:2379–2404

Kickulies M (2005) Fuel cell power for maritime applications. Fuel Cell Bull 9:12–15

Kim HN, Choi BC (2008) Effect of ethanol-diesel blend fuels on emission and particle size distribution in a common-rail direct injection diesel engine with warm-up catalytic converter. Renew Energy 33:2222–2228

Kim S, Shimpalee S, Van Zee JW (2004) The effect of stoichiometry on dynamic behavior of a proton exchange membrane fuel cell (PEMFC) during load changes. J Power Sources 135:110–121

Kim KH, Choi BC, Park SH, Kim EG, Chiaramonti D, Emission characteristics of compression ignition (CI) engine using diesel blended with hydrated butanol. Fuel 257:116037

Koscher GA, Kordesch K (2003) Chapter on alkaline methanol/air power devices. In: Vielstich W, Gasteiger HA, Lamm A (eds) Handbook of fuel cells—fundamentals, technology and applications, vol 4. Fuel cell technology and applications. Wiley, pp 1125–1129. ISBN 0-471-49926-9

Kovarik B, Charles F (1994) Kettering and the development of tetraethyl lead in the context of alternative fuel technologies. SAE technical paper 941942

Kumar D, Valera H, Gautam A, Agarwal AK (2021) Simulations of methanol fueled locomotive engine using high pressure co-axial direct injection system. Fuel 295:120231

Kunimatsu M, Shudo T, Nakajima Y, Murase I (2001) Efficiency analysis in a direct methanol fuel cell with a measurement of methanol concentration. SAE technical paper 2001-01-0237

Lapuerta M, Ramos A, Barba J, Fernández-Rodríguez D (2018) Cold- and warm-temperature emissions assessment of n-butanol blends in a Euro 6 vehicle. Appl Energy 218:173–183

Lee K, Gore CM, Wachsman ED (2012) Feasibility of low temperature solid oxide fuel cells operating on reformed hydrocarbon fuels. J Mater Chem 22:22405–22408

Lee SM, Kuan YD, Sung MF (2013) Diaphragm air-liquid micro pump applicable to the direct methanol fuel cell. J Power Sources 238:290–295

Lee J, Lee S, Han D, Gwak G, Ju H (2017) Numerical modeling and simulations of active direct methanol fuel cell (DMFC) systems under various ambient temperatures and operating conditions. Int J Hydrogen Energy 42(3):1736–1750

Lee KS, Choi BC, Lee CB, Oh KC (2020) Effects of SiO_2/Al_2O_3 ratio, reaction atmosphere and metal additive on de-NO_x performance of HC-SCR over Cu-based ZSM-5. J Ind Eng Chem 90:132–144

Lee KS, Choi BC, Kim CH, Lee CB, Oh KC (2021) De-NO_x characteristics of HC-SCR system employing combined Ag/Al_2O_3 and CuSn/ZSM-5 catalyst. J Ind Eng Chem 93:461–475

Leighty W, Ogden JM, Yang C (2012) Modeling transitions in the California light-duty vehicles sector to achieve deep reductions in transportation greenhouse gas emissions. Energy Policy 44:52–67

Li M, Liang J, Liu C, Sun G, Zhao G (2009) Effects of anode flow field design on CO_2 bubble behavior in μDMFC. Sensors 9(5):3314–3324

Liang MS (2021) Assessing emission and power tradeoffs of biodiesel and n-Butanol in diesel blends for fuel sustainability. Fuel 283:118861

Liu H, Qin J, Ji Z, Guo F, Dong P (2021) Study on the performance comparison of three configurations of aviation fuel cell gas turbine hybrid power generation system. J Power Sources 501:230007

Luiz CSSJ, Vítor PF, Julio AMS, Ednildo AT, Iuri MP (2018) Oxidized biodiesel as a cetane improver for diesel–biodiesel–ethanol mixtures in a vehicle engine. J Braz Soc Mech Sci Eng 40:79

Ma B, Yao A, Yao C, Wanga B, Gao J, Chen C, Wu T (2019) Experimental study on energy balance of different parameters in diesel methanol dual fuel engine. Appl Therm Eng 159:113954

Machiele P (1987) Flammability and toxicity tradeoffs with methanol fuels. SAE technical paper 872064

Markel A, Bailey B (1998) Modeling and cold start in alcohol-fueled engines. National Renewable Energy Laboratory

Marriott CD, Wiles MA, Gwidt JM, Parrish SE (2009) Development of a naturally aspirated spark ignition direct injected flex-fuel engine. SAE Int J Engines 1(1):267–295

Masdar MS, Dedikarni Zainoodin AM, Rosli MI, Kamarudin SK, Daud WRW (2017) Performance and stability of single and 6-cell stack passive direct methanol fuel cell (DMFC) for long-term operation. Int J Hydrogen Energy 42(14):9230–9242

Masimalai SK (2014) Influence of methanol induction on performance, emission and combustion behavior of a methanol-diesel dual fuel engine. SAE technical paper 2014-01-1315

Masiol M, Harrison RM (2014) Aircraft engine exhaust emissions and other airport-related contributions to ambient air pollution: a review. Atmos Environ 95:409–455

Maslan NH, Rosli MI, Goh CW, Masdar M (2015) Computational fluid dynamics simulation of the flow field of direct methanol fuel cells. IJMME-IJENS 15(5):114–125

MKC (2018) Public final report: methanol as an alternative fuel for vessels. Maritime Knowledge Centre, TNO Innovation for Life and Delft University of Technology, Nethelands

Mourad M, Mahmoud KRM (2018) Performance investigation of passenger vehicle fueled by propanol/gasoline blend according to a city driving cycle. Energy 21:741–749

Mourad M, Mahmoud K (2019) Investigation into SI engine performance characteristics and emissions fuelled with ethanol/butanol-gasoline blends. Renew Energy 143:762–771

Nielsen KV, Blanke M, Eriksson L, Vejlgaard-Laursen M (2018) Marine diesel engine control to meet emission requirements and maintain maneuverability. Control Eng Pract 76:12–21

Ning L, Duan Q, Chen Z, Kou H, Liu B, Yang B, Zeng K (2020) A comparative study on the combustion and emissions of a non-road common rail diesel engine fueled with primary alcohol fuels (methanol, ethanol, and n-butanol)/diesel dual fuel. Fuel 266:117034

Oh Y, Park J, Lee JT, Seo J, Park S (2016) Development strategies to satisfy corporate average CO_2 emission regulations of light duty vehicles (LDVs) in Korea. Energy Policy 98:121–132

Olabi AG, Wilberforce T, Abdelkareem MA (2021) Fuel cell application in the automotive industry and future perspective. Energy 214:118955

Olmer N, Comer B, Roy B, Mao X, Rutherford D (2017) Greenhouse gas emissions from global shipping, 2013–2015. Report by International Council on Clean Transportation (ICCT)

Onion G, Bodo LB (1983) Oxygenate fuels for diesel engines: a survey of world-wide activities. Biomass 3(2):77–133

Pachauri RK, Chauhan YK (2015) A study, analysis and power management schemes for fuel cells. Renew Sust Energ Rev 43:1301–1319

Pan P, Sun Y, Yuan C, Yan X, Tang X (2021) Research progress on ship power systems integrated with new energy sources: a review. Renew Sust Energ Rev 144:111048

Park SH, Kim SH, Lee CS (2009) Mixing stability and spray behavior characteristics of diesel–ethanol–methyl ester blended fuels in a common-rail diesel injection system. Energy Fuels 23:5228–5235

Paul A, Panua R, Debroy D (2017) An experimental study of combustion, performance, exergy and emission characteristics of a CI engine fueled by diesel-ethanol-biodiesel blends. Energy 141:839–852

Pavlenko N, Searle S (2021) Assessing the sustainability implications of alternative aviation fuels. In: International council on clean transportation (ICCT) working paper 2021-11, pp 1–17

Pearson RJ, Turner JWG (2012) GEM ternary blends: testing iso-stoichiometric mixtures of gasoline, ethanol and methanol in a production flex-fuel vehicle fitted with a physical alcohol sensor. SAE technical paper 2012-01-1279

Policarpo NA, Frutuoso FS, Cassiano DR, Cavalcante FSA, Araújo RS, Bertoncini BV, Oliveira MLM (2018) Emission estimates for an on-road flex-fuel vehicles operated by ethanol-gasoline blends in an urban region, Brazil. Urban Clim 24:111–120

Prasad BSN, Pandey JK, Kumar GN (2020) Impact of changing compression ratio on engine characteristics of an SI engine fueled with equi-volume blend of methanol and gasoline. Energy 191:116605.

Qi DH, Chen H, Geng LM, Bian YZ, Ren XC (2010) Performance and combustion characteristics of biodiesel–diesel–methanol blend fuelled engine. Appl Energy 87:1679–1686

Qian Y, Guo J, Zhang Y, Tao W, Lu X (2018) Combustion and emission behavior of N-propanol as partially alternative fuel in a direct injection spark ignition engine. Appl Therm Eng 144:126–136

Rakopoulos DC, Rakopoulos CD, Giakoumis EG, Papagiannakis RG, Kyritsis DC (2014) Influence of properties of various common bio-fuels on the combustion and emission characteristics of high-speed DI (direct injection) diesel engine: vegetable oil, bio-diesel, ethanol, n-butanol, diethyl ether. Energy 73:354–366

Ravi S, Karthikeyan A (2021) Assessment of performance and emission characteristics of diesel engine supplied with waste plastic oil propanol and ethylhexyl nitrate blends. Mater Today Proc 44:3642–3646

Regulation (EU) 2017/2400 (2018) Regulation (EU) 2017/2400 of 12 December 2017 implementing regulation (EC) No 595/2009 of the European Parliament and of the Council as regards the determination of the CO_2 emissions and fuel consumption of heavy-duty vehicles and amending Directive 2007/46/EC of the European Parliament and of the Council and Commission Regulation (EU) No 582/2011

Saxena MR, Maurya RK (2017) Effect of premixing ratio, injection timing and compression ratio on nano particle emissions from dual fuel non-road compression ignition engine fueled with gasoline/methanol (port injection) and diesel (direct injection). Fuel 203:894–914

Sayin C (2010) Engine performance and exhaust gas emissions of methanol and ethanol–diesel blends. Fuel 89:3410–3415

Sayin C, Balki MK (2015) Effect of compression ratio on the emission, performance and combustion characteristics of a gasoline engine fueled with isobutanol/gasoline blends. Energy 82:550–555

Sayin C, Ozsezen AN, Canakci M (2010) The influence of operating parameters on the performance and emissions of a DI diesel engine using methanol-blended-diesel fuel. Fuel 89:1407–1414

Schulze Lohoff A, Günther D, Hehemann M, Müller M, Stolten D (2016) Extending the lifetime of direct methanol fuel cell systems to more than 20,000 h by applying ion exchange resins. Int J Hydrogen Energy 41(34):15325–15334

Scott DS, Rogner HH, Scott MB (1993) Fuel cell locomotives in Canada. Int J Hydrogen Energy 18(3):253–263

Seyam S, Dincer I, Agelin-Chaab M, Development and assessment of a cleaner locomotive powering system with alternative fuels. Fuel 296:120529

Shanmugam R, Murugesan P, Guye GG, Duraisamy B (2021) Effect of additives on the stability of ethanol-diesel blends for IC engine application. Environ Sci Pollut Res 28:12153–12167

Sharaf OZ, Orhan MF (2014) An overview of fuel cell technology: fundamentals and applications. Renew Sust Energ Rev 32:810–853

Short D, Vu D, Durbin TD, Karavalakis G, Asa-Awuku A (2015) Particle speciation of emissions from iso-butanol and ethanol blended gasoline in light-duty vehicles. J Aerosol Sci 84:39–52

Shrivastava K, Thipse SS, Patil ID (2021) Optimization of diesel engine performance and emission parameters of Karanja biodiesel-ethanol-diesel blends at optimized operating conditions. Fuel 293:120451

Sileghem L (2015) A study of the combustion of alcohol-gasoline blends in internal combustion engines. Ghent University

Sileghem L, Vancoillie J, Demuynck J, Galle J, Verhelst S (2012) Alternative fuels for spark-ignition engines: mixing rules for the laminar burning velocity of gasoline-alcohol blends. Energy Fuels 26(8):4721–4727
Sileghem L, Wallner T, Verhelst S (2015) A quasi-dimensional model for SI engines fueled with gasoline-alcohol blends: Knock modeling. Fuel 140:217–226
Singh SB, Dhar A, Agarwal AK (2015) Technical feasibility study of butanol–gasoline blends for powering medium-duty transportation spark ignition engine. Renew Energy 76:706–716
Singh B, Ghosh S, Aich S, Roy B (2017) Low temperature solid oxide electrolytes (LT-SOE): a review. J Power Sources 339:103–135
Singhal SC, Kendal K (2003) High temperature and solid oxide fuel cells, fundamentals, design and applications. 1st edn. Elsevier. ISBN 9780080508085
Stambouli AB, Traversa E (2002) Solid oxide fuel cells (SOFCs): a review of an environmentally clean and efficient source of energy. Renew Sust Energ Rev 6:433–455
Staples MD, Malina R, Suresh P, Hileman JI, Barrett SRH (2018) Aviation CO_2 emissions reductions from the use of alternative jet fuels. Energy Policy 114:342–354
Stratton RW, Wolfe PJ, Hileman JI (2011) Impact of aviation non-CO_2 combustion effects on the environmental feasibility of alternative jet fuels. Environ Sci Technol 45:10736–10743
Strazza C, Del Borghi A, Costamagna P, Traverso A, Santin M (2010) Comparative LCA of methanol-fueled SOFCs as auxiliary power systems on-board ships. Appl Energy 87:1670–1678
Tang Q, Jiang P, Peng C, Chang H, Zhao Z (2021) Comparison and analysis of the effects of spark timing and lambda on a high-speed spark ignition engine fuelled with n-butanol/gasoline blends. Fuel 287:119505
The Government of the Republic of Korea (2020) 2050 carbon neutral strategy of the Republic of Korea, pp 73–80, https://unfccc.int/sites/default/files/resource/LTS1_RKorea.pdf, Dec 2020
Tse H, Leung CW, Cheung CS (2015) Investigation on the combustion characteristics and particulate emissions from a diesel engine fueled with diesel-biodiesel-ethanol blends. Energy 83:343–350
Turner JWG, Pearson RJ, Bell A, de Goede S, Woolard C (2012) Iso-stoichiometric ternary blends of gasoline, ethanol and methanol: investigations into exhaust emissions, blend properties and octane numbers. SAE Int J Fuels Lubr 5(3):945–967
Turner JWG, Pearson RJ, Dekker E, Iosefa B, Johansson K, Bergstrom K (2013) Extending the rol of alcohols as transport fuels using iso-stoichiometric ternary blends of gasoline, ethanol and methanol. Appl Energy 102:72–86
Vali RH, Wani MM (2021) The effect of mixed nano-additives on performance and emission characteristics of a diesel engine fuelled with diesel-ethanol blend. Mater Today Proc 43:3842–3846
Van TC, Ramirez J, Rainey T, Ristovski T (2019) Global impacts of recent IMO regulations on marine fuel oil refining processes and ship emissions. Transp Res Part D-Transp Environ 70:123–134
Vancoillie J (2013) Modeling the combustion of light alcohols in spark-ignition engines. Ghent University
Verhelst S, Turner JWG, Sileghem L, Vancoillie J (2019) Methanol as a fuel for internal combustion engines. Prog Energy Combust Sci 70:43–88
Vogler F, Würsig G (2010) New developments for maritime fuel cell systems. In: 18th world hydrogen energy conference, pp 445–453
Wagner TO, Russum LW (1973) Optimum octane number for unlead gasoline. SAE technical paper 730552
Walker M, Chance R (1983) Corrosion of metals and the effectiveness of inhibitors in ethanol fuels. SAE technical paper 831828
Wallner T, Ickes A, Lawyer K (2012) Analytical assessment of C2–C8 alcohols as spark-ignition engine fuels. In: Proceedings of the FISITA 2012 world automotive congress, pp 15–26
Wang L, He M, Hu Y, Zhang Y, Liu X, Wang G (2015) A "4-cell" modular passive DMFC (direct methanol fuel cell) stack for portable applications. Energy 82:229–235

Waqas M, Naser N, Sarathy M, Fejis J, Morganti K, Nyrenstedt G, Johansson B (2017) Auto-ignition of iso-stoichiometric blends of gasoline-ethanol-methanol (GEM) in SI, HCCI and CI combustion modes. SAE technical paper 2017-01-0726

Warnatz J, Maas U, Dibble R (1996) Combustion: physical and chemical fundamentals, modeling and simulation, experiments, pollutant formation. Springer, Berlin. ISBN 9783540259923

Wilberforce T, Hassan ZE, Ogungbemi E, Ijaodola O, Khatib FN, Durrant A, Thompson J, Baroutaji A, Olabi AG (2019) A comprehensive study of the effect of bipolar plate (BP) geometry design on the performance of proton exchange membrane (PEM) fuel cells. Renew Sust Energ Rev 111:236–260

Wong CW, Zhao TS, Ye Q, Liu JG (2006) Experimental investigations of the anode flow field of a micro direct methanol fuel cell. J Power Sources 155:291–296

Wu Y, Gen C, Huang Z (2011) Influence of cetane number improver on performance and emissions of a common-rail diesel engine fueled with biodiesel-methanol blend. Front Energy 5:412–418

Xiao H, Guo F, Li S, Wang R, Yang X (2019) Combustion performance and emission characteristics of a diesel engine burning biodiesel blended with n-butanol. Fuel 258:115887

Xing H, Spence S, Chen H (2020) A comprehensive review on countermeasures for CO_2 emissions from ships. Renew Sust Energ Rev 134:110222

Yang J, Rotha P, Zhu H, Durbin TD, Karavalakis G (2019) Impacts of gasoline aromatic and ethanol levels on the emissions from GDI vehicles: part 2. Influence on particulate matter, black carbon, and nanoparticle emissions. Fuel 252: 812–820

Yang H, Zhao TS (2005) Effect of anode flow field design on the performance of liquid feed direct methanol fuel cells. Electrochim Acta 50:3243–3252

Yanju W, Shenghua L, Hongsong L, Rui Y, Jie L, Ying W (2008) Effects of methanol/gasoline blends on a spark ignition engine performance and emissions. Energy Fuels 22:1254–1259

Yates A, Bell A, Swarts A (2010) Insights relating to the autoignition characteristics of alcohol fuels. Fuel 89(1):83–93

Yilmaz N, Vigil FM, Donaldson AB, Darabseh T (2013) Investigation of CI engine emissions in biodiesel ethanol diesel blends as a function of ethanol concentration. Fuel 115:790–793

Yuan Z, Zhang Y, Fu W, Li Z, Liu X (2013) Investigation of a small-volume direct methanol fuel cell stack for portable applications. Energy 51(1):462–467

Yuan W, Wang A, Ye G, Pan B, Tang K, Chen H (2017) Dynamic relationship between the CO_2 gas bubble behavior and the pressure drop characteristics in the anode flow field of an active liquid-feed direct methanol fuel cell. Appl Energy 188:431–443

Yuen PKP, Villaire W, Beckett J (2010) Automotive materials engineering challenges and solutions for the use of ethanol and methanol blended fuels. SAE technical paper 2010-01-0729

Yusri IM, Mamat R, Najafi G, Razman A, Awad OI, Azmi WH, Ishak WFW, Shaiful AIM (2017) Alcohol based automotive fuels from first four alcohol family in compression and spark ignition engine: A review on engine performance and exhaust emissions. Renew Sust Energ Rev 77:169–181

Yusuf AA, Inambao FL (2021) Effect of low bioethanol fraction on emissions, performance, and combustion behavior in a modernized electronic fuel injection engine. Biomass Conv Biorefin 11:885–893

Zacharof N, Fontaras G, Ciuffo B, Tansinilker A, Prado-Rujas I (2021) An estimation of heavy-duty vehicle fleet CO_2 emissions based on sampled data. Transp Res Part D Transp Environ 94:102784

Zaharin MSM, Abdullah NR, Masjuki HH, Ali OM, Najafi G, Yusaf T (2018) Evaluation on physicochemical properties of iso-butanol additives in ethanol-gasoline blend on performance and emission characteristics of a spark-ignition engine. Appl Therm Eng 144:960–971

Zhang ZH, Tsang KS, Cheung CS, Chan TL, Yao CD (2011) Effect of fumigation methanol and ethanol on the gaseous and particulate emissions of a direct-injection diesel engine. Atmos Environ 45:2001–2008

Zhang ZH, Cheung CS, Yao CD (2013) Influence of fumigation methanol on the combustion and particulate emissions of a diesel engine. Fuel 111:442–448

Zhao H, Ge Y, Tan J, Yin H, Guo J, Zhao W, Dai P (2011) Effects of different mixing ratios on emissions from passenger cars fueled with methanol/gasoline blends. J Environ Sci 23(11):1831–1838
Zhao W, Yan J, Gao S, Lee TH, Li X (2022) The combustion and emission characteristics of a common-rail diesel engine fueled with diesel, propanol, and pentanol blends under low intake pressures. Fuel 307:121692
Zhen X, Wang Y, Liu D (2020) Bio-butanol as a new generation of clean alternative fuel for SI (spark ignition) and CI (compression ignition) engines. Renew Nergy 147:2494–2521
Zhu L, Wang S, Sui PC, Gao X (2021) Multiscale modeling of an angled gas diffusion layer for polymer electrolyte membrane fuel cells: performance enhancing for aviation applications. Int J Hydrogen Energy 46(39):20702–20714

Chapter 6
Advances in the Use of Ethers and Alcohols as Additives for Improving Biofuel Properties for SI Engines

Samuel Eshorame Sanni and Babalola Aisosa Oni

Abstract Commercialization of biofuels as alternative fuels to conventional diesel fuel for application as transport-fuels for diesel engines is fast becoming attainable owing to the merits offered by the inclusion of significant quantities of an alcohol (ethanol) and a member of the "ether" group (dimethyl ether) as additives or property-improvers for biofuels obtained from biomass. These additives are fuels in kind, but have lower viscosities, flash points, flammability etc., hence they infuse some measures of atomization and moderation in the densities and viscosities of biofuels towards ensuring their suitability for use in Internal Combustion Engines (ICEs). Biofuels need be improved in terms of fuel quality such as performance, emission and combustion characteristics to meet market specification. This then informs the need for suitable fuel-modifiers which must be tested for their compatibilities with different biofuel-sources before they are used as fuels in ICEs. The mixing ratio of the added components with the biofuels is also to be given utmost attention as an alcohol such as ethanol and an ether (dimethyl ether), are known for their high volatilities which in turn regulate the BTEs and combustion potentials of the fuels, all aimed at improving the cetane numbers or indices of the blended fuels. Owing to the relative abundance of bioresources as precursors for biofuels relative to other sources of ethers and alcohols, literature has it that some prospective alcohols and ethers have been admixed with biofuels as means of upgrading their properties towards ensuring their high suitability for diesel engines with little or no modifications; this then implies that there might be need to begin to look into reconfiguring some diesel engines in order to abate engine wear, fuel degradation as well as catalyst-poisoning towards ensuring/maintaining high engine-compatibilities with these fuels. Therefore, this chapter is proposed for inclusion in Book 1 "Engine and fuels for future transport", and its focus will be on the effects of using lone ethanol, dimethyl ether or biofuels as well as their blends for use as future transport fuels.

S. E. Sanni (✉) · B. A. Oni
Department of Chemical Engineering, Covenant University, Ota, Ogun State, Nigeria

B. A. Oni
Department of Chemical Engineering, China University of Petroleum, Changping, Beijing, People's Republic of China

A. V. Agarwal and H. Valera (eds.), *Potential and Challenges of Low Carbon Fuels for Sustainable Transport*, Energy, Environment, and Sustainability,
https://doi.org/10.1007/978-981-16-8414-2_6

Keywords Biofuel · Dimethyl ether · Ethanol · Diesel Engine · Engine-fuel compatibility · Transport fuel

6.1 Introduction

Bioenergy is a renewable energy obtained from biomass. This includes biofuels for transportation (e.g., biodiesel and bioethanol), thermal/electrical energy from pellets and wood chips, and biogas from biological materials from industrial and municipal wastes (Keskin 2019). The depletion of fossil fuels, combined with rising energy demand, has prompted a search for alternative source of energy obtained from plants. Biofuels are derived from biomass to serve as alternative-fuels to fossil fuels (Ghadikolaei 2016). Biofuels stand alone or can be blended with fossil fuels such as diesel and petrol. Biofuels can be classified into first, second, third and fourth generation biofuels.

Biofuels are grouped into first, second, third and fourth generation fuels. Biofuels sourced from seed oils, and sugar/starch are tagged first generation biofuels (Aleiferis et al. 2013).

Vegetable/non-edible oils have negative consequences in engines due to their high flash points, densities, pour points, viscosities etc. Hence, it is important to convert the oils into bio-diesel to make them fit for use as alternative fuels for diesel. In the past, bioethanol has attracted a lot of attention (Reijnders and Huijbregts 2008; Atmanl et al. 2014).

Biofuels as transportation fuels account for the majority of the world's bioenergy. They are mainly obtained from food crops containing high amount of starch and sugar e.g., sugarcane/corn, and the end products are seed oils and ethanol, which are then trans esterified to biodiesel (Yusri et al. 2016). These first generation technologies were the first steps for the transition from traditional fossil fuels. Second generation biofuels are from non-edible biomass including residues from forestry and crops, energy crops (poplar, switchgrass, poplar, and miscanthus), lignocellulosic fractions from municipal and industrial solid wastes. Examples of the third generation biofuels is the algal biomass (Yusri et al. 2017).

More than two-thirds of bioenergy is derived from first-generation feedstocks, resulting in growing concerns for land, water, food, and fiber production, including other issues affecting land-use changes (Karavalakis et al. 2014). Therefore, the use of residues for the production of bioenergy has attracted more attention as they are readily available in most of the countries. The potentials of lignocellulosic biomasses differ and it is dependent on their relative abundance, types, biomass cost, feedstocks, patterns of energy demand and efficiencies of the technologies available (Awad et al. 2016a). The production of biofuel has increased rapidly in the last 10 years which provides about 3.4% of the world road transport/energy requirements.

Research has shown that about 40 million gross hectares are used to produce bioenergy crops especially for biogas, biodiesel, biofuel production, and bioethanol (Gravalos et al. 2013a).

There are some merits and demerits of biofuels in terms of social-economic and environmental sustainability. In other words, energy security, greenhouse gas (GHG) emission-reduction and the developments inherent in nonmunicipal settings are the most significant causes in the universal rise in energy demands for biofuels (Zhen and Wang 2015). Conversely, issues related to increased biofuel production include increase in: food commodity, risks associated with the GHG emissions via direct/indirect land-use acts arising from biofuel feedstocks, as well as the risks associated with ecosystem, forest, land degradation, and water resources (Balki and Sayin 2014a). The use of corn (first generation feedstock), has become an issue owing to high food competition and the use of arable land for fuel feedstocks which in turn serve as raw materials for biofuels (Calam et al. 2015a). The high demand of agricultural produce increases the risks associated with deforestation/use of lands with high biodiversity which in turn influences the demands for freshwater, fertilizers and pesticides which may impart positively or negatively on the environment (Qi and Lee 2016). Although some of the issues highlighted can be tackled via the use of second-generation feedstocks; however, there are still problems associated with the economic viability of biofuels from these sources, thus in the current economic context, this span of uncertainties are as a result of the drop in oil prices. Third-generation biofuels are derived from synthetic fibers, whereas fourth-generation (algal) biofuels are the most recent class of biofuels to emerge in order to emphasize the avoidance of food competition/land use since microalgae can be rapidly grown on non-arable lands, wastewaters, saline or brackish water etc. (Qi and Lee 2016a; Heywood 2018; Noor et al. 2014). The production of biofuels from microalgae is energy-intensive which is the reason such fuels are economically unviable.

6.2 Biofuel Additives

So far, 1st generation alcohol for SI application has mostly constituted gasoline-alcohol blends, where present, the standard fuel quality permits 5–10% blend of ethanol with gasoline. However, methanol has been employed as automobile fuel to replace gasoline for better performance in internal combustion engines (ICE) (Zhang et al. 2014).

Generally, alcohols of lower molecular weight, especially methanol and ethanol, comprise one group of alternative fuels which are attractive for ICEs. Alcohols has some merits which makes it an attractive alternative fuel for ICEs as compared to the fossil fuels (Balki et al. 2014). These advantages include:

- It decreases the release of harmful gases to the environment (Yusri et al. 2016).
- Combustion of alcohols in ICEs gives more higher combustion pressure compared to gasoline, thus improving the BTE and power output when compared to fossil fuel (Keskin 2010).

- Alcohol of lower molecular weight can be obtained from natural gas, biomass, coal etc., which are readily available and their precursors are less expensive (Reijnders and Huijbregts 2008).
- Alcohols produce more greenhouse gases (Zhen and Wang 2015)
- Alcohol having higher average octane rating can improve the efficiency of a fuel and power (Karavalakis et al. 2014).
- Generally, alcohols give lower evaporative emissions (Ghadikolaei 2016).
- Alcohols can generally be metabolized when absorbed on the ground (Keskin 2010).
- The combustion of alcohol in ICE releases small quantity of ash as a result of less carbon content present in alcohol (Calam et al. 2015a).

6.2.1 Methanol

Methanol (CH_3OH) being the first alcohol in the group comprises of one carbon-atom/ molecule. CH_3OH is a colourless and harsh liquid commonly called "wood spirit" (Balki et al. 2014). It gives several advantages as a substitute fuel source over gasoline. Methanol is relatively cheap, which can be obtained in different forms for example gasification of coal, component of synthesis gas by steam reforming of natural gas, etc. (Pourkhesalian et al. 2010). Methanol also gives low emission. Due to its low boiling point, faster evaporation takes place in the fuel for engine combustion, thus releasing low HC emissions when used as fuel. Additionally, the oxygen content of the fuel and its chemical properties can enhance engine performance with low resultant-emissions from SI engines (Balki et al. 2014).

Methanol can be synthesized by various methods; (a) from biomass (b) Catalytic synthesis.

(a) Biomass synthesis.

Methanol from biomass can be synthesized from animal waste, plants and fruits via anaerobic bacteria. MeOH can be obtained as by-product during the fermentation of ethanol. In countries like South Africa and China, MeOH can be produced from coal (Noor et al. 2014). Due to cost and less energy consumption, methanol from biomass is relatively cheap. Figure 6.1 shows the schematic of methanol production from biomass.

(b) Catalytic synthesis

In this method, a mixture of carbon dioxide, hydrogen and carbon (II) oxide form methanol and H_2O as a byproduct. The stoichiometric reactions and their enthalpies are shown in Eq. (6.1) and Eq. (6.2).

$$CO + 2H_2 \leftrightarrow CH_3OH \quad \Delta H_{298.15\,K,\,5\,MPa} = -90.7\,kJ/mol \tag{6.1}$$

$$CO_2 + 3H_2 \leftrightarrow CH_3OH + H_2O \quad \Delta H_{298.15\,K,\,5\,MPa} = -90.7\,kJ/mol \tag{6.2}$$

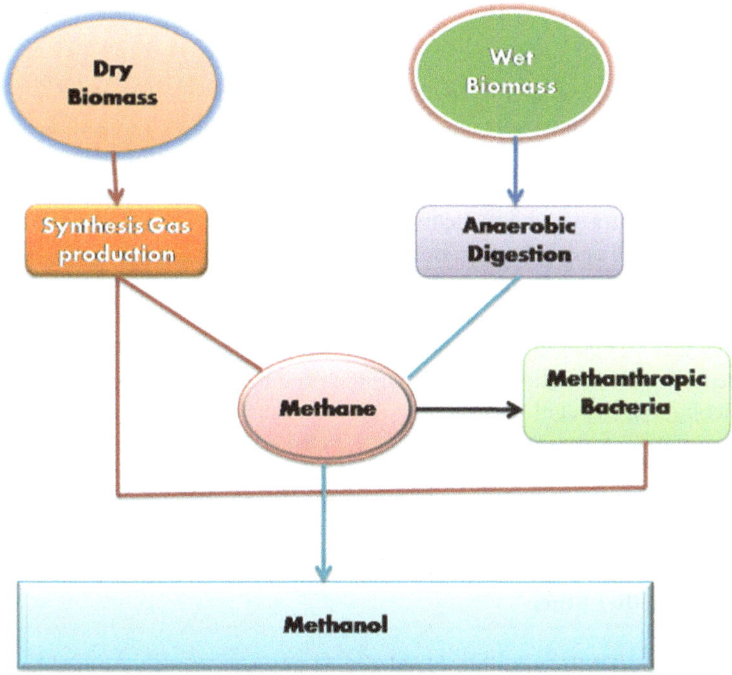

Fig. 6.1 Production of methanol from biomass. Methanol from Biomass. European Biofuels Technology Platform. 2011. Available from: www.biofuelstp.eu (Gu et al. 2012)

6.2.2 Ethanol

Bioethanol is another member of the alcohol family. It is a renewable fuel from biological materials through biomass fermentation (Awad et al. 2016). Bioethanol has been used in ICEs and it is still in use. The fuel is used as an additive in gasoline for octane rating and engine combustion enhancement in ICEs. Research has shown that alcohol blending (ethanol and methanol) decrease the emissions from engine compared to those of gasoline (Ghadikolaei 2016; Balki et al. 2014; Hagos et al. 2017). Furthermore report shows that low engine emissions were as a result of the O_2 present in both fuels. The physical and chemical properties of fuels are different when compared to gasoline, in terms of their heating values which are lower than that of gasoline. This further shows that ICEs will require higher volumes of methanol/ethanol blends for better combustion compared to gasoline standing alone as fuel. Alcohols (CH_3OH and C_2H_5OH) have higher octane rating compared to the gasoline with lower boiling points/vapour pressures (Hagos et al. 2017; Gu et al. 2012).

6.2.3 Butanol

Biobutanol can serve as fuel in an unmodified spark ignition (SI)-engine. It is relatively miscible with most solvents (Yusri et al. 2016). Butanol is relatively similar to gasoline owing to its low oxygen content, hydrocarbon chain-length, and relative heating value compared to those of ethanol and methanol (Nithyanandan et al. 2016). Biobutanol has attracted more attention in recent times, which has better characteristics than ethanol and methanol. Butanol also has high tolerance to water contamination and can be used as transport fuel delivered through distribution pipelines (Gravalos et al. 2013a). It serves as an agent of blending for SI engines (Zhuang and Hong 2013). Fusel oil is another by-product produced from alcohol. The oil is produced after fermentation by distilling the product of fermentation of natural or amyl alcohols (Balki et al. 2014).

6.2.4 Ethers

6.2.4.1 Dimethyl Ether

Dimethyl ether (DME)-gasoline blend for SI-engines appear a possible way of enhancing thermal efficiency and combustion properties under normal conditions (Siwale et al. 2014). DME can be synthesized from coal, crude oil, biomass, natural gas, residual oil and some waste materials (Pourkhesalian et al. 2010). DME is a prospective additive for biofuels and can serve as alternative fuel for future use in ICEs. Furthermore, several investigations have ensued as regards ascertaining the extent of desirability of DME for use as additive in fuels used in spark-ignited ethanol engines as a result of its high thermal efficiency, low combustion-noise, soot-free combustion and low emission. Addition of dimethyl ether to biofuels is seen as a feasible method that will enhance the overall performance of SI engines (Hagos et al. 2017; Gu et al. 2012; Nithyanandan et al. 2016). When DME is used as an additive, it helps to reduce engine emissions and economical characteristics in ethanol- gasoline blends for SI engine (Icingur and Calam 2012).

Dimethyl ether (DME) has been found to improve the ignition characteristics of internal combustion engines as well as the cetane number of diesels. DME can reduce NOx and smoke emissions when used as fuel-additive. DME is a gaseous fuel that may require major engine-modifications in order to ensure its adaptability to fuel-combustion (Zhang et al. 2013).

Diethyl ether (DEE) fuel is less volatile when compared to DME, DEE has a very high cetane number i.e., more than 125. DEE's latent heat of vaporization is 356 kJ/kg for which it can also provide charge cooling which helps in NO_x emissions reduction (Canakci et al. 2013). DEE also has a 21.6% oxygen content, which can greatly improve engine performance with reductions in CO/HC-emissions (Noor et al. 2014).

Due to their high-octane numbers of alcohol/ether-fuels, they are usually employed as octane- boosters/substitutes for tetraethyl-lead (TEL) in gasoline. As a result, the United States began to do away with adopting TEL infusion in gasoline by mandating the use of other viable additives, such as ethanol, which has become more appealing as an octane enhancer (Hagos et al. 2017). Leaded gasoline was outlawed in the United States in the 1970s, whereas in Italy, it has been outlawed since 2000, even though newer cars used unleaded fuel in the 1990s. Several countries have currently banned the use of leaded fuels (Keskin and Gürü 2011). Additionally, MTBE and methanol have been substituted with TEL as additives in gasoline owing to the fact that TEL in gasoline makes gasoline engines more prone to engine knocks. Emissions are emitted from the combustion process and transported through vehicle exhausts, thus posing serious health risks (Zhang et al. 2013; Canakci et al. 2013; Keskin and Gürü 2011). Since the replacement of TEL in the 1970s, MTBE is primarily being used as octane improver (4–8% by volume) (Gravalos et al. 2013b). In the late 1980s, the use of MTBE as an octane booster increased steadily, followed by its use in the production of cleaner-burning gasoline in some countries. Its use increased by 11% in 1981, rising to 15% in 1988. From 1980 to 1986, the adoption of MTBE as a gasoline additive gained a significant rise of about 50% per year (Heywood 2018).

High rating octane-fuels are recalcitrant to engine-knock and thus appropriate for use in spark-ignition engines. Ethanol reduces the tendencies for engine-knock, hence it can be integrated to compensate for high compression ratios which also culminates in increased engine power/efficiency compared to pristine/unblended gasoline (Yusri et al. 2017; Liang et al. 2012). Another incredibly interesting and advantageous characteristic of alcohols, particularly methanol, is their relatively high molar expansion, which leads to increased pressures resulting from the chemical interactions that occur without any additional heat. Alcohol and ether fuels have heating values that are nearly 30% lower than gasoline (Wang et al. 2013). The lower heating value (LHV) of alcohols (C_1–C_4) relative to gasoline is an indication that the engine will require more fuel, such that the fueling system of the engine will need to be redesigned to allow higher fueling-rates (Zhang et al. 2013; Ozsezen and Canakci 2011a). In addition, when compared to ethers, alcohols have the highest combustion efficiencies. Furthermore, alcohol and ether fuels help to enhance SI engine performance while lowering emissions (Çelik et al. 2011). Moreover, the prices of some alcohol fuels are lower than the prices of fossil fuels. In recent times, a large amount of literature has been devoted to studies that concern engine performance, emissions, and combustion properties from the use of alcohol fuels. From 1992 to 2013, there are a number of car models running on non-gasoline (Daniel et al. 2014).

6.2.4.2 Methyl Tert-Butyl Ether (MTBE)

Since 1985, an oxygenated and volatile organic fuel known as MTBE ($C_5H_{12}O$), a member of the ether family, has been tested with gasoline to improve octane value and reduce engine emissions from ICEs. It is produced by reacting isobutylene with methanol. Large quantities i.e., over 200,000 barrels/day were recorded in the United States as at 1999 for use as fuel-additive in gasoline engines (Wang et al. 2013; Ozsezen and Canakci 2011a; Çelik et al. 2011; Daniel et al. 2014). Owing to its inexpensive nature, simple means of production as well as its favorable transfer/blending properties, MTBE currently ranks the most widely adopted ether-based additive (Kiani et al. 2010). When introduced in gasoline it controls the expulsion of harmful emissions (Zhen and Wang 2015). Since 1979, in the US, MTBE has been adopted as additive in gasoline; its primary function is to serve as an octane booster (i.e., an antiknock agent) that replaces lead as an additive (Christensen et al. 2011). Previously, the use of MTBE had increased significantly to approximately 408,000–453,592 tons. Agarwal (Anonymous 2013) demonstrated that blending gasoline and MTBE enhances engine-performance with evidential lowering of the carbon monoxide emissions. Table 6.1 shows the physicochemical characteristics of some gasoline, alcohols, and ethers.

High-cetane-number fuel additives can measurably improve a SI engine's performance by improving the charges released by combustion inside the engine cylinder (Çelik et al. 2011). Furthermore, additives with high oxygen contents can have the additional effect of lowering the resulting emissions, such as the amount of CO and smoke emitted by SI engines. Chemical compounds such as DME, DEE, and others

Table 6.1 Physicochemical characteristics of some gasoline, alcohols, and ethers

Parameter	Gasoline	Butanol	Methanol	Ethanol	MTBE	DME
Formula	$C_{5\text{-}10}\,H_{12\text{-}22}$	$C_4\,H_{10}\,O$	CH_3OH	$C_2\,H_5\,OH$	$C_2\,H_{12}\,O$	CH_3OCH_3
Molecular weight (%)	106.2	74.1	32.0	46.7	88.2	46.1
Carbon mass (%)	87.5	64.9	37.5	52.2	66.1	52.2
H_2 mass (%)	12.5	13.5	–	34.7	13.7	13
Density (g/mL)	0.7	0.8	0.8	0.8	0.7	0.7
Boiling point, (°C)	27–225	117.3	78	78.3	52.2	– 25.1
RVP, kPa	53–60	18.6	32.4	17	54.47	–
RON	90–100	98	108.7	108.6–110	118	–
MON Calorific value (MJ/kg)	82–92 45	78 29.2	86.6 22.7	92 29.9	102 10.56	– 29
Freezing point, (°C)	– 40	–	– 97.5	– 114	– 108	–
Viscosity (mm^2/s)	0.5–0.6	–	0.6	1.2–1.5	0.4	–
Flash point (°C)	– 45 to – 13	–	11	12–20	– 25.5	–
Auto ignition temperature, (°C)	257	385	423	425	435	253
LHOV kJ/kg	349	707.9	920	923	320	–

**LHOV* latent heat of vaporization, *MON* motor octane number, *RON* research octane number, *RVP* reid vapour pressure can also provide such advantages (Daniel et al. 2012). Table 6.2 shows the properties and effects of some ether and alcohol fuels.

6.2.5 *Alcohol and Ether in Biofuels*

Furthermore, C_2H_5OH is isometric with DME. The number of O_2 atoms in each molecule, particularly in alcohol (methanol or ethanol), contributes to the reduction in CO/HC-emissions when used as engine fuel (Liang et al. 2011). In lieu of their high-octane induction, other characteristics of alcohol fuels used in SI-engines are related to their relative volatilities. Alcohol fuels have Latent heat of vaporization (LHOV) that range from 4 to 6 times that of gasoline. This LHOV affects the capability of a SI-cold-start engine, resulting in increased volumetric efficiency and a drop-in intake manifold temperature. Methanol, for example, has approximately 2.5 times the LHOV of gasoline (Icingur and Calam 2012; Çelik et al. 2011). Ether and alcohol fuels' heating value is 30–45% less than that recorded for unleaded gasoline, thus it requires 1.2–1.4 times additional alcohol-fuels to attain comparable energy output. Simultaneously, alcohol and ether low heating value fuels results in lower engine emissions and low exhaust temperatures (Kiani et al. 2010). The vapor pressure of engine fuel is an important property. Furthermore, vapor pressure can have an impact on proper cold engine ignition. Additionally, vapor pressure/ RVP is an important parameter to be considered in meeting the stipulated requirements for evaporative emissions. The Reid vapor pressure (RVP) of alcohol fuels generally ranges from 18 to 59 kPa, which is lower than that of gasoline, which ranges from 50 to 65 kPa (Agarwal 2007a).

6.2.5.1 Properties of Alcohols and Ethers Used in Biofuels

(i) Octane number

The octane number of a fuel is a measure of its proclivity to ignite prior combustion/during its compression stroke in the Otto cycle of ICEs. The octane number of gasoline, also known as RON/MON, determines its ignition quality (Agarwal 2007a). Many engines will typically fail in less than 50 h when subjected to heavy knock, implying that the damage may be cumulative. To avoid engine knocks, fuel with a high-octane number should be employed (Awad et al. 2018a). Without additional reforming, unleaded straight-run gasoline is not compatible with high compression engines. Owing to the fact that tetraethyl lead (TEL) is a good anti-knock additive, in the 1920s, engine manufacturers used TEL to improve the performance of manufactured engines with considering their compression ratios. Furthermore, Qi and Lee (Awad et al. 2018b) blended TEL with gasoline to increase its octane number and chances of being used at high compression ratios. Following that, there was essentially a universal agreement on the need to eliminate the use of lead as an anti-knock component in gasoline. Many researchers around the world have conclusively

Table 6.2 Ether and alcohol on fuel properties and their effects

Alcohol	Conc. of % alcohol	Octane number	Reid vapor pressure	Density	Heating value	Kinematic viscosity (mm^2/s)	References
Ethanol	5– 25% (E5, 10, 20, 30)	Increased by 1.36–5.56%	Increased by 10.33–10.99%	Increased by 0.21–0.49%	Decreased by 4.76–14.70%	-	Reijnders and Huijbregts (2008)
MBTE	15–20% (M5, M10, E15, M20)	Increased by 1.41–4.65%	Increased by 6.36–7.27%		Decreased by 1.88–3.83%	–	Kiani et al. (2010)
Methanol	5–10(M5, M10			Increased by 0.15–0.44%	Decreased by 1.88–3.83%	Increased by 7.09–10.32%	Christensen et al. (2011)
Ethanol	5–10% (E5, E10,)	Increased by 15–5.56%	Decreased by 2.67–5.47%	Increased by 7.61–15.69%			Anonymous (2013)
Butanol	B35			Increased by 7.63%	Decreased by 3.54%		Daniel et al. 2012)
Ethanol	10, 20% (E10, E20)	Increased by 2.47–4.16%		Increased by 0.67–1.20%	Decreased by 3.93–7.83%		Gu et al. (2012)
n-Butanol	10–20% (n-B10, n-B20)	Decreased by 0.34–0.67%		Increased by 0.80			Noor et al. 2012)

established its negative neurological effects, particularly on humans (Calam et al. 2015b).

When alcohol-ether fuels are mixed with gasoline, the octane number of gasoline gains a rapid rise due to the induced high RON/MON (Yusri et al. 2017).

Over the years, ethanol production has received a lot of attention as a booster of octane number (Çelik et al. 2011).

In addition to ethanol and MTBE, fusel oil, a byproduct of some fermented agricultural products has some form of semblance with alcoholic fuel. These characteristics include a high RON of about 106, high MON of about 103, high amount of O_2 and uniform boiling point which are indicative of its use as alternative fuel for SI engines (Table 6.2). As a result, the fuel has a high-octane rating, making it an excellent octane-improver. Calam et al. (2015a) studied the properties of various fusel oils such as F5, F10, F20, F30 and F500 on the performance and emissions of a SI-engine. As the amount of fuel additive increased, the fuel's octane number increased. Higher octane-rating fuels are also desirable because they allow for improved engine efficiencies and decreased BSFC due to higher compression ratios.

(ii) Calorific value

A fuel's calorific value is the quantity of heat expelled per unit mass of the fuel during combustion (Zhuang and Hong 2013). A fuel's heating value has an impact on the engine's BSFC. Furthermore, the engine's power-output and BTE are affected by the fuel's energy value as well as the flow rate of air in the engine. The energy content of a fuel is an important consideration when selecting appropriate fuels. As shown in Table 6.1, alcohols and ethers typically have heating values 30% lower than that of gasoline. Several factors influence the heating values of fuels/biofuels, they include the amount of O_2, H_2, C and H_2O as well as their free fatty acid contents, which influence the heating value of biofuels (Zhen and Wang 2015; Qi and Lee 2016b). Furthermore, some researchers established a link between fuel properties such as density, viscosity and the heating values of the fuels. Furthermore, biofuels, particularly alcohols, contain less carbon but more O_2 which makes them low heating fuels. Sarathy et al. (2009) reported that a fuel's heating value is related to its O_2 content. The LHV of a fuel decreases as the percent-oxygen and water increase, whereas it increases as the H_2 and sulfur contents increase. The carbon content of a fuel is one of its most influential parameters that imparts on its LHV because the fuel's percentage-carbon is connected to its heating value, this then informs the reason why gasoline has higher heating value than biofuels. Hu et al. (2007) reported that the amount of water in a fuel has the capacity to reduce its heating value. The presence of high amount of moisture in a fuel negatively affects the performance and combustion properties of an engine (Aleiferis et al. 2013; Song et al. 2006). Najafi et al. (2009) showed that the heating value of fuels increase with carbon chain-length for saturated esters and acids. According to Costa and Sodré (2010), increasing the heating value of fuels causes a resultant increase in its percent-H_2 and carbon. Hence, using experimental correlations, attempts have been made to predict the high heating values, amount of water as well as the amount of carbon-hydrogen ash of some fuels.

(iii) Boiling point

Generally, alcohol has a lower boiling point relative to gasoline (see Table 6.2). One of the important parameter of a fuel is the boiling point and it is related to a fuels volatility. At low boil point, fuels evaporate faster, with little or no spillage. Ethers and alcohols boil at a single temperature which is largely responsible for their increased energy-release (Irimescu 2012). At low boiling point, DME evaporates faster when liquefied-DME is injected into the cylinder of an engine. At a completely low temperature, of an alcohol- fuel evaporates as, this process is called the Volatility. When vaporization does not take place at an available temperature range required by an engine or its components, the injector or carburetor may not deliver the appropriate air–fuel mixture thus leading to incomplete combustion of the fuel. This will certainly increase the emissions of HC and CO emissions and wasted fuel (Sarathy et al. 2009). Flash point is another factor to be considered, which is also related to a fuel's volatility where an evaporative compound such as gasoline ignites within the confines of an open flame or spark (Hu et al. 2007).

(iv) Auto ignition temperature (AIT)

The auto-ignition temperature range for gasoline lies between 257 and 221 °C. That of ethanol is 329 °C; for the AIT of other fuels see Table 6.1. AIT is the point of spontaneous ignition of vapor within an enclosed space. Consequently, vapor and gas-leaks may accumulate within the space such that any hot surface becomes a potential ignition-source. Another effect is that of turbulence, at increased turbulence, the AIT also increases (Hu et al. 2007). Generally, alcoholic fuels have comparable ignition and combustion characteristics (Najafi et al. 2009). The AIT of methanol is higher than that of gasoline fuel; more so, generally, the AITs of alcohols are higher compared to that of pure gasoline, thus it is safer to store and transport alcohols (Liang et al. 2011).

(v) Latent heat of vaporization (LHOV)

For a substance undergoing phase transition from liquid to vapor, absorption of heat is required before the intended change can occur; the energy required is termed the LHOV (Costa and Sodré 2010). To undergo such a change, ethanol needs about 923 kJ/kg (at 77 °F/1 atm), whereas for gasoline the value is approximately 349 kJ/kg. The relative proportion, which is a factor of 2.68, is the basis for which gasoline has a lower LHOV and cannot absorb nearly as much heat as ethanol. Since gasoline has a lower LHOV and cannot absorb nearly as much heat as ethanol, the heat is merely transported as waste-heat to the engine's coolant. The ethanol additive, which has a high LHOV, is responsible for the reduction in NOx emissions (Gravalos et al. 2013b; Calam et al. 2015; Najafi et al. 2009). In a Direct Injection SI engine, LHOV is seen to exhibit an indirect anti-knock property by cooling the air–fuel mixture as it evaporates at the intake stroke and thus lengthens the fuel's ignition under knocking-conditions; this is known as "cooling effect" of the fuel (Costa and Sodré 2010). As a result of the higher heat requirement to vaporize the blinding fuel/gasohol, ethanol and methanol are seen to exhibit higher LHOVs compared to gasoline. Due to the amount of oxygen, high latent heat/lower heating values of gasohols, the combustion temperature is reduced thus resulting in lower NOx emissions (Calam et al. 2015b; Sarathy et al. 2009).

(vi) Oxygen content

High amount of oxygen in alcohols may result in complete/cleaner combustion which results in a decrease in the in-cylinder temperature (Hagos et al. 2017). Methanol has higher oxygen content relative to ethanol and MTBE. As a result, methanol provides lesser dilution relative to other oxygenates and according to Yasin et al. (2015), engine-knock is related to the amount of oxygen in fuels. This explains why knock occurs at lower loads when DMF is used as an alternative. Alcohols outperform Dimethyl formamide (DMF) in terms of combustion efficiency and this is primarily due to their relative amounts of oxygen which limits the latter's consumption performance. Furthermore, the amount of O_2 in an alcohol contributes immensely to its improved combustion-completeness (Christensen et al. 2011). Gasoline has a higher calorific value than ether and alcohol-fuels due to the high amounts of carbon/oxygen in alcohols and ethers (Awad et al. 2016). The influence of oxygen on some fuels' heating values is depicted in Fig. 6.2. The higher the oxygen content of the fuel, the lower its heating value as opposed to its carbon content. Although DME and ethanol

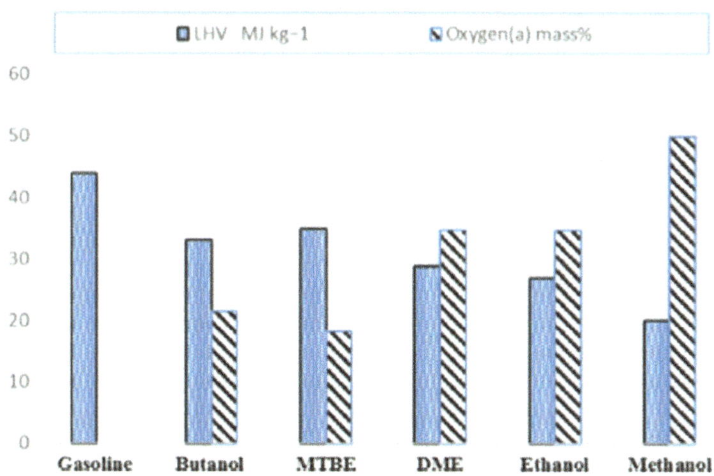

Fig. 6.2 Effects of O_2 on the heating values of alcohols and ethers. Reprinted from Awad et al. (2016a; Pourkhesalian et al. 2010) with copyright permission obtained from Elsevier

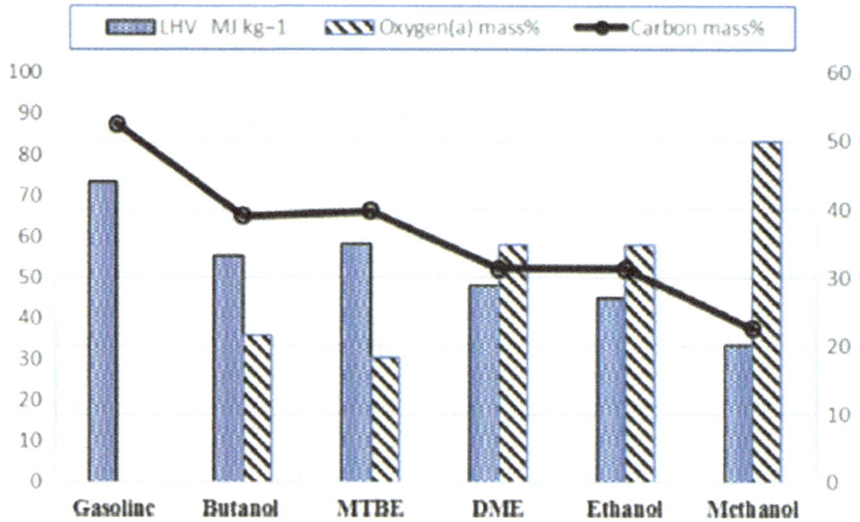

Fig. 6.3 Variation of CO and oxygen with heating value of some alcohols and ethers. Reprinted from Awad et al. (2016a; Pourkhesalian et al. 2010) with copyright permission obtained from Elsevier

have the same oxygen content, their heating values differ slightly; the same is true for butanol and MTBE. This may be as a result of the O:C ratio in the fuels (Fig. 6.3).

(vii) Phase Separation Temperature (PST)

Because of the hygroscopic nature of the alcohols, gasohols are usually three-component systems comprising of an alcohol/gasoline and water (Pereira et al. 2014). The most significant issue with using gasohols as engine fuels is the possibility of them separating into two liquid phases due to a variety of factors such as temperature, amount of water, the nature of the alcohol in the blend, as well as the gasoline-content. According to Roberto da SilvaTrindade et al. (2017), the PST of water-entrained ethanol-gasoline fuel-blends poses serious technical issues (Balki et al. 2014). Due to the poor solubility of water in gasoline, it is often not a desirable performance-concern. The presence of water in the fuel tank of a vehicle, results in its accumulation at the tank/channel's bottom. Water in fuels can influence the corrosion of engine plugs, engine-parts as well as fuel-tank corrosion. Thus, in cold temperatures, the water content of fuels may freeze, halting fuel flow. High amount of dissolved water in a fuel may reduce its heating value, but more importantly, for gasohols (ethanol-gasoline blends), when high amount of water is present, a separate aqueous phase results which drives/strips ethanol from gasoline (Gu et al. 2012). The alcohol-gasoline mixture should be dry enough to prevent splitting the mixture into water, alcohol and hydrocarbons at low temperatures, according to ASTM D4814-08b. As a result, when evaluating a gasohol blend as potential fuel for SI engines, the first thing to do is to tackle the PST problem so as to limit the measure of heterogeneity. Higher aliphatic alcohols e.g., tertiary butyl alcohol as well as benzyl alcohol have been tested and confirmed compatible with gasoline when mixed as fuels with very minimal tendencies for phase separation (Balki and Sayin 2014b; Calam et al. 2015b; Costa and Sodré 2010; Chen et al. 2014; Irimescu 2012). The solubility of water in gasoline for every gasohol mix is highly dependent on the type of alcohol used. The recorded PSTs reveal that an increase in fuel's hazy appearance increased linearly as the fuel's water content increased (Roberto da Silva Trindade et al. 2017). Some alcohols are adopted as co-solvents due to their abilities to reduce the PST of gasohols with resultant increase in the stability of the mix. Jiang et al. (2014) established the impact of introducing fusel oil as blending agent for a mixture of gasoline and methanol.

6.3 Ethanol-Ether Admixed Biofuel-Properties and Engine Compatibility

The blending of alcohol and gasoline has piqued the interest of researchers in a number of countries such as Brazil where blends of E5–E85 (5–85% C_2H_5OH), Germany (3% CH_3OH), South Africa (12–1% alcohol blend), and parts of the United States (10–1% C_2H_5OH) (Balki and Sayin 2014b; Mofijur et al. 2016). Without any further modification, alcohols can be mixed in any proportion with gasoline for use as fuel in a variety of ICEs. Methanol-gasoline blend can be in various proportions in SI engines (Yasin et al. 2015). Gravalos et al. (2013a) investigated the emission characteristics of high to low molar mass alcohol-gasoline blends with methanol, butanol, ethanol, and propanol as the alcohol component. Based on their findings, it was discovered that when running the engine on gasohol-blends, the resulting CO/HC-concentrations decreased at different engine-loads, whereas the NO emissions of the gasohols were higher than those of the pristine gasoline. There are property-specs that must be met in order for gasohols to run properly in SI engines. Alcohol addition to gasoline alters certain key properties of the individual fuels, particularly their heating value, blend-viscosity, octane rating and stability. Corrosiveness and material compatibility are two other factors to consider. As a result, properties that jeopardize the safety of the engine should be prioritized in any fuel evaluation protocol; this then means that flash point and flammability of the fuel are critical. In summary, as the alcohol concentration in the blend increased, so did the fuel properties (Agarwal et al. 2014; Fenard and Vanhove 2021; Lehn et al. 2021; Liu et al. 2006). Increasing the concentration of alcohol in the blend reduced the heating value of the fuels with resultant increase in the kinematic viscosity and density of the blended fuels. Moreover, with the exception of butanol, the fuels' octane numbers increased and this property played an important role in the performance and emission characteristics of the SI engine (Zhang et al. 2013; Chen et al. 2014).

6.4 The Mechanism of Performance of Alcohol Admixed Biofuel towards Ensuring High Engine Compatibility and Engine Performance

Mechanism for (i) methanol-gasoline mix; ethanol gasoline mix; butanol-gasoline mix etc.

$$CH_3OH + \text{Gasoline} \rightarrow \text{R - Gasoline - OH (Gasohol)} \qquad \text{(Scheme 1)}$$

R can range from methyl to ethyl to butyl etc. and is the displaced alkyl group from the alcohol

$$RCHO + \text{Gasoline} \rightarrow \text{R - Gasoline - CHO (Gasoldehyde)} \qquad \text{(Scheme 2)}$$

Again, R is any alkyl group ranging from methyl to ethyl, propyl, butyl etc.

Considering the product sides of the stoichiometric reactions displayed in Schemes 1 and 2, the resulting gasohol and gasoldehyde are projected to have better or improved fuel properties relative to their pristine counterparts (Karavalakis et al. 2014).

Their findings indicate that at the time of combustion of the DMF and at a pressure of 8.5 bar, the IMEP engine load was found to be exerted within the regions characterized by 1 and 4° crank angles lower relative to those of ethanol and gasoline.

6.5 Characteristics of Alcohol and Ether-Admixed Biofuels for High Engine Performance: Source, Composition and Properties

The type of fuel used has a direct impact on the performance characteristics of SI engines. Torque, engine power, specific fuel consumption (SFC), and brake thermal efficiency (BTE) are among these properties (Noor et al. 2014). Tables 6.3, 6.4 and 6.5 show the effects of alcohol and ether-induced fuels on the performance of SI engines. Several engine types, fuel-blends, operating conditions, as well as engine-torque, power, specific fuel consumption (SFC) and brake thermal efficiencies (BTE) for different scenarios have been documented.

6.5.1 Effect on Torque and Power

According to some studies, a slight increase in engine-torque caused a corresponding increase in the fuel consumption of the alcohol-based fuels owing to the fact that alcohols have lower energy-densities per unit volume relative to gasoline. The power

Table 6.3 Engine power, torque and thermal efficiencies of alcohol and ether fuels in spark ignition engines

Type of alcohol	Blending ratio	Type of SI engine	Test condition	Effect on toque, BTE and engine power	References
Ethanol and Methanol	E100 and M100	4C, 4S, WC engine	Various engine speeds and loads	Higher BTE was found in Alcohols as compared to gasoline due to in cylinder temperature and lower pressure	Qi et al. (2005)
Ethanol and DMF	DMF100 and E100		Under 2500 rpm speed, AFR and various loads conditions	ISFC accelerated when DMF and ethanol were used because DME and ethanol and have lower heating values than gasoline	Yucesu et al. (2006)
Butanol	B35	4S, 4C, engine with low BSFC 14.44 (MJ/kW h)	Speeds at (3000–8000 rpm)	Engine 100% load torque doubled clearly for all speeds below 7000 rpm but constant speeds above rpm of 700	Yüksel and Yüksel (2004)
DME	DME at 100	SC, 4S engine	Various engine speeds and loads	Comparing with gasoline, DME increases the BTE by 20%. With DME addition, the flame development reduced	Sezer and Bilgin (2008)
MTBE	MTBE10, 20, 30	4S, Hydra engine, SC	Engine speed varying from 2500 to 5000 rpm	When compared gasoline, the torque increased by 1.24%, 1.96%, and 2.2% for the F10, 20, and F30	Kelly and Konstantinidis (2011)

and torque delivered by an engine are heavily influenced by the engine's in-cylinder pressure. Calam et al. (2015b) studied the performance of a 4-stroke, SI Hydra engine, single cylinder, operated at 1500, 2500, 3500 and 5000 rpm, test-fuels (fusel oil admixed with unleaded gasoline at different concentrations 10, 20, 30%) (F0, F10, F20 and F30) and engine loads (25, 50, 75, and 100%). The engine-torques increased at all loads and engine speeds for increased quantities of the fusel oil (Ozsezen and Canakci 2011b). The F30 gave the highest engine torque at a speed of 2500 rpm. Wang et al. (2013) tested ethanol, gasoline, and DMF on a four-stroke DISI research engine, spray guided, single cylinder, at a speed of 1500 rpm, air–fuel ratio (AFR), and engine-loads in the range of 3–8.5 bar, it was observed that a mean effective pressure (IMEP) at optimum spark timing was obtained. Their findings indicate that

Table 6.4 NOx and hydrocarbon emissions effect of alcohol and ether fuels on spark ignition engines

Type of alcohol	Blending ratio	Type of SI engine	Test condition	Effect on NOx and hydrocarbon emissions	Ref
Ethanol	E5, 10 15	4S, WC, SC	Various engine CR, torque, and speeds	NOx emissions recorded 16.18% increase as the blending ratio increased. While HC emissions decreased substantially 41.15%	Roberto da Silva Trindade et al. (2017)
Methanol ethanol	E10 and M10	air cooled, SC 4S,	Various engine CR, torque, and speeds	E10 showed lower CO emission as compared to pure gasoline	Uygun (2018)
Methanol	E5, 10, 15	SC, 2C	Various engine loads and speeds	Oxides of carbon emissions reduced with ethanol blends	Yüksel and Yüksel (2004)
MTBE	MTBE0, MTBE5, and MTBE11	1.6L, 4S,	Various engine speeds between (1000–2500 rpm) and (0–20 HP) engine power	13 w/w percent MTBE was added to gasoline, HC emissions were reduced at high engine loads	Zhang et al. (2021)
Isobutanol	iB0, iB7, iB10	4S 2 valves, SC,	Various engine loads and speeds	At 2900 rpm or lower, butanol-gasoline blends had lower HC emissions. However, at speeds greater than 2500 rpm, gasoline produced lower HC emissions than the blends	Polat (2016)

(continued)

Table 6.4 (continued)

Type of alcohol	Blending ratio	Type of SI engine	Test condition	Effect on NOx and hydrocarbon emissions	Ref
MTBE	MTBE0, MTBE10, MTBE15, and MTBE20	2.96L, 6C, 4S, l SI engine	Various engine loads and speeds	MTBE blends resulted in increased NOx emissions stoichiometric blends were used. At idle speed, the addition of MTBE had no discernible effect on NOx emissions, which were already very low. While the HC emission decreased as the MTBE ratio increased	Sarathy et al. (2014)
DME	DME (0–30%)	1.6L, 4C, 4S, SI engine port fuel infection system	Various engine loads and speeds	NOx and HC emissions clearly reduced with increasing DME amount	Yasunaga et al. (2010)
Bio-ethanol	E0, E10 E15 E20, E30 and E85	4C, 4S, SI engine	Various engine speeds and engine loads	With E85, the maximum reduction in NOx and HC emissions was about 15% and 20%, respectively	Eble et al. (2017)
n-butanol	Bu0, Bu10, Bu30, Bu40, and Bu100	3C,4S, port fuel injection, with CR 9.6	Various engine loads and speeds	Butanol and ethanol addition to gasoline reduced NOx emissions across the CR range. HC emission also decreased across the CR range	Markides and Mastorakos (2011)

Table 6.5 CO and CO_2 emissions effect of alcohol and ether fuels on spark ignition engines

Type of alcohol	Blending ratio	Type of SI engine	Test condition	Effect on CO and CO_2 emissions	Ref
Methanol ethanol	E10 and M10	SC 4S, air cooled	Various injection timing and engine loads	E10 had lower CO emission compared to gasoline	Sezer and Bilgin (2008)
Isobutanol	iB0, iB7, iB10	SC 4S, air cooled	Various engine speeds	CO emission decreased at low speeds while increasing at high speeds. While CO_2 levels reduced	Yasunaga et al. (2010)
Ethanol	E5, E10 and E15	SC 4S, WC, dual Fuel	Various engine speeds, torque and CR	CO and CO_2 emissions decreased by ethanol addition about 45 and 7.5%	Ozsezen and Canakci (2011b)
n-butanol	Bu0, Bu10, Bu30, Bu40, and Bu100	3C,4S, port fuel injection, with CR 9.6	Various spark timings and EGR rates	n-butanol-gasoline reduces CO emission	Agarwal et al. (2014)
Methanol	E5, E10 and E15	SC 2S	Various speeds and engine loads	Ethanol blends reduce CO and CO_2 emissions	Najafi et al. (2009)
Bio-ethanol	E0, E10 E15 E20, E30 and E85	4C, 4S, SI engine	Various engine speeds and engine loads	15% reduction in CO emission was observed by E85	Milpied et al. (2009)
DME	DME (0–30%)	1.6L, 4C, 4S, SI engine port fuel infection system	1400 rpm engine speed	CO emission first reduced and then increased with increase in concentration of DME	Agarwal (2007b)
Ethanol and Methanol	E5, E10, M5, and M10	4C, 4S, engine with MPFJS	–	Emission of CO_2 decreased	Markides and Mastorakos (2011)
MTBE	MTBE10, MTBE20 and MTBE30	SC, 4S DISI engine	Various speeds and engine loads	using MTBE, CO_2 emission increased	Hu et al. (2017)

(continued)

Table 6.5 (continued)

Type of alcohol	Blending ratio	Type of SI engine	Test condition	Effect on CO and CO_2 emissions	Ref
MTBE	MTBE10, MTBE20 and MTBE30	SC, 4S DISI engine	Various speeds and engine loads	using MTBE, CO_2 emission increased	Yüksel and Yüksel (2004)
MTBE	MTBE0, MTBE2, MTBE5, MTBE8 and MTBE11	1.6L, 4C, 4S, Opel SI engine	Various speeds between 1000 and 4500 rpm	CO emission decreased at high engine load with the addition of MTBE	Çelik et al. (2011)
MTBE	MTBE0, MTBE10, MTBE15, and MTBE20	2.96L, 6C, 4S, 1 SI engine	Various speeds between 1000 and 3500 rpm	CO emission decreased with increasing MTBE ratio. Also, leaded fuel caused higher CO emission compared to MTBE blend	Liang et al. (2011)
Methanol and ethanol	M100 and E100	4C, 4S, WC engine	CR at (6–16)	Emission of CO_2 decreased	Liu et al. (2006)
Ethanol	E25, E50, E75 and E100	4C, 4S, high CR engine	Various CR at constant load and speed	Emission of CO decreased	Yusri et al. (2016)
Methanol and ethanol	M100 and E100	4C, 4S, naturally aspirated Engine	Various engine speeds with CR of 8.6	Emission of CO_2 decreased	Noor et al. (2014)

at the time of combustion of the DMF and at a pressure of 8.5 bar, the IMEP engine load was found to be exerted within the regions characterized by 1 and 4-degree crank angles lower relative to those of ethanol and gasoline. The in-cylinder peak pressure obtained for DMF was higher than that of C_2H_5OH. However, the C_2H_5OH and DMF displayed greater ant-knocking potentials compared to that of gasoline as a result of its higher octane-rating. Also, C_2H_5OH had the highest thermal efficiency of about 38.5% at 8.5 bar IMEP, while those of gasoline and DMF were 36 and 37%, respectively (López et al. 2015). Efforts have been made to account for the ability of a fuel's octane rating to increase the compression ratio (CR) in SI engines so as to induce high torque, engine efficiency and power (Chen et al. 2014; Roberto da Silva Trindade et al. 2017; Lehn et al. 2021; Agarwal 2007b). High LHOV and octane values of alcohols make them possible candidates for high compression ratio engines with high power output. A high-octane rating allows for a higher CR and thus better engine-performance. Yucesu et al. (2006) investigated the effects of gasoline, E10, 20, 40, and 60 fuels on the performance of a SI engine with a CR ranging from 8:1 to 13:1. It was discovered that at higher CR with ethanol in the fuel, the engine torque and power increased. Heywood (2018) focused on the influence of CR on the performance of a 5.3L 8-cylinder engine running at 2000 rpm (full load). The ignition timing and fuel–air mixture equivalence were altered at the time of the experiment. The results showed that increasing the CR gave a corresponding rise in the mean effective pressure (MEP) up to a certain point after which it then decreased. As a result, the engine power was reduced. Yüksel and Yüksel (2004) studied the impacts of a butanol–gasoline mix (B35) on the performance of an engine and discovered that the engine's full load torque increased significantly for speeds lower than 7000 rpm for the B35-fuel blend relative to gasoline. The full load engine torque was maintained at the same value at engine speeds greater than 7000 rpm. Moreover, Balki et al. (2014) found that adding 10% ethanol to gasoline improves brake power by 5% and increases the octane rating by 5% for every 10% addition of C_2H_5OH. However, in some studies, it has been reported that engine torque and power are lower when using alcohol fuels versus gasoline. Sezer and Bilgin (2008) studied engine efficiency within fuel–air equivalence ratios ranging from 0.8 to 1.2 on a 4-stroke single cylinder engine using a B30 gasoline-butanol blend. They discovered a 7% decrease in power for the fuel blend when compared to gasoline. Consequently, Kelly and Konstantinidis (2011) discovered that when fuel oil had a lower heating value and high amount of water, the engine power decreased.

6.5.2 *Effect on Fuel Consumption (FC)*

Many factors influence fuel economy, including vehicle weight, aerodynamics, transmission type, engine design, fuel supply system, and size. For some vehicles, these factors remain constant (Hu et al. 2017). However, several factors influence fuel economy one of which is the heating value of alcohol in gasoline Daniel et al. (2012), as well as the road/weather-conditions. In a study, the influence of heating value on the combustion of a fuel in relation to its alcohol and oxygen contents in a SI engine was investigated. Since the heating value of alcohol and ether fuels is approximately 30% less than that of gasoline, fuels with high proportions of alcohol and ether must produce the same measure of power in terms of fuel consumption (Sakai et al. 2017). Calam et al. (2015a) conducted an experiment using a Hydra brand, 4-cylinder SI engine with a CR of 11:1 and various gasoline-fuel oil blends including B0, F5, 10, 20, 30, and 50 fuels at full load conditions with maximum engine torque of 3500 Nm. It was observed that introducing fusel oil in gasoline helped to reduce the heating value of the fusel oil thus causing a reduction in the calorific value of the test-fuels. As a result, as the amount of fuel oil in the blend increases, so is the amount of fuel taken into the cylinder thus resulting in an increase in the brake specific fuel consumption (BSFC) (Tang et al. 2017; Serinyel et al. 2018). When the F10, 20, and 30 fuels were used, it was observed that the SFC and maximum torque-speed increased by 2.4, 2.7, and 3.1%, for the fuels relative to gasoline. Zhang et al. (2014) adopted methanol-gasoline blends as fuels for operating an engine. Engine-performance tests were carried out with the throttle wide open (WOT) and at various engine speeds in the range of 1000–2500 rpm, using various methanol gasoline blends. With the exception of the E15 blend, the BSEC increased at all engine speeds and blending ratios in comparison to that of the pristine gasoline. Javed et al. (2017) showed that BSFC of gasoline can increase to about 70% when hydrous ethanol is mixed with gasoline. Ogura et al., (2007) monitored the combustion properties of a 3-cylinder engine fueled with E10 and 20 ethanol–gasoline blends at 2500 rpm at several engine- loads. The blended BTE values were similar with those of gasoline with higher BSFC than gasoline.

6.5.3 Effect on NOx and HC Emissions

Table 6.4 shows the recorded NOx/HC emissions from alcohol and ether fuels; the emissions are lower in most cases compared to that recorded for pure gasoline. The operating conditions, on the other hand, had an impact on the NOx and HC emissions. When alcohol was used to blend gasoline, the peak in-cylinder temperature decreased due to its high heat of vaporization, lowering NOx emissions (Wurmel et al. 2007). Zervas et al. (2003) investigated how fuel composition and air–fuel ratio affect NOx emissions. They asserted that for stoichiometric quantities of alcohol blends of $(CH_3)_2CHOH$, C_2H_5OH, CH_3OH, and MTBE with 20% v/v of gasoline can result in a 60% reduction in NOx emissions. Najafi et al. (2009) observed the effect of ethanol–gasoline blends on CO and HC- emissions and compared them with those of gasoline. According to the findings, a blend with more than 20% ethanol reduced CO and HC emissions significantly relative to that of gasoline. Zhang et al. (2018) also confirmed the reduction in NOx emissions when the fusel oil used had comparable properties with gasoline. Kisenyi et al. (1994) determined the effects of 15% v/v MTBE as additive in gasoline on the performance of SI engines; the analysis revealed the ability of MTBE to decrease HC emissions by 10–20%, NOx emissions by 1.0–1.7%, and CO emissions by 10–15%. Al-Baghdadi investigated the effect of introducing DME on the idle performance of a SI ethanol engine under idle and stoichiometric conditions. The resulting HC emissions decreased with an increase in the proportion of added DME. A 4-cylinder SI engine fueled by different ethanol-gasoline blends which run at various CR and speeds was used by Koç et al. (2009). Their result showed a reduction in HC emissions at 1500 and 5000 rpm at both CRs with the addition of ethanol.

6.5.4 *Effect of on Carbon Dioxide and Carbon Monoxide Emissions*

Despite the fact that carbon dioxide is non-toxic and not classified an engine pollutant, it is responsible for the rising global warming (Noor et al. 2014). At complete combustion, HC fuels produce CO_2. CO is released from as exhaust gas as a result of incomplete burning of the fuel. Undoubtedly, if the combustion temperature is insufficient to achieve complete combustion, the conversion of CO to CO_2 may not occur (Pourkhesalian et al. 2010). Furthermore, the (C–H) ratio in fuels influences CO_2 formation. Moreover, oxygen content is important in improving engine combustion which in turn increases the CO_2 levels and as a result, reduces the amount of CO released. From Table 6.5, increased concentrations of alcohol and ether fuels reduced CO emissions while increasing CO_2 emissions relative to that seen for pure gasoline. The higher O_2 content of alcohol and ether fuels could explain this. For a gasoline engine, Ghadikolaei (2016) studied the effects of several MTBE mixing ratios i.e., 10, 15, and 20% with gasoline on engine performance and pollution. They reported higher CO_2 exhaust emissions for the MTBE–gasoline mix, which is evidential of better combustibility relative to pure gasoline; this may have resulted due to the associated lower carbon number for the mix as compared to that of the pristine gasoline which is usually between 5 and 8 carbon atoms, respectively. According to Aleiferis et al. (2013), about 2.51% increment in the amount of CO_2 released occurred as a result of using pure ethanol as fuel in an ICE engine.

6.6 Recent Advances in the Use of Ether and Ethanol Admixed Biofuels for Improved Engine Performance

To properly apply ether and alcohol as alternative fuels, two conditions must be achieved. First, redesigning the engine is required in order to fully benefit from the properties of alcohol fuel. Also, multiple mixtures of alcohol must be blended to improve their properties.

6.6.1 *Engine Modifications*

Certain adjustments in engine management and fuel-system properties are required in automobile/SI engines. Alcohol compression ratios should be increased to take full advantage of higher octane values. The following are the key modifications needed to convert current automobile engines to vehicles that run on high-level alcohol blends:

- To improve fuel efficiency, CRs or variable-boost turbochargers must be used; this portends well for smaller and lighter engines (Kiani et al. 2010).
- Improved cold-starting methods must be permitted to allow for appropriate reduction or elimination of gasoline content (Atmanl et al. 2014).
- Due to methanol's higher-octane rating and faster flame speed, spark retard reductions must be encouraged (Hagos et al. 2017).

6.6.2 *Fuel Property-Enhancement*

Alcohols are potential alternative fuels that can give high engine efficiencies when their properties are modified by mixing them with gasoline such that a measure of compatibility is induced in the base alcohol fuel as well as high heating values (Awad et al. 2018b; Qi and Lee 2016b; Costa and Sodré 2010). Corrosion-inhibiting additives influence material incompatibility thus affecting the physicochemical composition of the base fuels (Keskin 2010; Burke et al. 2014). An additive should be easily removed from fuel, must not increase emissions and leave any residue. Additive should be relatively cheap and not disrupt compliance requirements. Some alcohols including methanol are not miscible with gasoline. As a result, mixing them with gasoline may cause phase separation issues. Additives such as non-ionic surfactants, fusel oil, isopropanol, and 1-butanol, can be used to avoid this problem. The effect of adding small amount of fusel oil as a blending agent to pure gasoline- methanol blends was investigated by Sivasankaralingam et al. (2016) they discovered that the PST of the blends decreased as the percentage of fusel oil increased.

6.7 Concluding Remarks

Alcohols and ethers would play important roles in achieving energy demands in the automobile sector if used as additives in gasoline. In this chapter, knowledge on different alcohol types such as ethanol, butanol, and methanol, as well as DME, MTBE and fusel oil additives has been provided in terms of their effects on SI engines; the conclusion reached thus far is that they have the ability to improve gasoline or improve the fuel properties. However, before recommending the use of an alternative fuel for existing engine designs on a large scale, a number of factors must be considered. Among the major factors mentioned are:

- Due to their high-octane numbers, alcohol fuels with high CRs should be used to improve the performance of spark ignition engines.
- The quantity of water in fusel oil and alcohols should be minimized before being used as fuel.
- Fusel oil may be mixed with CH_3OH to limit/curb phase separation.

Finally, alcohols may be adopted for use in high proportions in their mixed forms in SI engines, nevertheless, engine modifications would be required to ensure the fuels' high compatibility with the engines. Because alcohols and ethers have been shown to have significant environmental impacts, these additives must be kept within permissible and tolerable limits when used in fuel blends.

References

Agarwal AK (2007a) Biofuels (alcohols and biodiesel) applications as fuels for internal combustion engines. Prog Energy Combust Sci 33:233–271

Agarwal AK (2007) Biofuels (alcohols and biodiesel) applications as fuels for internal combustion engines. Progress Energy Combust Sci 33:233–271

Agarwal AK, Karare H, Dhar A (2014) Combustion, performance, emissions and particulate characterization of a methanol–gasoline blend (gasohol) fuelled medium duty spark ignition transportation engine. Fuel Process Technol 121:16–24

Aleiferis PG, Serras-Pereira J, Richardson D (2013) Characterization of flame development with ethanol, butanol, iso-octane, gasoline and methane in a direct-injection spark-ignition engine. Fuel 109:256–278

Anonymous (2013) Market supply information in ethanol Turkish tobacco and alcohol market regulatory authority

Atmanl A, Ileri E, Yüksel B (2014) Experimental investigation of engine performance and exhaust emissions of a diesel engine fueled with diesel–n-butanol—vegetable oil blends. Energy Convers Manage 81:312–21

Awad OI, Mamat R, Ali OM, Yusri IM, Abdullah AA, Yusop AF et al (2016) The effect of adding fusel oil to diesel on the performance and the emissions characteristics in a single cylinder CI engine. J Energy Inst. https://doi.org/10.1016/j.joei.2016.04.004

Awad OI, Ali OM, Hammid AT, Mamat R (2018b) Impact of fusel oil moisture reduction on the fuel properties and combustion characteristics of SI engine fueled with gasoline-fusel oil blends. Renew Energy 123:79–91

Awad OI, Ali OM, Mamat R, Abdullah A, Najafi G, Kamarulzaman M et al (2016a) Using fusel oil as a blend in gasoline to improve SI engine efficiencies: a comprehensive review. Renew Sustain Energy Rev

Awad OA, Mamat R, Ibrahim TK, Kettner M, Kadirgama K, Leman AM (2018a) Effects of fusel oil water content reduction on fuel properties, performance and emissions of SI engine fueled with gasoline—fusel oil blends. Renew Energy 118:858–869

Balki MK, Sayin C (2014b) The effect of compression ratio on the performance, emissions and combustion of an SI (spark ignition) engine fueled with pure ethanol, methanol and unleaded gasoline. Energy 71:194–201

Balki MK, Sayin C, Canakci M (2014) The effect of different alcohol fuels on the performance, emission and combustion characteristics of a gasoline engine. Fuel 115:901–906

Balki MK, Sayin C (2014a) The effect of compression ratio on the performance, emissions and combustion of an SI (spark ignition) engine fueled with pure ethanol, methanol and unleaded gasoline. Energy 71:194–201

Burke U, Somers KP, O'Toole P, Zinner CM, Marquet N, Bourque G et al (2015) An ignition delay and kinetic modeling study of methane, dimethyl ether, and their mixtures at high pressures. Combust Flame 162(2):315–330

Burke SM, Metcalfe W, Herbinet O, Battin-Leclerc F, Haas FM, Santner J et al (2014) An experimental and modeling study of propene oxidation. Part 1: Speciation measurements in jet-stirred and flow reactors. Combust Flame 161(11):2765–2784

Burke SM, Burke U, Mc Donagh R, Mathieu O, Osorio I, Keesee C et al (2015) An experimental and modeling study of propene oxidation. Part 2: Ignition delay time and flame speed measurements. Combust Flame 162(2):296–314

Calam A, İçingür Y, Solmaz H, Yamık H (2015a) A comparison of engine performance and the emission of fusel oil and gasoline mixtures at different ignition timings. Int J Green Energy 12:767–72

Calam A, Solmaz H, Uyumaz A, Polat S, Yilmaz E, İçingür Y (2015b) Investigation of usability of the fusel oil in a single cylinder spark ignition engine. J Energy Inst 88 258–265

Canakci M, Ozsezen AN, Alptekin E, Eyidogan M (2013) Impact of alcohol–gasoline fuel blends on the exhaust emission of an SI engine. Renew Energy 52:111–117

Çelik MB, Özdalyan B, Alkan F (2011) The use of pure methanol as fuel at high compression ratio in a single cylinder gasoline engine. Fuel 90:1591–1598

Chen Z, Wu Z, Liu J, Lee C (2014) Combustion and emissions characteristics of high n-butanol/diesel ratio blend in a heavy-duty diesel engine and EGR impact. Energy Convers Manage 78:787–795

Christensen E, Yanowitz J, Ratcliff M, McCormick RL (2011) Renewable oxygenate blending effects on gasoline properties. Energy Fuels 25:4723–4733

Cooney CP, Worm JJ, Naber JD (2009) Combustion characterization in an internal combustion engine with ethanol– gasoline blended fuels varying compression ratios and ignition timing. Energy Fuels 23:2319–2324

Costa RC, Sodré JR (2010) Hydrous ethanol versus gasoline-ethanol blend: engine performance and emissions. Fuel 89:287–293

Daniel R, Tian G, Xu H, Shuai S (2012) Ignition timing sensitivities of oxygenated biofuels compared to gasoline in a direct-injection SI engine. Fuel 99:72–82

Eble J, Kiecherer J, Olzmann M (2017) Low-temperature autoignition of diethyl ether/ O_2 mixtures: mechanistic considerations and kinetic modeling. Z Phys Chem 231(10):1603

Fenard Y, Vanhove G (2021) A mini-review on the advances in the kinetic understanding of the combustion of linear and cyclic oxymethylene ethers. Energy Fuels 35(18):14325–14342. https://doi.org/10.1021/acs.energyfuels.1c01924

Ghadikolaei MA (2016) Effect of alcohol blend and fumigation on regulated and unregulated emissions of IC engines—a review. Renew Sustain Energy Rev 57:1440–1495

Gravalos I, Moshou D, Gialamas T, Xyradakis P, Kateris D, Tsiropoulos Z (2013a) Emissions characteristics of spark ignition engine operating on lower–higher molecular mass alcohol blended gasoline fuels. Renew Energy 50:27–32

Gravalos I, Loutridis S, Moshou D, Gialamas T, Kateris D, Tsiropoulos Z et al (2013b) Detection of fuel type on a spark ignition engine from engine vibration behavior. Appl Therm Eng 54:171–175

Gu X, Huang Z, Cai J, Gong J, Wu X, Lee C (2012) Emission characteristics of a spark ignition engine fuelled with gasoline-n-butanol blends in combination with EGR. Fuel 93:611–617

Hagos FY, Ali OM, Mamat R, Abdullah AA (2017) Effect of emulsification and blending on the oxygenation and substitution of diesel fuel for compression ignition engine. Renew Sustain Energy Rev 75:1281–1294

Heywood JB (2018) Internal combustion engine fundamentals, 2nd edn. McGraw-Hill Education. ISBN: 9781260116106

Hu T, Wei Y, Liu S, Zhou L (2007) Improvement of spark-ignition (SI) engine combustion and emission during cold start, fueled with methanol/gasoline blends. Energy Fuels 21:171–175

Hu E, Chen Y, Zhang Z, Chen J-Y, Huang Z (2017) Ab initio calculation and kinetic modeling study of diethyl ether ignition with application toward a skeletal mechanism for CI engine modeling. Fuel 209(Supplement C):509–520
Icingur Y, Calam A (2012) The effects of the blends of fusel oil and gasoline on performance and emissions in a spark ignition engine. J Faculty Eng Archit Gazi Univ 27:143–149
Irimescu A (2012) Performance and fuel conversion efficiency of a spark ignition engine fueled with iso-butanol. Appl Energy 96:477–483
Javed T, Ahmed A, Lovisotto L, Issayev G, Badra J, Sarathy SM et al (2017) Ignition studies of two low-octane gasolines. Combust Flame 185:152–159
Jiang Y, Liu J, Jiang W, Yang Y, Yang S (2014) Current status and prospects of industrial bio-production of *n*-butanol in China. Biotechnol Adv 33(7):1493–1501
Karavalakis G, Short D, Vu D, Villela M, Asa-Awuku A, Durbin TD (2014) Evaluating the regulated emissions, air toxics, ultrafine particles, and black carbon from SI-PFI and SI-DI vehicles operating on different ethanol and iso-butanol blends. Fuel 128:410–421
Kelly JM, Konstantinidis D (2011) Mechanics of rubber bearings for seismic and vibration isolation. Wiley
Keskin A (2010) The influence of ethanol–gasoline blends on spark ignition engine vibration characteristics and noise emissions. Energy Sour Part a: Recov, Utilization, Environ Effects 32:1851–1860
Keskin A, Gürü M (2011) The effects of ethanol and propanol additions into unleaded gasoline on exhaust and noise emissions of a spark ignition engine. Energy Sour Part A Recov, Utilization Environ Effects 33:2194–2205
Kiani MKD, Ghobadian B, Tavakoli T, Nikbakht A, Najafi G (2010) Application of artificial neural networks for the prediction of performance and exhaust emissions in SI engine using ethanol-gasoline blends. Energy 35:65–69
Kisenyi J, Savage C, Simmonds A. The impact of oxygenates on exhaust emissions of six European cars. SAE Technical Paper; 1994.
Koç M, Sekmen Y, Topgül T, Yücesu HS (2009) The effects of ethanol–unleaded gasoline blends on engine performance and exhaust emissions in a spark-ignition engine. Renew Energy 34:2101–2106
Liang C, Ji C, Liu X (2011) Combustion and emissions performance of a DME-enriched spark-ignited methanol engine at idle condition. Appl Energy 88:3704–3711
Liang C, Ji C, Gao B, Liu X, Zhu Y (2012) Investigation on the performance of a spark ignited ethanol engine with DME enrichment. Energy Convers Manage 58:19–25
Liu HF, Zhang QC, Huang H, Huang JC (2006) The application of fuels methanol and dimethyl ether (DME) on engines. Energy Convers Manage 25:13–16
Liu H, Wang Z, Wang JX (2014) Methanol-gasoline DFSI (dual-fuel spark ignition) combustion with dual-injection for engine knock suppression. Energy 73:686–693
López AF, Cadrazco M, Agudelo AF, Corredor LA, Vélez JA, Agudelo JF (2015) Impact of n-butanol and hydrous ethanol fumigation on the performance and pollutant emissions of an automotive diesel engine. Fuel 153:483–491
Markides CN, Mastorakos E (2011) Experimental investigation of the effects of turbulence and mixing on autoignition chemistry. Flow Turbul Combust 86:585–608
Milpied J, Jeuland N, Plassat G, Guichaous S, Dioc N, Marchal A et al (2009) Impact of fuel properties on the performances and knock behaviour of a downsized turbocharged DI SI engine–focus on Octane numbers and latent heat of vaporization. SAE Int J Fuels Lubr 2:118–126
Mofijur M, Rasul M, Hyde J, Azad A, Mamat R, Bhuiya M (2016) Role of biofuel and their binary (diesel–biodiesel) and ternary (ethanol–biodiesel–diesel) blends on internal combustion engines emission reduction. Renew Sustain Energy Rev 53:265–278
Najafi G, Ghobadian B, Tavakoli T, Buttsworth DR, Yusaf TF, Faizollahnejad M (2009) Performance and exhaust emissions of a gasoline engine with ethanol blended gasoline fuels using artificial neural network. Appl Energy 86:630–639

Nithyanandan K, Zhang J, Li Y, Wu H, Lee TH, Lin Y et al (2016) Improved SI engine efficiency using acetone–butanol–ethanol (ABE). Fuel
Noor MM, Wandel AP, Yusaf T (2012) A review of MILD combustion and open furnace design consideration. Int J Autom Mech Eng 6:27
Noor MM, Wandel AP, Yusaf T (2014) MILD combustion: The future for lean and clean combustion technology. Int Rev Mech Eng 8:251–257
Ogura T, Miyoshi A, Koshi M (2007) Rate coefficients of H-atom abstraction from ethers and isomerization of alkoxyalkylperoxy radicals. PCCP 9(37):5133–5142
Ozsezen A, Canakci M (2011b) Performance and combustion characteristics of alcohol–gasoline blends at wide-open throttle. Energy 36:2747–2752
Ozsezen AN, Canakci M (2011a) Performance and combustion characteristics of alcohol–gasoline blends at wide-open throttle. Energy 36:2747–2752
Pereira LP, Dias MO, Junqueira TL, Pavanello LG, Cavalett MFCO, Filho RM, Bonomi A (2014) Butanol production in a sugarcane biorefinery using ethanol as feedstock. Part II: integration to a second-generation sugarcane distillery Chem Eng Res Des 92:1452–1462
Polat S (2016) An experimental study on combustion, engine performance and exhaustemissions in a HCCI engine fuelled with diethyl ether–ethanol fuel blends. Fuel Process Technol 143(Supplement C):140–150
Pourkhesalian AM, Shamekhi AH, Salimi F (2010) Alternative fuel and gasoline in an SI engine: a comparative study of performance and emissions characteristics. Fuel 89:1056–1063
Qi D, Liu SQ, Zhang CH, Bian YZ (2005) Properties, performance, and emissions of methanol-gasoline blends in a spark ignition engine. Proc Inst Mech Eng Part D J Autom Eng 219:405–412
Qi D, Lee C (2016) Combustion and emissions behaviour for ethanol–gasoline-blended fuels in a multipoint electronic fuel injection engine. Int J Sustain Energy 35:323–338
Rakopoulos DC, Rakopoulos CD, Papagiannakis RG, Kyritsis DC (2011) Combustion heat release analysis of ethanol or n-butanol diesel fuel blends in heavy-duty DI diesel engine. Fuel 90:1855–1867
Reijnders L, Huijbregts M (2008) Biofuels for road transport: a seed to wheel perspective. Springer Science & Business Media
Roberto da Silva Trindade W, Rogério Gonçalves dosSantos R (2017) Review on the characteristics of butanol, its production and use as fuel in internal combustion engines. Renew Sustain Energy Rev 69(Mar 2017):642–651
Sakai Y, Herzler J, Werler M, Schulz C, Fikri M (2017) A quantum chemical and kinetics modeling study on the autoignition mechanism of diethyl ether. P Combust Inst 36(1):195–202
Sarathy SM, Thomson MJ, Togbé C, Dagaut P, Halter F, Mounaim-Rousselle C (2009) An experimental and kinetic modeling study of n-butanol combustion. Combust Flame 156:852–864
Sarathy SM, Tingas E-A, Nasir EF, Detogni A, Wang Z, Farooq A et al (2019) Three-stage heat release in n-heptane auto-ignition. Proc Combust Inst 37(1):485–492
Sarathy SM, Oßwald P, Hansen N, Kohse-Höinghaus K (2014) Alcohol combustion chemistry. Prog Energy Combust Sci 44(Suppl C):40–102
Serinyel Z, Lailliau M, Thion S, Dayma G, Dagaut P (2018) An experimental chemical kinetic study of the oxidation of diethyl ether in a jet-stirred reactor and comprehensive modeling. Combust Flame 193:453–462
Sezer I, Bilgin A (2008) Effects of methyl tert-butyl ether addition to base gasoline on the performance and CO emissions of a spark ignition engine. Energy Fuels 2008:1341–1348
Sivasankaralingam V, Raman V, Mubarak Ali MJ, Alfazazi A, Lu T, Im H et al (2016) Experimental and numerical investigation of ethanol/diethyl ether mixtures in a CI engine. SAE Int
Siwale L, Kristóf L, Bereczky A, Mbarawa M, Kolesnikov A (2014) Performance, combustion and emission characteristics of n-butanol additive in methanol–gasoline blend fired in a naturally-aspirated spark ignition engine. Fuel Process Technol 118:318–326
Song C-L, Zhang W-M, Pei Y-Q, Fan G-L, Xu G-P (2006) Comparative effects of MTBE and ethanol additions into gasoline on exhaust emissions. Atmos Environ 40:1957–1970

Tang Z, Zhang L, Chen X, Tang G (2017) Improved kinetic mechanism for diethyl ether oxidation with a reduced model. Energy Fuel 31(3):2803–2813

Tran L-S, Herbinet O, Li Y, Wullenkord J, Zeng M, Bräuer E et al (2018) Low-temperature gas-phase oxidation of diethyl ether: Fuel reactivity and fuel-specific products. Proc Combust Inst

Uygun Y (2018) Ignition studies of undiluted diethyl ether in a high-pressure shock tube. Combust Flame 194:396–409

vom Lehn F, Cai L, Pitsch H (2019) Impact of thermochemistry on optimized kinetic model predictions: auto-ignition of diethyl ether. Combust Flame 210:454–466

vom Lehn F, Cai L, Tripathi R, Broda R, Pitsch H (2021) A property database of fuel compounds with emphasis on spark-ignition engine applications. Appl Energy Combust Sci 5:100018. https://doi.org/10.1016/j.jaecs.2020.100018

Wang C, Xu H, Daniel R, Ghafourian A, Herreros JM, Shuai S et al (2013) Combustion characteristics and emissions of 2-methylfuran compared to 2, 5-dimethylfuran, gasoline and ethanol in a DISI engine. Fuel 103:200–211

Wurmel J, Simmie JM, Curran HJ (2007) Studying the chemistry of HCCI in rapid compression machines. Int J Veh Des 44(1–2):84–106

Yasin MHM, Paruka P, Mamat R, Yusop AF, Najafi G, Alias A (2015) Effect of low proportion palm biodiesel blend on performance, combustion and emission characteristics of a diesel engine. Energy Procedia 75:92–98

Yasunaga K, Gillespie F, Simmie JM, Curran HJ, Kuraguchi Y, Hoshikawa H et al (2010) A multiple shock tube and chemical kinetic modeling study of diethyl ether pyrolysis and oxidation. J Phys Chem A 114(34):9098–9109

Yucesu HS, Topgul T, Cinar C, Okur M (2006) Effect of ethanol–gasoline blends on engine performance and exhaust emissions in different compression ratios. Appl Therm Energy

Yüksel F, Yüksel B (2004) The use of ethanol–gasoline blend as a fuel in an SI engine. Renew Energy 29:1181–1191

Yusri I, Mamat R, Azmi W, Najafi G, Sidik N, Awad OI (2016) Experimental investigation of combustion, emissions and thermal balance of secondary butyl alcohol-gasoline blends in a spark ignition engine. Energy Convers Manag 123:1–14

Yusri IM, Mamat R, Najafi G, Razman A, Awad OI, Azmi WH et al (2017) Alcohol based automotive fuels from first four alcohol family in compression and spark ignition engine: a review on engine performance and exhaust emissions. Renew Sustain Energy Rev 77:169–181

Zervas E, Montagne X, Lahaye J (2003) Emissions of regulated pollutants from a spark ignition engine. Influence of fuel and air/fuel equivalence ratio. Environ Science Technol 37:3232–3238

Zhang J, Niu S, Zhang Y, Tang C, Jiang X, Hu E et al (2013) Experimental and modeling study of the auto-ignition of n-heptane/n-butanol mixtures. Combust Flame 160:31–39

Zhang Y, El-Merhubi H, Lefort B, Le Moyne L, Curran HJ, Kéromnès A (2018) Probing the low-temperature chemistry of ethanol via the addition of dimethyl ether. Combust Flame 190:74–86

Zhang X, Zou J, Cao C, Chen W, Yang J, Qi F, Li Y (2021) Exploring the low-temperature oxidation chemistry of 1-butene and i-butene triggered by dimethyl ether. Proc Combust Inst 38:289–298

Zhang Z, Wang T, Jia M, Wei Q, Meng X, Shu G (2014) Combustion and particle number emissions of a direct injection spark ignition engine operating on ethanol/gasoline and n-butanol/gasoline blends with exhaust gas recirculation. Fuel 130:177–88

Zhen X, Wang Y (2015) An overview of methanol as an internal combustion engine fuel. Renew Sustain Energy Rev 52:477–493

Zhuang Y, Hong G (2013) Primary investigation to leveraging effect of using ethanol fuel on reducing gasoline fuel consumption. Fuel 105:425–431

Chapter 7
Influence of Oxygenated Fuel and Additives in Biofuel Run Compression Ignition Engine

Debangsu Kashyap, Samar Das, and Pankaj Kalita

Abstract Effort for reduction of global emission level is currently one of the prime areas of concern for the research community across the globe. Particularly due to more stringent standards of emission control, the prevailing diesel engines are on the verge of losing their permission to operate. Engine fuel modification technologies are reported to improve engine combustion and reduce engine emission levels. The enhancement in liquid fuel using oxygenated additives can be a sustainable and cost effective solution to address the issues of the existing diesel engine emission. Among the existing vehicular fuel improvements technologies, the use of biodiesel, alcohols viz. methanol, ethanol, propanol, butanol, and ethers improves performance as well as emission characteristics of engine significantly. As found in open source, the application of biodiesel-alcohol blended fuels can reduce carbon emission by 50–60% and hence considered as a possible conventional fuel substitution for engine applications. These fuels can be applied either completely or as blends with diesel in diesel engines. Moreover, water emulsification with different blends of biofuels can further restrict the engine emission levels particularly the NOx and smoke emission levels up to 25%. Therefore, the current chapter delivers a critical analysis of the use of oxygenated additives for running diesel engines. The improvements in the physiochemical properties of biodiesel/diesel-alcohol blended fuels and their influence on the engine emission characteristics are discussed in the chapter.

Keywords Biodiesel · Alcohols · Emulsification · CI engine · Performance · Emission

D. Kashyap · S. Das · P. Kalita (✉)
Fuel and Combustion Lab, Centre for Energy, Indian Institute of Technology Guwahati, Guwahati, Assam 781039, India
e-mail: pankajk@iitg.ac.in

A. V. Agarwal and H. Valera (eds.), *Potential and Challenges of Low Carbon Fuels for Sustainable Transport*, Energy, Environment, and Sustainability,
https://doi.org/10.1007/978-981-16-8414-2_7

Abbreviations

ABE	Acetone-butanol-ethanol
BFO	Borassus Flabellifer Oil
BMEP	Brake Mean Effective pressure
BSEC	Brake Specific Energy Consumption
BSFC	Brake Specific Fuel Consumption
BTDC	Before Top Dead Centre
BTE	Brake Thermal efficiency
CFPP	Cold Filter Plugging Point
CI	Compression Ignition
CN	Cetane Number
CNT	Carbon Nano Tubes
CO	Carbon monoxide
CP	Cloud Point
CR	Compression Ratio
CV	Calorific Value
DEE	Di-ethyl Ether
DME	Di-methyl Ether
DOC	Diesel Oxygen Catalyst
EGR	Exhaust Gas Recirculation
EGT	Exhaust gas Temperature
FIE	Fuel Injection Equipment
FP	Flash Point
HC	Hydrocarbon
HLB	Hydrophile-Lipophile Balance
HRR	Heat Release Rate
IP	Injection Pressure
IT	Injection Timing
JME	Jatropha Methysl Ester
KME	Karanja Methyl Ester
KOH	Potassium Hydroxide
KSOME	Koroch Seed Oil Methyl Ester
LGO	Lemon Grass Oil
LPO	Lemon Oil Water
MOBE	Mahua Oil Butyl Ester
MOEE	Mahua Oil Ethyl Ester
MOME	Mahua Oil Methyl Ester
Nox	Oxides of Nitrogen
PCP	Peak Cylinder Pressure
PM	Particulate Matter
POME	Palm Oil Methyl Ester
PP	Pour Point
PPME	Pongamia Pinnata Methyl Ester

Sox Oxides of Sulphur

7.1 Introduction

Global rise in continuous energy demand in the different sectors along with the automobile sector has resulted in declining of fossil fuel reservoirs. In addition to that, the harmful pollutant emission from the internal combustion engines are causing serious concerns of environmental pollutions. The Twenty-first session of the Paris conference held in December 2015 discussed finding out ways to curb pollutant emissions from all sources including automobiles (Mondal and Mandal 2019). Diesel engines are the most extensively used engines for both light and heavy duty vehicles. The higher compression ratio (CR) in diesel engines compared to petrol engines makes it suitable in terms of engine thermal efficiency. Over the past, diesel engines have undergone vast improvements such as electronically controlled direct fuel injection, high pressure common rail engine system, piezo-injectors, trapping of particulate and lean oxides of nitrogen (NOx), and so on. These modern and highly developed electronic control system in diesel engines have resulted in improved engine performance from 45% to around 55–63% along with reduced emissions (Hagos et al. 2017). However, inspite of these positive improvements, the drawbacks of high emission levels of particulate matter (PM), NOx, CO_2, SOx (oxides of sulphur), HC, CO in diesel engines are causing serious concerns of atmospheric pollution. Along with this, the stringent emission norms imposed by governments for curbing the adverse effects of harmful pollutants has lead the researchers to develop cleaner engine technologies. The EURO 6 regulations for diesel transport released in September 2014 recommended a substantial cutback in the emission NOx and PM from 0.25 g/km and 0.025 g/km (as per EURO 4, January 2005) to 0.08 g/km and 0.005 g/km respectively (Yahaya Khan et al. 2014).

Conventional diesel engine operates with diesel fuel through compression ignition mechanism. Based on the type of engine, the specifications of the fuel also vary. The influential characteristics of diesel fuel in diesel engines include density, cetane number, viscosity, sulphur content, low-temperature properties, aromatic content, boiling range, and volatility (Debnath et al. 2015). The ignition process of diesel engines has a vital role in the combustion of diesel and air mixture which are influenced by the above properties of diesel fuel. For improved functioning of the engine, the short ignition delay is recommended. However, the key challenges are PM and NOx emission which are dependent on the quality of ignition taking place in the combustion chamber. Injection of fuel after ignition delay results in prompt burning of fuel and air mixture in the combustion chamber under a rich fuel–air ratio zone. As a consequence, some droplets of injected fuel do not vaporize completely and instead will be emitted as particles of solid carbons. In addition, the increase in engine speed and load also increases the volume of injected fuel into the combustion chamber with decreased combustion time which conduct high formation of PM (Hagos et al.

2017). Again, the high temperature reactions in the chamber lead to the NOx formation, which is a combination of NO (nitric oxide), NO_2 (nitrogen dioxide), and N_2O (nitrous oxide). Among these, NO and N_2O are non-toxic gases. Moreover, NO also reacts with oxygen to form NO_2, which is a reactive and poisonous gas and highly contributes to atmospheric pollution. The NOx formation in diesel engines seems to be high during slightly lean and stoichiometric condition where the combustion temperature inside the chamber attains a higher value. As such researchers have focussed on newer technologies to reduce the PM and NOx emission levels with least drawback on engine performance (Pulkrabek 2003).

The modern technologies involve either hardware modification of fuel side improvement of the existing diesel engines. The hardware side modification includes, injector nozzle modification, fuel injection equipment (FIE) for improving atomization of fuel, Exhaust gas recirculation (EGR) for utilizing the exhaust gas for diluting intake air thereby reducing the NOx formation, and different post-treatment processes for exhaust cleaning are some of the modern diesel engine technologies to abide by the emission regulations. However, these types of modifications are costly and hence uneconomical (Hagos et al. 2017). Further, the rapid diminution of fossil fuels and -continuous rise in unwanted pollutants has led researchers to focus on the use of alternative sources of energy. The use of biodiesel and alcohols in CI engines is the most effective way for fuel side modification of CI engines without undergoing any engine hardware improvements along with replacement of conventional diesel in CI engines (Demirbas 2008).

In addition to the engines modification techniques, the type of engines viz. Low temperature combustion engine (LTC), partially premixed combustion ignition engines (PPCI), Homogeneous charge compression ignition engines (HCCI) and Reactivity controlled compression ignition engine (RCCI) plays a crutial role in as far as application of oxygenated fuels are concerned. These advanced engines offers ultra-low NOx and smoke emission comparing the existing diesel engines. Additionaly, higher levels of CO and HC emission have been reported in in these technologies. However different after treatment process such as EGR can be applied to the engines in overcoming the HC and CO emission levels (Noh and No 2017; Tziourtzioumis and Stamatelos 2017).

Therefore, the current chapter aims at fuel side development of CI engine, specifically focussing on the improvement of the performance and emission characteristics without any modifications on the hardware side of the engine. The chapter vividly depicts the utilization and characteristics of different oxygenated biofuels viz, biodiesels, alcohols (methanol, ethanol, propanol, and butanol) along with their blending and emulsification techniques for substitution of fossil diesel in existing CI engines. Further, the chapter critically discusses the use of these alternative fuel technologies for enhancing the CI engine performance and emission characteristics.

7.2 Fuel Improvements Techniques in CI Engine

Improvements in the engine performance, combustion, and exhaust emission through different improvement techniques such as engine and fuel modification are accompanied by different challenges. However, the fuel side modification technique is found to be suitable as this type of modification can be done with minimum alteration in the design side of the engine. Also, with the continuous diminution of fossil resources, using alternative fuels in CI engines is the need of the hour. Considering the improvements in fuel based technologies, the biodiesel, alcohols, and addition of water in blends or emulsified along with diesel is found to be suitable for enhancing engine performance and lowering the emission. Blending in general, is a mixing process used for two or more constituents irrespective of their phase, miscibility, and other physical properties. It can be between substances such as solid–solid, solid–liquid, immiscible liquids, miscible liquids and also miscible-immiscible liquids. The homogeneous mixing depends upon the shear force between the substances. In addition, for effective blending, macro level mixing structure is important, and the surface tension effect defines the stability of individual substances (Hagos et al. 2017). On the other hand, in emulsification, a liquid is dispersed a clearly immiscible or partially miscible incessant liquid with the use of ultrasonic vibrators, motorised stirring in the presence of surfactants (emulsifiers). Emulsions are classified as macro, micro and nano emulsion depending on the droplet size of the dispersed liquid. Among these, micro-emulsion is reported to be responsible for engines performance and emission improvements (Watanabe and Okazaki 2013; Hagos et al. 2011; Mattiello et al. 1992). However, one of the major drawback of oxygenated fuel application in CI engine is the formation of deposits in the engine. Some of the prominent areas of deposits which significantly effects the performance of the engine includes combustion chamber, inlet valves, fuel injectors. This is because blending of oxygenated fuel with diesel changes its polarity and solubility which in turn results in deposit formation. Hence, suitable deposit control additives need to added to oxygenated fuels while blending with diesel in CI engines (Stępień et al. 2021).

The following subsection gives a detailed analysis of different fuel improvements techniques in CI engines.

7.2.1 Biodiesel Application in CI Engines

Biodiesel is one of the alternative fuel which could be potentially utilized in diesel engines to solve the growing issue of energy and environmental concerns. Biodiesel is produced through transesterification of fatty acid methyl ester or ethyl esters processed from renewable agricultural resources like vegetable oils, animal fats and waste oils, etc. with the help of a catalyst. The transesterification reaction process is shown in Fig. 7.1 (Giuffrè et al. 2017). There are wide ranges of vegetable oils that are favourable for biodiesel production mainly categorized as edible and non-edible oils.

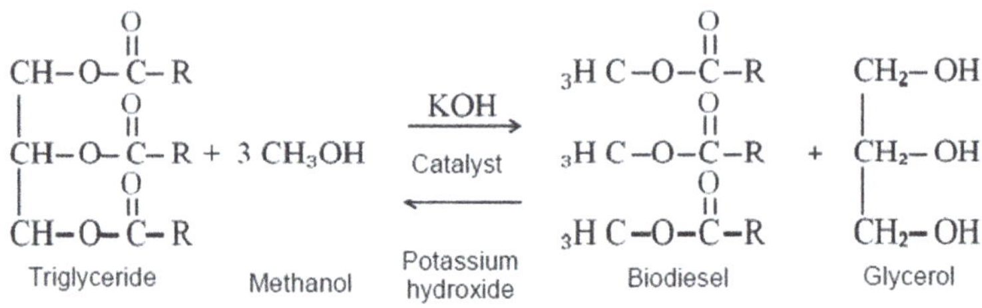

Fig. 7.1 Transesterification reaction process (Giuffrè et al. 2017)

However, food versus fuel controversy has restricted the usage of edible resources in a long term and promote the use of non-edible resources as a potential raw materials for biofuel generation. Additionally, the non-edible sources known as second generation feedstock can be farmed on either non-agricultural or marginal land (Sakthivel et al. 2018). Some of the non-edible oil seeds used in biodiesel production are shown in Fig. 7.2. This includes non-edible seeds such as Jatropha, karnaja, Mahua, Neem, Jojoba, thevetia peruviana. According to SWOT (Strength, Weakness, Opportunities, and Threats) analysis, biodiesel as an alternative fuel compared to conventional diesel has better ignition quality. Biodiesel having higher cetane number and its ability to use directly in diesel engine with little adjustments (Tamilselvan et al. 2017).

Some of the advantages of biodiesel application are (Murugesan et al. 2009).

- It is non-toxic in nature and degradation is faster than diesel.
- It is locally available and produce from renewable sources.
- It has high blending property with diesel irrespective of any blending agent in any proportion.
- Safe storage due to higher flash point.
- It can be applied directly to engines without any substantial engine alteration.

Fig. 7.2 Different non-edible oil seeds

- Biodiesel has no sulphur content.
- It is considered as CO_2 neutral with the ability to reduce more CO_2 emissions than fossil diesel.

Along with these advantages, one of the major drawbacks of biodiesel is its higher combustion temperature that leads to higher NOx emission. Another drawback is the higher viscosity of biodiesel which forms larger size of fuel droplets leading to poor atomization. This in turn affects the ignition delay with higher time to evaporate causing decreased droplet evaporation rate (Rahman et al. 2010). On the other hand, the oxygenated property of biodiesel along with negative sulphur content leads to a significant reduction in the formation of PM in exhaust emission. Hence to achieve a trade-off between NOx and PM emissions it is preferred to use biodiesel as blends along with diesel in CI engines (Dorado et al. 2003).

7.2.1.1 Characteristics of Biodiesel

There are various factors that defines the quality of biodiesel such as composition of raw materials, oil extraction method, synthetization and refining processes. With the use of formulated standards, the quality of biodiesel can be assessed. Standards prescribed by American Society for Testing and Materials (ASTM 6751), European Standard (EN14214), Indian standard (IS-15607), etc. are followed to characterize all biodiesel fuels prior to use in engine. Density, viscosity, calorific value, flash, cloud and pour point, cetane number, acid value etc. are some of the crucial properties that influences the engine operation (Kullolli et al. 2016). Quantity of injected fuel through the injection system depends on the density of the fuel to provide proper combustion. It is also crucial from injector nozzle design point of view because it can directly influence the fuel atomization which is directly related to the thermal efficiency of the engine (Atabani et al. 2012). Viscosity also estimates the quality of spray atomization and spray penetration (Kullolli et al. 2016). The viscosity relies on the composition of the fatty acid of the oil from which biodiesel is made, as well as on the extent of oxidation and polymerization of biodiesel. Moreover, kinematic viscosity is useful for monitoring the fuel quality of biodiesel during storage (Atabani et al. 2012). Cold filter plugging point (CFPP) is also important as it describes the limit of filterability of fuels. The ignition temperature of vapour of the volatile fuel can be determined using Flash point (FP). Biodiesel possesses (150 °C) higher FP than petro diesel (50–65 °C). It signifies safe storing capacity and transit for biodiesel than conventional diesel. Cloud point (CP) and Pour point (PP) are the major criterion set for controlling the low temperature functioning of biodiesel. CP estimates the lowest possible temperature at which the wax present in the fuel starts crystalizing, where PP defines the flow ability of the fuel. Generally, biodiesel possesses higher CP and PP in comparison with conventional diesel (Sakthivel et al. 2018). Cetane number (CN) directly influences the ignition delay period. High CN indicates the auto ignition capability of the fuel after injection into the combustion chamber. Whereas, fuel with low CN promotes engine knock, exhaust emission and unnecessary deposits because of improper combustion. Biodiesel having high oxygen content possess high CN which enhance combustion efficiency. Another important combustion factor of fuel is the calorific value (CV) or heating value of the fuel. It is precisely defined as the amount of energy that can be received from burning one unit of fuel. Hence, biodiesel with higher CV is favourable to be used in the internal combustion engine. Due to the presence of oxygen, biodiesel accounts lower CV than conventional diesel fuel. The acid number is vital which sets the lubricant degradation criterion of the fuel during its operation. Corrosion inside the fuel supply system are mainly caused by higher acid content (Atabani et al. 2012; Srivastava et al. 2018). The oxidation of biodiesel is an another parameter that is used for the qualitative assessment of fuel. This defines the degree of oxidation, potential reactivity with air, and the need for antioxidants (Kullolli et al. 2016). In comparison with diesel fuel, biodiesel has more oxygen, lower volatility, and higher viscosity, but has a narrow boiling range. The physicochemical properties of a fatty acid methyl ester molecule are influenced by the chain length and branching of the alkyl group, the number position, geometric configuration of double bonds, and the presence of additional functional groups in the fatty acid chain, which all relate to fuel feedstock composition (He 2016). The physio-chemical properties of biodiesel obtained from different feedstocks are illustrated in Table 7.1.

Researchers have found suitable physiochemical properties with different non-edible seeds for biodiesel generation. Patil and Deng used potassium hydroxide catalyst for transesterification process of non-edible seeds; Jatropha curcas, pongamia, canola, and corn and observed yield of 90–95%, 80–85%, 80–95%, and 85–96% biodiesel respectively (Patil and Deng 2009). Ali et al. have tested the different properties of neem biodiesel such as density, kinematic viscosity, and calorific value and found to be within the ASTM standards (Ali et al. 2013). Similarly, Atbani et al. also estimated the physiochemical properties of biodiesel produced from jatropha curcas, pongamia pinnata, madhuca indica and found to be within the limits of ASTM and DIN EN specificationswere analyzed by Atabani et al. (2013). Similar analysis (Ana and Udofia 2015) of biodiesel production from seeds of *Thevetia peruviana* also met the standards of ASTM and DIN EN specifications.

7.2.1.2 Application of Biodiesel and Its Blends in CI Engine

Biodiesel derived from locally available natural resources is one of the treading research topics around the globe for application in a conventional diesel engine. Due to comparable physiochemical properties of biodiesel with conventional diesel, along with its higher miscibility makes biodiesel suitable for application in diesel engines. In addition, the applicability of biodiesel in CI engine without any modification of engine design makes it more economical and cost effective. However, researchers have opted to use biodiesel as blends along with diesel due to its higher viscosity and combustion temperature. Various studies have been reported with different biodiesel blends obtained from various non-edible oil seeds along with fossil diesel for analyzing the CI engine performance, combustion, and exhaust emission

Table 7.1 Fuel properties of different non-edible biodiesel and diesel (Onoji et al. 2016; Cai et al. 2021)

Fuels	Kinematic Viscosity at 40 °C (mm^2/s)	Density at 40 °C (kg/m^3)	Flash Point (°C)	Cloud point (°C)	Pour point (°C)	Calorific value (kJ/kg)	Cetane number
Rubber seed oil (Hevea brasiliensis)	3.7–8.02	860–892	110–154.6	− 6 to 5	− 8 to 4.8	36,500–41,070	37–66.2
Jatropha (Jatropha Curcus L.)	3.7–5.8	864–880	163–238	10	5–6	38,500–42,000	46–55
Karanja (Pongamia Pinnata L)	4.37–9.60	876–890	163–187	13–15	− 3 to 5.1	36,000–38,000	38–52
Polanga (calophyllum inophyllum)	4–5.34	888.6–610	151–170	13.2–14	4.3	39,250–41,300	57.3
Mohua (madhuca indica)	3.9–5.8	904–916	127–129	3–5	1–6	39,400–39,910	51–52
Rapeseed (Brassica napus)	4.50	879	169	− 3	− 10	37,600–41,100	53.7
Yellow Oleander (*Thevetia Peruviana*)	4.12–5.2	868–880	108–175	3–6	− 2 to 4	39,100–41,500	47–58
Palm oil (Elaeis guineensis)	4.61	873	163	14	13	37,700–40,700	61.9–65.8
Cotton seed	2.2–4.9	874–911	120–243	1.7	− 10 to − 15	39,500–40,100	41.2–59.5
Jojoba oil (Simmondsia chinensis)	19.2–25.4	863–866	61–75	6–16	− 6 to 6	42,760–47,380	63.5
Tobacco oil (Nicotiana tabacum L.)	3.5–4.23	860–888.5	152–165.4		− 12	38,430–39,810	49–51.6
Neem (Azadirachta indica)	5.2–48.5	860–965	34–165	8	4	33,700–39,800	51–55.31
Linseed oil (Linum usitatissimum)	16.2–36.6	865–950	108	1.7	− 4 to − 18	37,700–39,800	28–35

(continued)

Table 7.1 (continued)

Fuels	Kinematic Viscosity at 40 °C (mm^2/s)	Density at 40 °C (kg/m^3)	Flash Point (°C)	Cloud point (°C)	Pour point (°C)	Calorific value (kJ/kg)	Cetane number
Moringa oleifera	5.05	859.6	150.5	19	19	40,050	56.6

characteristics. Soo-Young No (No 2011) utilized seven different biodiesels obtained from non-edible seeds of mahua, karanja, jatropha, neem, rubber, linseed and cotton in CI engine. The researcher reported lower emission of CO, HC and PM while NOx emission was found to be higher. They have also stated that the diesel engine was successfully operated with blending of 20–80% vegetable oil and diesel without any modification. Sukumar Puhan et al. (Puhan et al. 2005a) conducted experiments with mahua oil methyl ester (MOME), mahua oil ethyl ester (MOEE), mahua oil butyl ester (MOBE) and observed lower emission of CO and NOx and higher CO_2 emission compared to diesel. Based on the overall outcomes, the authors recommended the use of MOME as a possible substitute for diesel comparing to the other esters. As per the observations made by Chauhan et al. (Puhan et al. 2005a) lower calorific value of karanja biodiesel was the major reason behind lower BTE (3–5%) than diesel. In addition, the researchers explained the presence of oxygen in biodiesel to be the primary reason behind lower emission of unburned HC, CO and smoke than diesel due to improved combustion of fuel. However, the NOx emission was found to be higher for karanja biodiesel and its blend compared to diesel. NOx emission is a function of temperature of exhaust gas. The better combustion of biodiesel tends to increase the exhaust gas temperature which consequently increases the NOx emission. In a similar investigation, Chauhan et al. (Chauhan et al. 2012) reported with a lower BTE and higher BSFC for Jatropha biodiesel and its blends than diesel,. They reported NOx emission to be higher and smoke, CO and HC for jatropha biodiesel and its blends to be lower than diesel fuel. Some experimental results of biodiesel run diesel engines are illustrated in Figs. 7.3, 7.4, 7.5, 7.6, 7.7 and 7.8.

In another investigation, Balusamy and Marappan (2009) analyzed the performance and emission characteristics of neem oil, mahua oil, thevetia Peruviana seed oil, pongamia oil and jatropha oil, and in the CI engine. The fuel blends were prepared separately in blends of 20:80 with diesel. The experimental outcomes revealed that 20% methyl ester of thevetia peruniana oil showed enhanced engine performance and reduced emission characteristics comparing other biodiesel blends. Similarly, B10 and B20 blends of Jatropha methyl esters were utilized by Mofijur et al. (2013) and found 4.67% and 8.86% reduction in BP respectively compared to B0. BSFC was reported to be increased with the addition of biodiesel in the blend. HC, CO and NOx emission was reported to be reduced for B10 and B20 compared to that of B0. Yadav et al. (2018) used the hydrodynamic cavitation method for biodiesel production from *Thevetia peruviana* seeds for analyzing the performance and emission behaviour of the engine. The experiments were conducted with 10, 20, and 30% fuel blends at

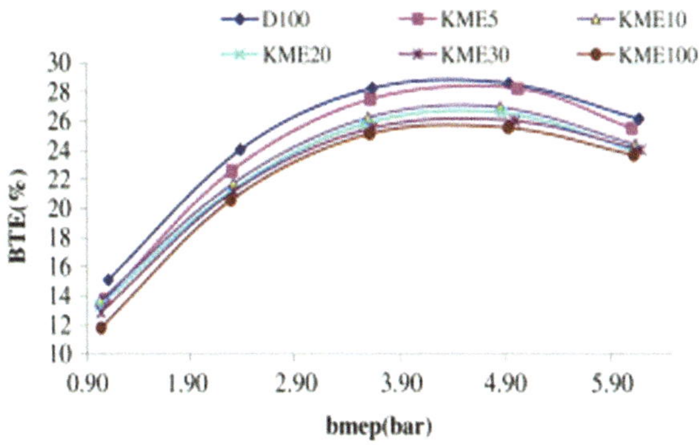

Fig. 7.3 Variation of BTE at different biodiesel blends (Puhan et al. 2005a)

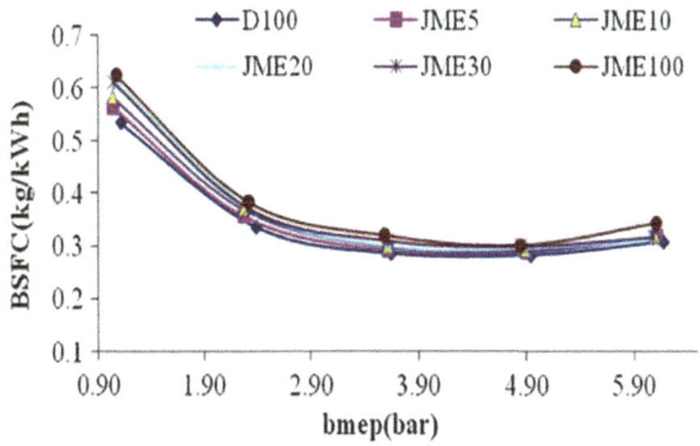

Fig. 7.4 Variation of BSFC at different biodiesel blends (Chauhan et al. 2012)

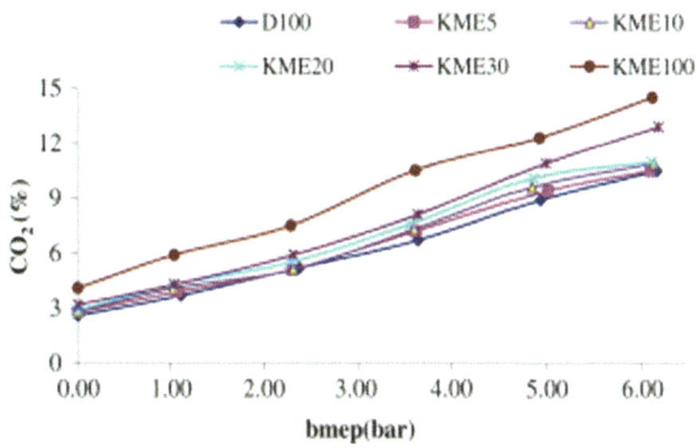

Fig. 7.5 Variation of CO_2 at different biodiesel blends (Puhan et al. 2005a)

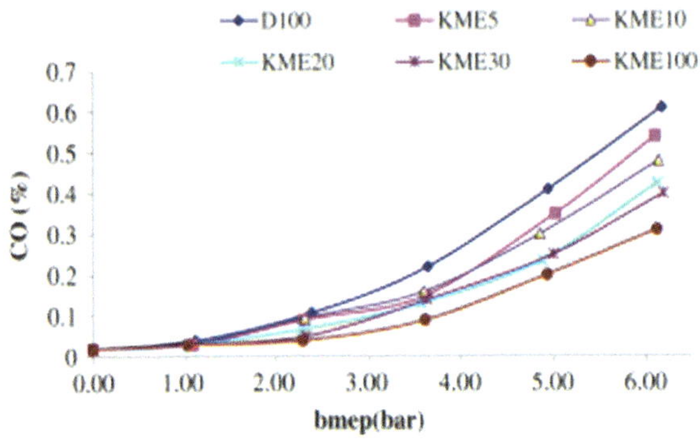

Fig. 7.6 Variation of CO at different biodiesel blends (Puhan et al. 2005a)

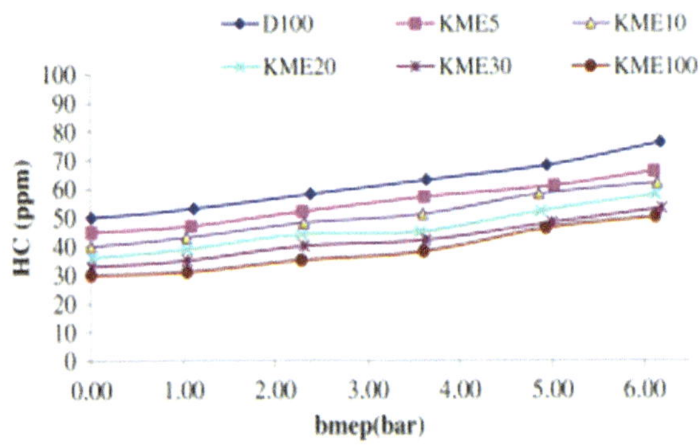

Fig. 7.7 Variation of HC at different biodiesel blends (Puhan et al. 2005a)

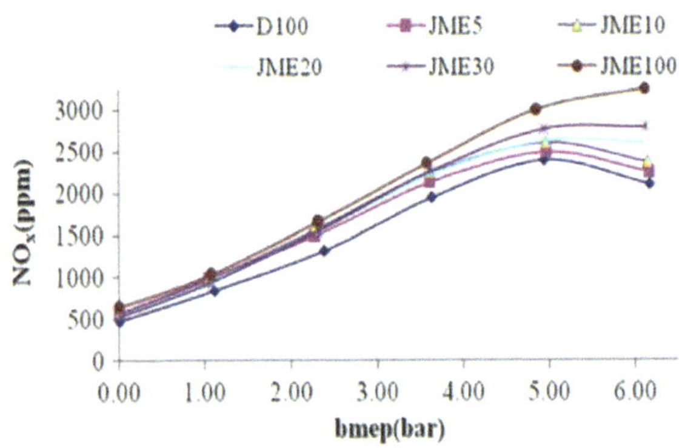

Fig. 7.8 Variation of NOx at different biodiesel blends (Chauhan et al. 2012)

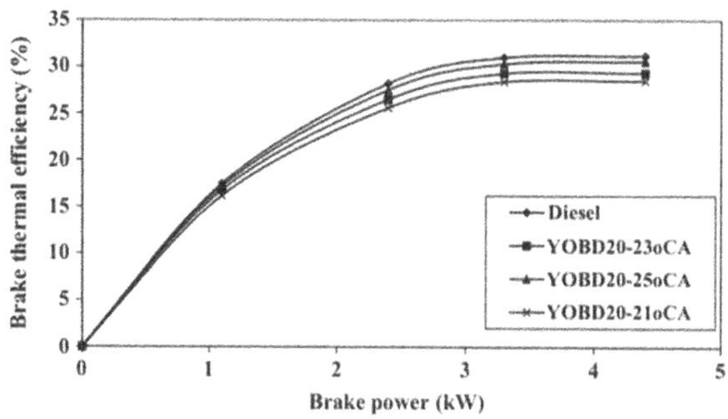

Fig. 7.9 Variation of BTE at different IT (Rajan and Senthil Kumar 2020)

different engine speeds. The found hydrodynamic cavitation method to be efficient and less time consuming. On the other hand, 20% blends of biodiesel resulted in optimum engine performance with significantly lower engine emission.

The engine operating parameters viz. injection timing (IT), CR, injection pressure (IP), engine load, different biodiesel mixture blends, plays a vital role in the engine performance and exhaust emission analysis of CI engines. Rajan and Senthil Kumar (2020) investigated the effect of different IT in diesel engines operated with 20% yellow oleander biodiesel. IT of 21°BTDC, 23°BTDC and 27°BTDC were considered for the analysis. The experimental data indicated that at advanced IT of 27°BTDC, the BTE improved by 1.24% and BSFC reduced by 9% as shown in Fig. 7.9. The reason is due to fact that longer ignition delay (ID) allows more time to undergo combustion. This improves the degree of combustion and results in improvement of BTE. On the other hand, BTE reduced by 0.896% at retarded IT of 21°BTDC compared to the standard IT (23°BTDC). Further, the NOx emission increased while CO, HC emission reduced at advance IT, with opposite trends being observed at retarded IT. The advancement of IT promotes better oxidation of fuel and thereby improves combustion. This decreases the CO and HC emission. In addition, the prolonged ID extends the time for fuel to burn which increases the temperature of fuel, as a result the NOx emission rises. The variation of CO and NOx emission are shown in Figs. 7.10 and 7.11. According to the Jindal (2011), neat karanja methyl ester (KME) resulted with higher BSFC and BTE at CR 18 and IP 250 bar in comparison with the standard values. With increase in CR and IP the fuel droplets break down and provides large surface area. This ensures better air–fuel mixing and resulted in better combustion as shown in Figs. 7.12 and 7.13. In addition, as illustrated in Figs. 7.14, 7.15 and 7.16, compared to neat KME and diesel, reduced HC, NOx, smoke and EGT was observed at CR 18 and IP of 250 bar. The researchers achieved an overall improvement in CI engine operation with higher CR and IP fuelled with neat KME. Jindal et al. (2010) conducted experiments with jatropha methyl ester (JME) and observed 10% and 8.9% improvements in BSFC and BTE at IP of 250 bar and CR of 18 compared to standard CR and IP. They have also concluded that higher value of CR and IP lowers the emission of HC, smoke and

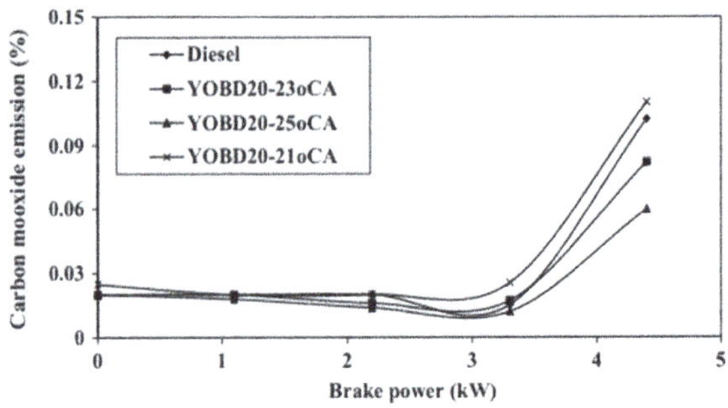

Fig. 7.10 Variation of CO at different IT (Rajan and Senthil Kumar 2020)

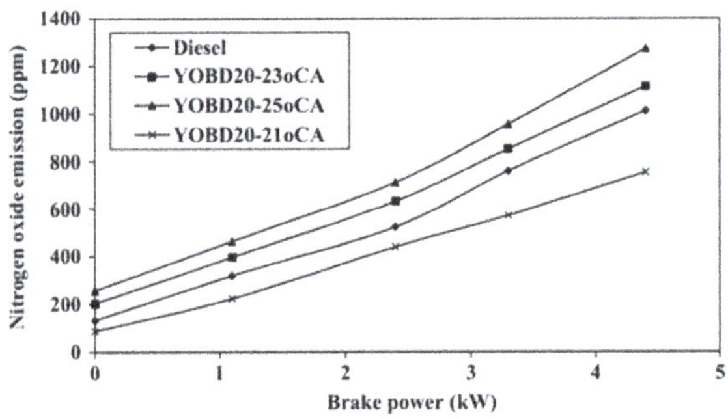

Fig. 7.11 Variation of NOx at different IT (Rajan and Senthil Kumar 2020)

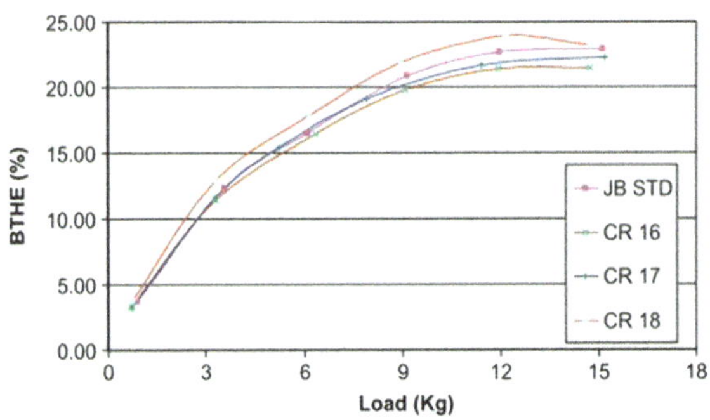

Fig. 7.12 Variation of BTE at different CR (Jindal 2011)

EGT for biodiesel operation. Saravanan et al. (2010), Raheman and Ghadge (2008) analyzed the effect of methyl esters of mahua biodiesel and pure mahua biodiesel respectively in CI engine. Compared to diesel, mahua ester application resulted with 26% and 20% lower CO and HC emission. Saravanan et al. (2010) reported 4% lower NOx emission for mahua methyl ester than diesel. On the other hand, BSFC of pure

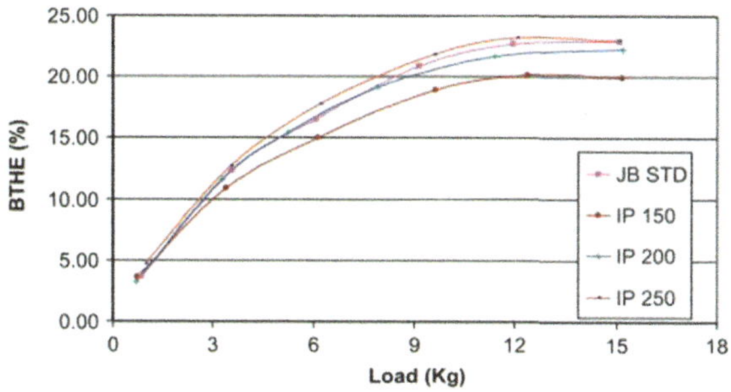

Fig. 7.13 Variation of BTE at different IP (Jindal 2011)

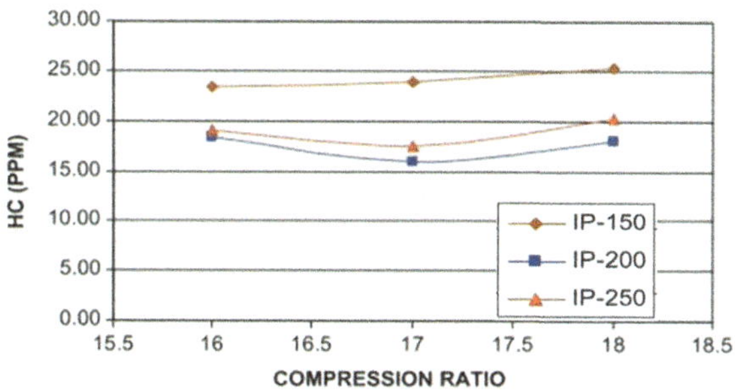

Fig. 7.14 Variation of HC at different CR (Jindal 2011)

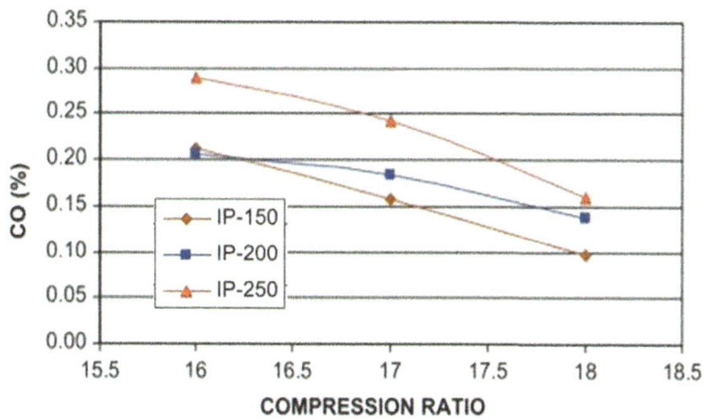

Fig. 7.15 Variation of CO at different CR (Jindal 2011)

mahua biodiesel was reduced, while EGT and BTE increased with the increasing load and CR with advanced IT. Again, Raheman and Ghadge (Nalgundwar et al. 2016) recommended the use of pure mahua biodiesel in the Ricardo engine at CR 20 and IT 40 without hampering the engine performance. Nalgundwar et al. prepared different blending combination of palm and jatropha oil with diesel for conducting experiments in a diesel engine. The share of palm and jatropha oil were varied from

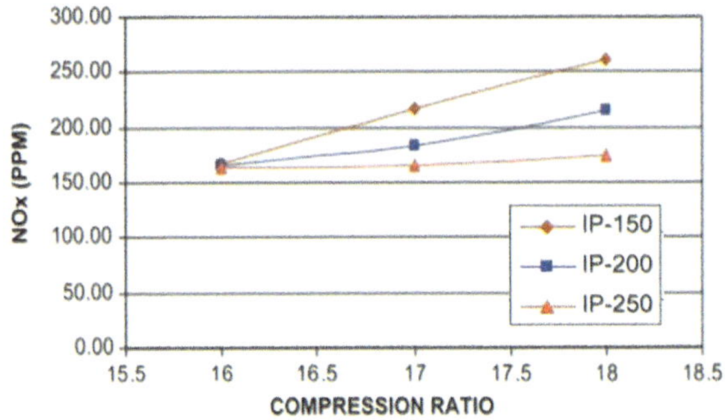

Fig. 7.16 Variation of BTE at different CR (Jindal 2011)

5 to 50% for better analysis of the engine. Blended fuel with combination of 60% diesel, 20% jatropha and 20% palm resulted with 2.55% reduction in BSFC and 15.4% increment in BTE. A reduction up to 14.5% and 42.3% was noticed for CO and CO_2 respectively, while NOx emission increased up to 28% compared to standard diesel experiment.

A summary of CI engine performance using various non-edible biodiesel and biodiesel blend is presented in Table 7.2.

7.2.2 Application of Alcohol as Fuel in CI Engine

Alcohol and ethers are oxygenated fuels which are produced from renewable resources as well as synthetically for industry application. The former sources of alcohols and ethers serves as excellent octane enhancers. The growing interest in using alcohols as a substitute for diesel is due to its oxygen enrichment property that enhances the premixed combustion stage and improves the diffusive combustion stage. This fact is the primary reason behind growing interest for using oxygenated fuels in IC engines. Methanol, ethanol, butanol, and propanol are the mostly used alcohols in diesel engines. However, the lower alcohol fuels viz. methanol and ethanol have certain complications for direct use in diesel engines because of low cetane number, low boiling and flash point, high latent heat of vaporization, and high resistance to auto-ignition. Moreover, the lower heating value, poor lubricating property, and poor miscibility with diesel restrict its use in diesel engines (Rajesh Kumar and Saravanan 2016). To counter the aforementioned limitations, several techniques have been proposed which include, pulverization of methanol/ethanol, diesel and methanol/ethanol blends, alcohol diesel emulsion, and alcohol fumigation. However lower cetane number of these alcohols encourages its application for low temperature condition combustion engines. Literatures on the application of alcohol-diesel blend suggested that upto 15% mixing of alcohol with diesel is technically feasible in existing CI engines without additional attachments (Hagos et al. 2017).

Table 7.2 Summary of biodiesel/biodiesel blends performance on CI engine

Investigator	Engine used	Biodiesel	Experimental condition	Research findings
Rao et al. (2008)	1-Cylinder, 4 stroke, WC, DI, CR: 16.5:1, P: 3.7 kW	Pongamia, Jatropha and Neem methyl esters	• Variable load, constant speed (1500 rpm)	• BTE of B10 and B20 are close to that of diesel fuel • Decrease in CO, HC and smoke biodiesel blend
Agarwal et al. (2007)	1-Cylinder, 4 stroke, WC, DI, CR: 17.5, RP: 7.4 kW	Jatropha oil	• Variable load, constant speed (1500 rpm)	• Increase in BSFC, EGT Decrease in Thermal efficiency for unheated jatropha oil • CO, CO_2, HC and smoke opacity of preheated blends are close to the diesel fuel
Arunprasad and Balusamy (2018)	1-Cylinder, 4 stroke, WC, DI, 3.5 kW	*Thevetia peruviana*, Jatropha, Pongamia and Azadirachta indica	Effect of injection pressure and injection timing	• Experiments were performed with mixed non edible oils • Increase in BTE with pressure from 25.7 to 28.1% • Increase in NOx and decrease in CO and HC emission with injection pressure • BTE increase upto 27.6% with IT • HC and CO decreased while NOx increased from 16.01 to 30.54%

(continued)

Table 7.2 (continued)

Investigator	Engine used	Biodiesel	Experimental condition	Research findings
EL-Kasaby and Nemit-Allah (2013)	1-Cylinder, 4 stroke, WC, DI, CR: 18	Jatropha oil biodiesel	• Variable speed (1000–2000 rpm)	• Increase in Engine power, torque and BTE with the addition of biodiesel • Increase in BSFC for biodiesel blends and • Increase in Peak cylinder pressure and shorter ignition delay for biodiesel blends • Decrease of CO emission and increase of NOx emission with the addition of biodiesel blends
Sureshkumar et al. (2008)	1-Cylinder, 4 stroke, WC, CR: 16.5:1, RP: 3.68 kW	Pongamia pinnata methyl ester (PPME)	• Variable load, constant speed (1500 rpm)	• Decrease in BSEC for B20 and B40 blends and decrease EGT for all PPME blends • Decrease in CO_2 for medium blends of PPME decrease in CO, HC and NOx with the addition of PPME
Investigator	**Engine used**	**Biodiesel**	**Experimental condition**	**Research findings**
Paul et al. (2014)	2-Cylinder, 4 stroke, WC, DI, RP: 7.35 kW	Jatropha biodiesel	• Variable load, constant speed (1500 rpm)	• BSFC increases and BTE decreases with the addition of jatropha biodiesel • Cylinder pressure for jatropha Biodiesel increase • NOx and CO_2 increases and PM and smoke emission decreases with the addition of biodiesel

(continued)

Table 7.2 (continued)

Investigator	Engine used	Biodiesel	Experimental condition	Research findings
Agarwal and Dhar (2013)	1-Cylinder, 4 stroke, WC, DI, CR: 17.5:1, RP: 7.4 kW	Karanja oil	• Variable load, constant speed (1500 rpm)	• BSFC and EGT increase and thermal efficiency decreases with the addition of Karanja oil • Decrease in cylinder pressure for Karanja oil blends • HC, Smoke emissions decreases for up to K50 and NO and CO_2 emission increases for Karanja oil blends
Gogoi and Baruah (2011)	1-Cylinder, 4 stroke, WC, DI, CR: 12–18, RP: 3.5 kW	Koroch seed oil methyl ester (KSOME)	• Variable load	• BSFC increases and thermal efficiency decreases with the addition of KSOME • Pressure crank angle is similar of KSOME blends are similar to diesel fuel • Earlier HRR occurred for KSOME blends
Lenin et al. (Haiter et al. 2012)	1-Cylinder, 4 stroke, WC, DI, CR: 17.5:1, RP: 5.2 kW	Mahua oil methyl ester	• Variable load, constant speed (1500 *rpm*)	• BSFC increases and BTE and EGT decreases with the addition of methyl ester • Decease in Peak cylinder pressure and HRR for methyl ester blends • CO, CO_2 and NOx increases and HC decreases for methyl ester blends

(continued)

Table 7.2 (continued)

Investigator	Engine used	Biodiesel	Experimental condition	Research findings
Puhan et al. (2005b)	1-Cylinder, 4 stroke, WC, DI, CR: 16.5:1, RP: 3.7 kW	Mahua oil ethyl ester (MOEE)	• Variable BMEP • Constant speed (1500 rpm)	• BSFC, EGT increases and BTE slightly increases for MOEE • CO, HC, NOx emission decreases and CO_2 emission increases for MOEE
Investigator	Engine used	Biodiesel	Experimental condition	Research findings
Godiganur et al. (2009)	6-Cylinder, WC, DI, CR: 17.6:1	Mahua oil methyl ester	• Variable load, constant speed (1500 rpm)	• BSFC for all blends and BTE decreases for the biodiesel blends, except B20 • For B100 at full load CO and HC decrease and EGT, NOx increases
Banapurmath et al. (2008)	1-Cylinder, 4 stroke, WC, DI, CR: 17.5:1, RP: 5.2 kW	Honge, Jatropha and Sesame oil methyl ester	• Variable load, constant speed (1500 rpm)	• Thermal efficiency decreases for all methyl ester blends • Ignition delay is longer and shorter premixed HRR for all biodiesels • Combustion duration increases for all the methyl esters • Increase in CO, HC and Smoke opacity for all methyl esters

(continued)

Table 7.2 (continued)

Investigator	Engine used	Biodiesel	Experimental condition	Research findings
Devan et al. (Devan and Mahalakshmi 2009)	1-Cylinder, 4 stroke, AC, DI, CR: 17.5:1, RP: 4.4 kW	Paradise oil methyl ester-eucalyptus oil blends	Variable load, constant speed (1500 rpm)	• BTE increases and BSEC decreases for Me50-Eu50 blend and for Me20-Eu80 blend EGT increases • Increase in cylinder pressure and HRR for eucalyptus oil in the methyl ester blends • CO, HC, smoke and NOx increases for Me50- Eu50 blend
Vallinayagam et al. (2014)	1-Cylinder, 4 stroke, WC, DI, CR: 17.5:1, RP: 5.2 kW	Pine oil-Kapok methyl ester (KME) blends	• Variable load, constant speed (1500 rpm)	• BSFC decreases for B25P75 blend • BTE and EGT increases for B25P75 and B50P50 lend • Increase in Peak HRR and in-cylinder pressure for B25P75 • Decrease in CO, HC and smoke by 18.9%, 8.1% and 12.5% respectively for B50P50 blend

On the other hand, higher alcohol fuels viz. butanol, propanol, pentanol have a higher energy density, better stability of blends, higher cetane number, and less hygroscopic nature as compared to lower alcohol fuels. In addition, the long-chained structure of alcohols improves the ignition quality of alcohol molecules. However, inspite of the suitable properties of alcohols, one of the major concerns of alcohol substitution in diesel engines is the economic viability as the cost of production of alcohol is higher. Another major concern of alcohol production is the utilization of food grains as raw materials which are having an added value in the global food market. The use of cellulose, hemicellulose, and lignin based materials has recently gained momentum for the production of alcohol (Kumar et al. 2013). Ethers like alcohols have similar functional groups, for example, butanol (C_4H_9OH) and Diethyl ether (DEE) ($C_4H_{10}O$) are almost similar in formula but different in properties because of the presence of the OH group in alcohols. Ether additives containing higher cetane number improves the combustion of fuel inside the combustion chamber, thereby improving the overall engine performance. Two of the most commonly used ethers are DEE and DME (dimethyl ether). However, DME requires major engine modification for undergoing combustion as it is gaseous at ambient temperature and pressure. Contrastingly, DEE having less volatility, lower auto-ignition temperature, high cetane number, oxygen content and miscibility makes it favourable for blending with mineral diesel (Hagos et al. 2017). Table 7.3 represents the physiochemical properties of alcohols and ethers in comparison to that of diesel fuel.

Table 7.3, indicates that higher alcohol group comparing to lower alcohols like methanol and ethanol have better physiochemical properties and have higher potential to replace diesel fuel either partially or wholly. Higher alcohols have high carbon content, low polarity and are less hygroscopic in nature which makes them suitable to maintain blend stability without the use of co-solvents or any emulsifying agents. Another important characteristic of higher alcohols is their low vapor pressure which results in lower evaporative emissions (Rajesh Kumar and Saravanan 2016). On the other hand, lower alcohols due to their higher oxygen content may lead to corrosion if used alone in an unmodified diesel engine. As such an optimum threshold blending percentage with diesel is specified while using in a diesel engine. However, factors such as the aromatic content of diesel, the percentage composition of diesel, and the amount of water in the blends significantly affect the diesel-alcohol miscibility (Kumar et al. 2013). To overcome such difficulties particularly in lower alcohols, emulsification is done to maintain the stability of diesel-alcohol blends. From Table 7.3 it can also be inferred that, unlike alcohols, ethers improve the fuel combustion in CI engines because of their higher cetane number. The succeeding sections give a comprehensive idea of different alcohols and their application in CI engines.

7.2.2.1 Methanol as Biofuel

Methanol containing single carbon atom is the simplest form of alcohol usually known as. methyl alcohol, is a very pale odor, colourless and tasteless liquid. Because of polar solvent characteristic it is miscible in water, alcohols, esters and other organic

Table 7.3 Physicochemical properties of alcohols and ethers compared to diesel (Hagos et al. 2017; Kumar et al. 2013)

Properties	Diesel	Methanol	Ethanol	Propanol	Butanol	DEE
Molecular formula	$C_{12}H_{26}$	CH3-OH	C2H5-OH	C3H7-OH	C4H10-OH	C4H9OH
Molecular weight (kg/kmol)	190–211.7	32.04	46.07	60.09	74.12	74.12
C (wt%)	86.13	37.48	52.14	59.96	64.82	13.6
H (wt%)	13.87	12.48	13.02	13.31	13.49	64.8
O (wt%)	0	49.93	34.73	26.62	21.59	21.6
Solubility (g/L)	Immiscible	Miscible	Miscible	Miscible	77	–
Lubricity	315	1100	1057	922	591	614
Cetane number	52	5	8	12	17	> 125
Self ignition temperature (°C)	254–300	463	420	350	345	160
Density (kg/m^3) at 15 °C	835	791.3	789.4	803.7	809.7	710
Viscosity at 40 °C (mm^2/s)	2.72	0.58	1.13	1.74	2.22	0.23
Lower heating (MJ/kg)	42.49	19.58	26.83	30.63	33.09	33.9
Latent heat of evaporation (kJ/kg)	270–375	1162.64	918.42	727.88	581.4	–
Vapor pressure (mmHg)	0.4	127	55	20	7	–
Boiling point (°C)	180–360	64.7	78.3	97.1	117.5	34.6
Flash point (°C)	> 55	11–12	17	11.7	35–37	– 45

solvents, while less miscible with fats and oils. Methanol is highly flammable and may blast once exposed to flame. Due to higher oxygen content, it is a hydroscopic liquid and directly absorbs water vapour from the air. This absorption of water results in dilution of methanol causes phase separation while preparing methanol-diesel blends (Kumar et al. 2013).

Methanol was earlier produced from wood before the development of modern technologies. However, with the advent of sustainable technologies, methanol production from biomass has become unfeasible. Generation of methanol is associated with cost-intensive chemical processes. Some of the viable methodologies for methanol production include, waste biomass such as crop residues, wood resources, forage, forest resource, early grown energy crops, grass, and lignocellulosic element of municipal wastes. For using as fuel in CI engines, methanol is derived from synthetic gas or biogas (Vasudevan 2005). Presently, the primary sources of methanol production is syngas, natural gas, coal, refinery off-gas, and petroleum along with partial oxidation reaction followed by catalytic transformation of wood waste or garbage into biomethanol. The addition of an ample amount of hydrogen to syngas can double the production of methanol from similar quantities of biomass. Inspite of the available technology for methanol production, renewable methanol is incapable of competing economically with other fuel resources (Agarwal et al. 2019). However, to overcome these financial barriers, some of the cost effective methanol production technologies include, biomass gasification, synchronized biomethanol production along with bioethanol using hydrogenation of CO_2 formed during the fermentation of sugar, methanol production using CO_2, and water using sunlight (Aasberg 2004).

Several studies have been reported for methanol use in diesel engines because of its similarity with petroleum in terms of storage and transportation competence. The usage of methanol either as blends with diesel or as an biodiesel additive have attained growing attention. Methanol with high octane number, oxygen content, negligible NOx emission, lack of sulphur compounds, lower ozone forming potential define its suitability for applying in CI engines. However, the major difficulty of methanol is the ignition of the air–fuel mixture during diesel engine operation, primarily because of high latent heat of vaporization and longer ID. Another drawback is the miscibility of methanol with diesel as discussed earlier, due to which it is important to maintain a lower methanol to diesel ratio for avoiding adverse effects on combustion (Kumar et al. 2013). Hence to minimize these problems and to use methanol in diesel engines appropriately, dual injection, alcohol fumigation, and emulsions of methanol-diesel have been investigated in different research studies. Azev et al. (1986) investigated the effect of higher alcohols, hydrocarbons, surfactants, the composition of fuels, addition of water and aromatic hydrocarbons on the miscibility of methanol with diesel. The results revealed that 5% methanol with 10% water formed an emulsion which sustained for several days, forming suspended droplets of methanol in diesel. The addition of 0.25% of surfactant (dibutyl-ester of sulfosuccinic acid) to the above emulsion resulted in a reasonably steady emulsion of water-fuel which lasted for 15 days. Further, it was found that stability of the four component system, reduced with increasing water content, unlike the use of higher alcohol that stabilizes the methanol-diesel blends. The advantage of methanol blended diesel is its ability to reduce emission levels when used in diesel engines. Huang et al. (2004) utilized various methanol-diesel blends in diesel engines for studying the emission characteristics and found that the HC, CO, and smoke emission decreased while NOx emission augmented with an increase in methanol content. Najafi and Yusuf (2009) examined three methanol-diesel blending ratios of 10:90, 20:80, and 30:70. The results revealed that BTE improved with all the methanol blends while comparing with diesel fuel under entire operating conditions, also 30% methanol blends resulted in the lowest BSFC compared with other methanol blends. Sayin et al. (Sayin 2010) utilized methanol-diesel (M5, M10) and ethanol–diesel (E5, E10) to assess the performance and exhaust gas emission of a diesel engine. Dodecanol was used to stabilize the blends. The exhaust emission of smoke, HC, and CO reduced while NOx emission increased when the blends were used. M10 methanol blend resulted in lowest BTE and higher BSFC compared to all other blends. Ethanol blends showed better engine performance compared to methanol blends. In another study Song et al. (Song and Wang 2008) studied the variation of engine load (0.35, 0.56, and 0.77 MPa) and engine speed (1600 and 200 rpm) with methanol use in a diesel engine. The primary objective of the study was to identify the optimal performance of dual fuel engine and study the influence of methanol on the performance, combustion, and emission behaviour of the dual fuel engine. Dual fuel engines produced similar or higher power with better fuel economy compared to diesel engine. Smoke emission reduced significantly and slight reduction in NOx emission was observed. Yet, HC and CO emission increased with addition of methanol.

Fuel injection systems such as injection timing and injection pressure play a significant role in reducing the emission to acceptable levels. Canakci et al. (2009) varied the injection pressure (180, 200, and 220 bar) at four different engine loads (5, 10, 15, and 20 Nm) to study the effects on a diesel engine. An increase in IP resulted in rise of NOx and CO_2 emission, while with lower load minimum NOx and CO_2 were reported. In addition, smoke, HC, and CO emission were reduced with injection pressure and the best results were obtained at a load of 20 Nm. The effect of varying methanol blends (5, 10, and 15%) along with three different injection timing (15°, 20°, and 25° CA BTDC) and four engine loads (5, 10, 15, and 20 Nm) were investigated by Sayin et al. (2009) for analyzing engine performance and emission characteristics. The increase in methanol blends improved the combustion and emission characteristics of the diesel engine. With the advancement of IT, smoke, HC and CO emissions reduced however emission of NOx and CO_2 increased. Under loading conditions of 20Nm, advancement of IT resulted in optimum HC and CO emission while retarding IT at 10 Nm resulted in minimum NOx and CO_2 emission. The variations of engine performance and emission are significantly impacted by injection parameters. With increase in engine load combustion temperature increases which promotes better combustion of fuel mixture. Similar is the case with higher CR, IT and IP. As a result, BTE of engine increases along with NOx and CO_2 level, in contrast to lower CO and HC emission. In addition to the above mentioned methanol-diesel blends, the engine performance and emission have impacted by methanol fumigation as reported by different researchers. Methanol fumigation has the potential to reduce NOx and CO_2 emission but on the other hand, it significantly reduces the HC, CO and unburned methanol emission. However, in applying fumigated methanol with diesel, oxygen catalyst has the potential to decrease HC, CO, NOx and smoke emission under different loading conditions. In addition, the use of methanol fumigation reduced BTE at low engine loads with increased BSFC. Therefore, it can be summarized that both methanol fumigation and methanol blending have similar impacts on an engine performance, with methanol fumigation showing slightly better results with the application of diesel oxygen catalyst.

7.2.2.2 Ethanol as Biofuel

Ethanol is two-carbon atom alcohol, flammable in nature, polar solvent and is soluble in water. It is heavier than air having a vapor density of 1.59 and as a result, unlike gasoline vapour, ethanol vapors do not rise. As mentioned in Table 7.3, the specific gravity of ethanol is 0.79, lower than that of water, however, it is soluble in water. Ethanol as compared to gasoline and methanol is less toxic and non-carcinogenic.

Ethanol under clean form burns without any smoke and under denatured form burns with almost negligible smoke. Due to the high cetane rating of ethanol, it serves as a proper fuel for use in CI engines (Hagos et al. 2017).

Ethanol is produced by simple fermentation of sugars (sugar beet, sugarcane, molasses), starches (wheat, corn), and cellulosic biomass (straw, wood, grass, corn stover) is one of the most popular alcohol-based fuels. Among the feedstocks, sugar cane or sugar beet are usually preferred followed by corn starch (Verma et al. 2000). The production of ethanol from lignocellulosic biomass is limited as the conversion of the crystal-like complicated structure of cellulose to sugars requires an expensive pre-treatment process. The use of gasohol as an alternative fuel is gaining much worldwide attention for several reasons. Bioethanol having a general formula of $(CH_2O)_n$ is prepared from wide range of carbohydrates. Commercial yeast (Saccharomyces cerevisiae) is used for the fermentation of sucrose. The chemical reaction starts with the enzymatic hydrolysis of sucrose prior to the fermentation of simple sugars (Baltz et al. 1982; Kim and Dale 2005). The invertase enzyme present in yeast works as a catalysis in hydrolysis of sucrose and converts it into glucose and fructose. Another yeast enzyme called zymase, converts glucose and fructose into ethanol. Enzymatic saccharification of cellulose to sugars is one of the key phase in ethanol processing followed by a small pre-treatment phase. Bioethanol contributes to around 94% of total biofuel production worldwide. Out of which 60% of global bioethanol is produced from sugarcane and the rest comes from other crops. This increase in bioethanol production can address several issues of conventional engines primarily minimizing air pollution, and limiting the use of fossil fuel consumption (Kumar et al. 2013).

Ethanol application in CI engines as diesel-ethanol blends has been the area of research since its first investigation in 1970s. Several ethanol blending technologies are relevant for CI engines application as reported by various researchers which include diesel-ethanol blends, fumigation, ethanol emulsions, and dual fuel injections. Ajav et al. (1999) investigated four ethanol–diesel blends (E5, E10, E15 and E20) in a CI engine and reported an increase of BSFC upto 9% for 20% blending. The increase in ethanol blends reduces the density and lowers the calorific value of the fuel blends compared to diesel. As a result the fuel consumption increases at higher blends also BTE decreases. With a higher percentage of ethanol reduced the temperature of exhaust gas without any substantial power reduction. Further, the NOx and CO emission reduced by 24% and 62% respectively. In another investigation Bilgin et al. (2002) varied ethanol–diesel blends (2, 4 and 6%) along with compression ratios (19, 21 and 23) of a diesel engine. The addition of 4% ethanol results in higher power output with increased efficiency over the range of CR. The best efficiency was obtained at CR of 21 with a 3.55 increment ratio. The variation of ethanol blends (E10. E10, E15) along with four different engine loads was reported by Rakopoulos et al. (2007) Smoke and CO emission were significantly reduced across all the ethanol blends, while NOx emission reduced slightly as compared to neat diesel mode. However, with an increase in the percentage of ethanol, NOx emission reduced significantly while a slight improvement in BTE was observed. On the contrary, HC emission increased with an increase in the percentage of ethanol.

The variations are shown in Figs. 7.17, 7.18 and 7.19. One of the key advantages of bioethanol in CI engines is that it can be used without any engine modification. However, one of the primary problems of using ethanol is its low cetane number, due to which its addition in diesel lowers the cetane number of diesel. Modern diesel engines require fuel having cetane number of 40 or higher for the suitable performance of the diesel engine. Addition of additives i.e. 2-ethylhexylnitrate for cetane number enhancement could be the remedy to such problem (Kumar et al. 2013). The extent of improvement with such enhancing additives varies from engine to engine

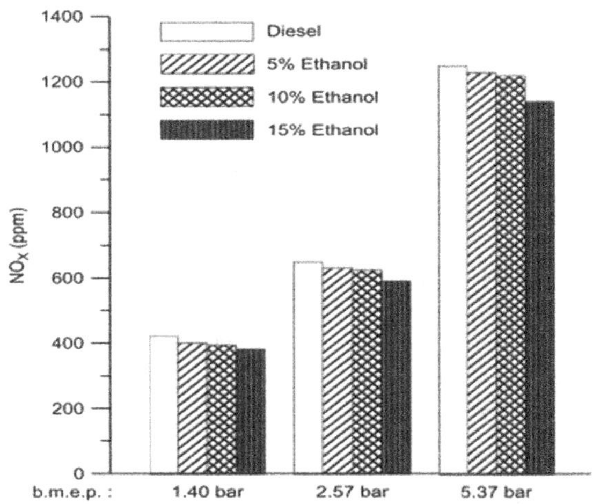

Fig. 7.17 Variation of NOx with different ethanol blends (Rakopoulos et al. 2007)

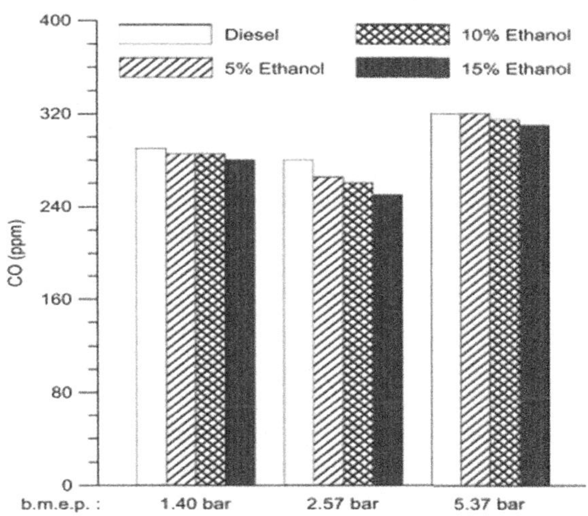

Fig. 7.18 Variation of CO with different ethanol blends (Rakopoulos et al. 2007)

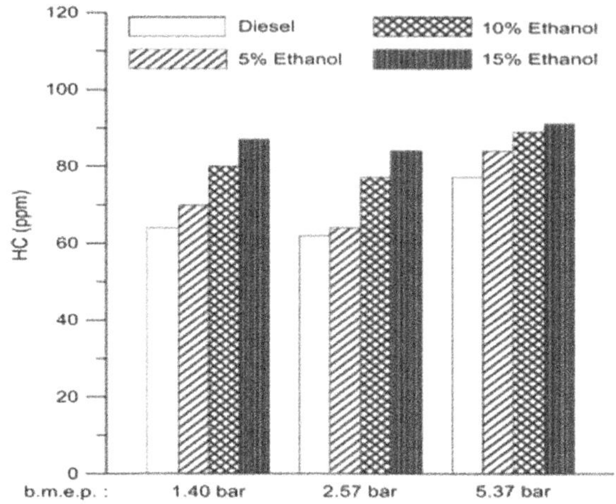

Fig. 7.19 Variation of HC with different ethanol blends (Rakopoulos et al. 2007)

depending upon its operational ranges. Another drawback of ethanol blended fuel is its low viscosity and lubricity which causes the degradation of lubricity of diesel fuel. Therefore, use of ignition enhancers and other additives are necessary to improve the ignition quality and life span of diesel engines while using blended ethanol–diesel fuels. The low miscibility of ethanol with diesel is a problem with the effective blend formation of ethanol with diesel resulting in phase separation. The mixing quality of ethanol depends on certain parameters such as the amount of hydrocarbon and wax content of base diesel, the amount of ethanol and the temperature of diesel fuel (Hansen et al. 2005). The reduction in aromatic content of diesel also lowers the solubility of ethanol. The solubility of ethanol also depends upon the amount of water content of ethanol. Anhydrous ethanol is readily miscible with that of diesel in bends of 0–30% and 70–100%. However, with higher water content in ethanol certain additives need to be used to maintain the homogeneity of the fuel under all temperatures and mixing conditions. In order to maintain the stability, two types of methods are normally applied; addition of surfactants for stability enhancement and addition of co-solvents for maintaining reliable blends (Kumar et al. 2013). Moses et al. (1980) used 2% of commercial surfactant in 5% ethanol blend with diesel. The combination formed an impetuous, clear, and thermodynamically stable blend. In another investigation, emulsified ethanol–diesel blend (30, 40, and 50%) was prepared using sorbitan mono oleate as a surfactant for using in a CI engine. It has been reported that 50% ethanol–diesel blend gave better results compared to pure diesel and other blends. There was rise in BTE and NOx emission, but smoke density and PM lowered. Can et al. (2005) studied the use of co- solvents in ethanol–diesel blends. The researchers utilized an unsaturated fatty acid based solvent and isooctyl nitrate as ignition improvers for blending ethanol- diesel in different ratios (10–30%). They found that engine exhaust emissions varied with ethanol content, engine operating conditions, additives, and ignition improver. Huang et al. (2009)

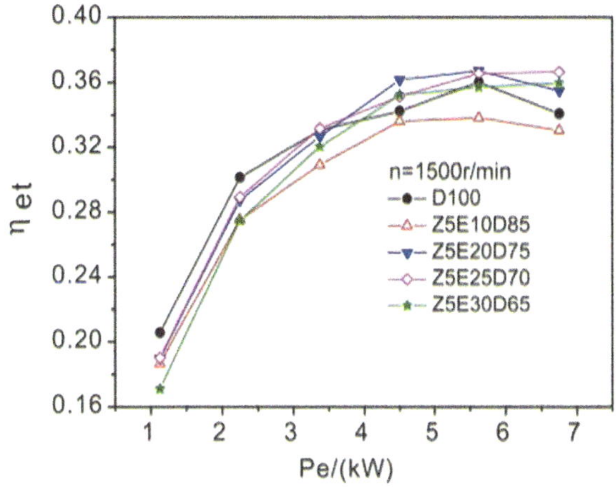

Fig. 7.20 Variation of BTE with different ethanol blends (Huang et al. 2009)

analyzed the diesel engine performance at different ethanol–diesel (10, 20, 25 and 30%) blends with and without the use of 5% n-butanol as additive. With an increase in the amount of ethanol BTE as well as smoke, emissions decreased compared to neat diesel fuel as shown in Fig. 7.20. CO emissions was less at and above half loads while HC emission was higher except at full loads. In addition, NOx emission was varied at different loads, speed, and blends. To improve the combustion phase analysis, different research investigations have been carried out with or without cetane number improver. Kim and Choi (2008) studied three different ultra-low sulphur fuels (pure diesel, ethanol–diesel blend and ethanol–diesel blend with CN improver). There was slight increment in HC and CO emission of ethanol–diesel blend while smoke emissions and PM decreased by more than 42% and 26% respectively. The effect of CN improver [2-methylhexyl nitrate (EHN) $C_{10}H_4C_{13}$–NO_2] on the engine performance, combustion and exhaust emission behaviour of a CI engine fuelled with ethanol–diesel blends was analysed by Liu et al. (2010) It was found that BTE, total and diffusive combustion was increased with CN improver whereas ignition delay reduced. The addition of ethanol and CN improver, diesel showed not much improvement in NOx emission, while PM and smoke emission decreased significantly for 30% ethanol–diesel blend, and decreased further with an increase in CN improver. The reason behind the variation is because the use of CN improver brings the ID of ethanol blends close to diesel. This results in lowering of heat release rates of the engine as shown in Fig. 7.21. As a result the NOx emission was observed to lower with addition of CN improver along with ethanol blends as shown in Fig. 7.22. In addition to above mentioned methods of blend improvement techniques, ethanol fumigation has been reported by different researchers. For fumigation, slight engine modifications is required so as to inject alcohol into the air with the help of low-pressure fuel injectors. Fumigation of Ethanol resulted in substantial reduction in

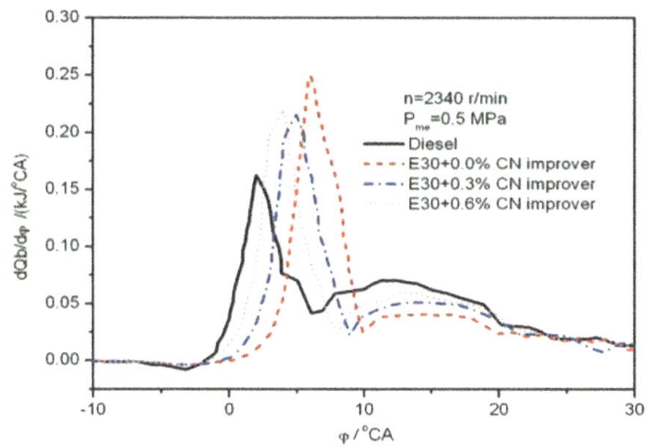

Fig. 7.21 Variation of heat release rates with and without CN improver (Liu et al. 2010)

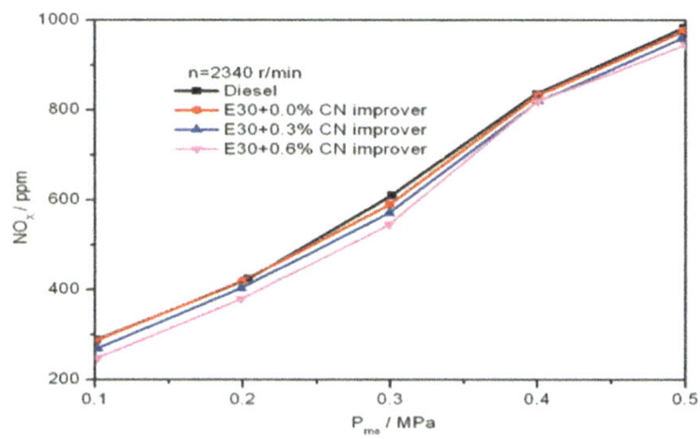

Fig. 7.22 Variation of NOx with and without CN improver (Liu et al. 2010)

NOx, smoke, and PM emission, while HC and CO emission increased with no significant changes in CO_2 emission. As far as BTE is concerned, a slight reduction was observed. Further, the use of diesel content catalyst reduces the CO and HC emission with higher exhaust gas temperature; NOx emission reduced drastically. Hence, it can be summarized that alcohol fumigation and can be used as an effective emission control technique in diesel engines (Tsang et al. 2010; Janousek 2010).

7.2.2.3 Propanol as Biofuel

Propanol, a three carbon structure is a straight chain alcohol having a higher cetane number and energy density compared to methanol and ethanol, which makes it a potential alternative fuel for blending with diesel in CI engines. Propanol can be produced in different ways, one of the most cost-effective processes is production from petrochemicals by oxo-synthesis process (Cornils 2004). Another method of propanol production is from the feedstocks of biomass or municipal wastes. Under this process, syngas (CO, H_2, and CO_2) is obtained through gasification of these feedstocks and then converted to biofuels using microbial catalyst such as Clostridium ljungdahlii and Clostridium ragsdalei (Liu et al. 2014). However, propanol production from clostridium through microbial production via threonine catabolism or form fermentation of yeast yielded very less quantities of 70 mg/l (Shen and Liao 2011). Therefore, researchers have opted for propanol production from bio-synthetic pathways from Escherichia coli, a less complex microorganism than clostridium species (Srirangan et al. 2014). Recently several modern biochemical, genetic and metabolic techniques have been adopted for enhancing the production of propanol. Among the 3-carbon fuels viz. propane, propanal, acetone and n-propanol, propanal showed the highest flame speeds while propane and n-propanol resulted in comparable laminar flame speeds. Propanol is used in diesel engines either as a blending component with raw diesel or as a diluting agent with vegetable oils for viscosity reduction. All diesel and propanol blends are found to be stable at temperatures above 2.5 °C (Rajesh Kumar and Saravanan 2016).

The use of propanol in diesel engines has been reported by different researchers for investigating the performance and emission characteristics of the diesel engine. Laza et al. (Laza and Bereczky 2009) used to different blends of propanol (10 and 20% v/v) with rapeseed oil and neem oil in CI engine. The results indicated that the use of propanol with rapeseed oil reduced smoke emission but increased HC, CO and NOx emission significantly. On the other hand, a propanol blend of 20% v/v with neem oil improved the engine's performance along with a considerable reduction in emission levels. In another study, propanol was used as a supplement in blends of (10, 20 and 30% v/v) with callophyllum vegetable oil in a four stroke, single cylinder diesel engine under six different engine loads. A significant reduction in HC, CO, and smoke emission with increased NOx was obtained with propanol-callophyllum blend compared to neat diesel operation. The use of propanol/ diesel blends (5, 10, 15 and 20%) as reported by Singh et al. (Ajav et al. 1999) reduced NOx emission significantly when compared with diesel fuel. With 15 and 20% blends, BTE of the engine also increased slightly with lower smoke emission for 20% blend at peak loads. The results of 4 and 8% propanol- diesel blends in a constant speed diesel engine showed enhanced performance of the engine (Balamurugan and Nalini 2014). Smoke and NOx emission decreased while HC emission increased with an increase in the percentage of propanol and also BTE increased upto 7.635% at 8% propanol-diesel blend. Therefore, from different studies of propanol use in a diesel engine, it can be summarized that propanol can be mixed with diesel upto 20% v/v without phase separation and with vegetable oils upto 30% v/v without the use of an emulsifying agent.

7.2.2.4 Butanol as Biofuel

Butanol is a four carbon atom alcohol that exists as four isomers: n-butanol(CH_3–$CH_2CH_2CH_2OH$), 2-butanol ($CH_3CH_2CHOHCH_3$), i-butanol $(CH_3)_2CH_2CHOH$, and t-butanol $(CH_3)_3COH$. All the four isomers have the same formulae and energy but each has a different manufacturing process and molecular structures, which affect their properties. Inspite of similar molecular weights and functional groups but their solubility differs. n-butanol is a colourless, slightly hydrophobic in nature, flammable with a sharp banana-like smell and distinct alcoholic odor. It is soluble with most common organic solvent but frugally soluble in water. Similar to n- butanol, iso-butanol is also a colourless fluid, slightly soluble in water with a sweet musty odor. 2-butanol is flammable, slight water soluble, and totally soluble in polar organic solvents. In contrast to these isomers, t-butanol is a clear fluid with a camphor like odor that is completely soluble in water and miscible with ethanol and diethyl ether (Kumar et al. 2013).

Butanol use as a sustainable engine fuel has been a growing interest since the late 1990s. For commercial and industrial usage, butanol is presently produced purely from petrochemical (Harvey and Meylemans 2011). However, enhanced biobutanol production through microbial fermentation processes is more economical compared to petrochemical based production. Acetone-butanol-ethanol (ABE), one of the oldest known fermentation processes is promoted by a bacteria of genus Clostrifium species, particularly acetobutylicum. This bacterial species secrets numerous enzymes that enable polymeric carbohydrates to break down to monomers. The above fermentation is capable of producing 66% of butanol was the only biotechnological method used until the first part of the twentieth century in industries (Qureshi et al. 2010). However, with the development of the petrochemical industries, the production through this fermentation process has declined due to the growing costs of biomass feedstock. The oil crisis at the beginning of 1970s has increased the revival of bio-industry with advanced researches in both biotechnological and molecular biological fields for improving the microbial tolerance of butanol toxicity. This led to the increase in ABE solvent production yield (Dürre 2007, 2008). A new butanol production technology developed by Taiwans industrial Technology Research Institute known as Butyfix process is capable of converting 94% of available carbon in biomass. Biobutanol is an excellent liquid having superior biofuel properties. The feedstocks for biobutanol production similar to bioethanol production which involves, wheat, corn, sugarcane, sugar beet, and cassava. Further, several researches has also reported that glycerol obtained as a by-product of biodiesel can be used for butanol production via an anaerobic fermentation process (Rajesh Kumar and Saravanan 2016).

Butanol blending with diesel has several possibilities to make diesel fuel technology compatible with alcohols for use in CI engines. n- butanol is one of the most effective fuel additives for use in CI engines. Unlike the lower alcohol fuels, butanol has the potential to surrogate gasoline and perform well in engines (Agarwal et al. 2017). The moderate range of cetane number also makes a suitable alternative biofuel for significant blending with that of diesel. Butanol has properties similar to that of diesel and can be used not only as blends but as a single fuel in engines in the near future. The production of biobutanol from biomass feedstocks with significant technological development can result in reducing the utilization of diesel engines (Harvey and Meylemans 2011). The use of butanol in diesel as diesel-butanol blend has been reported by different researchers. Rakopoulos et al. (2010)

studied the effect of n-butanol and diesel fuel blends (8%, 16%, and 24%) in a high-speed direct injection diesel engine for analyzing the performance and emission characteristics at three different loads. The increase in butanol blends marginally increased the specific fuel consumption with a minor rise in BTE. In addition, smoke density reduced significantly with slight decrease in NOx and CO emission for the butanol-diesel blends. However, HC emission increased with butanol use in diesel engine. Another research study analysed the effects of adding iso-butanol to diesel fuel (Al-Hasan and Al-Momany 2008). The test was performed at engine speed ranging from 375 to 625 rpm, different iso butanol-diesel blends (10, 20, 30 and 40% v/v) and under different engine loads. The results of the analysis revealed rise in the exhaust gas temperature and BSFC with the increase in engine speed,, while BTE and air–fuel ratio decreased. The addition of iso butanol blend upto 30% v/v to diesel showed desired engine performance while with 40% v/v blend deteriorated the engines performance. The use of iso butanol—diesel in different blends (5, 10, 15 and 20%) at different engine speed ranging from 1200 to 2800 rpm and under full load was studied by Karabektas and Hosoz (2009). All the blends resulted in lower BTE and break power with the highest BTE reported for 10% iso-butanol blend. The use of iso-butanol blends reduced CO and NOx emission while HC emission increased. The effect of variation of engine speed with BTE CO and NOx emission are shown in Figs. 7.23, 7.24 and 7.25.

However, Asfar et al. (2003) revealed that the addition of 5–10% of iso-butanol to 10% olive oil biodiesel further reduced smoke HC and CO emissions. In another investigation Rakopoulos et al. (Rakopoulos et al. 2011a) utilized two different fuels ethanol–diesel (5 and 10%) and n-butanol (8 and 16%) blends in a six-cylinder heavy-duty diesel engine. The emission results of ethanol and n-butanol blends were found to be similar, with the exception that smoke density decreased with ethanol blends while NOx emission decreased more with n-butanol blends. BTE and BSFC values for n-butanol blends (8 and 16%) were higher at speeds of 1200 and 1500 rpm compared to diesel fuel. One of the drawbacks of low CN fuels is that is can cause

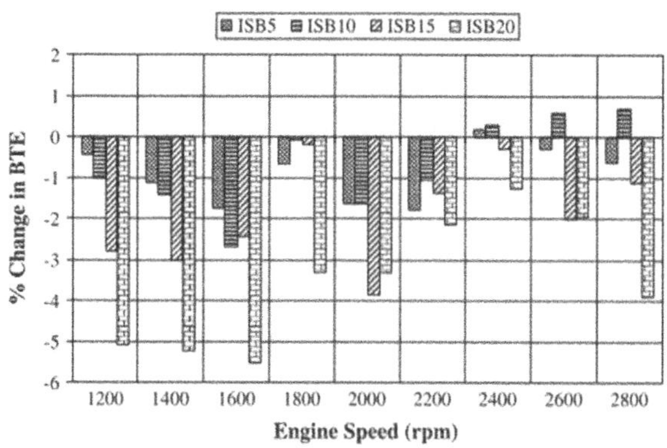

Fig. 7.23 Variation of BTE at different engine speeds (Al-Hasan and Al-Momany 2008)

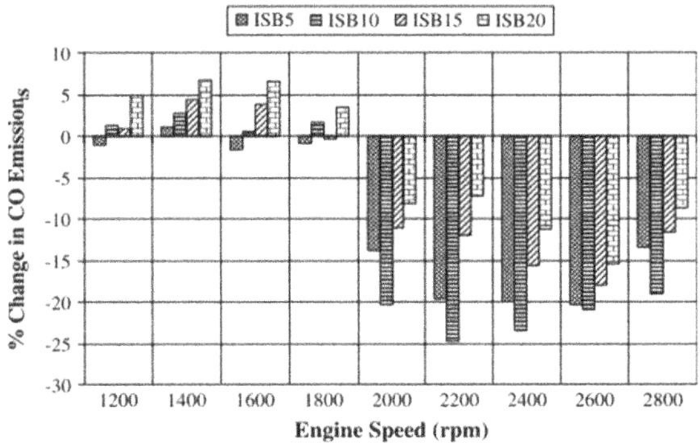

Fig. 7.24 Variation of CO at different engine speeds (Al-Hasan and Al-Momany 2008)

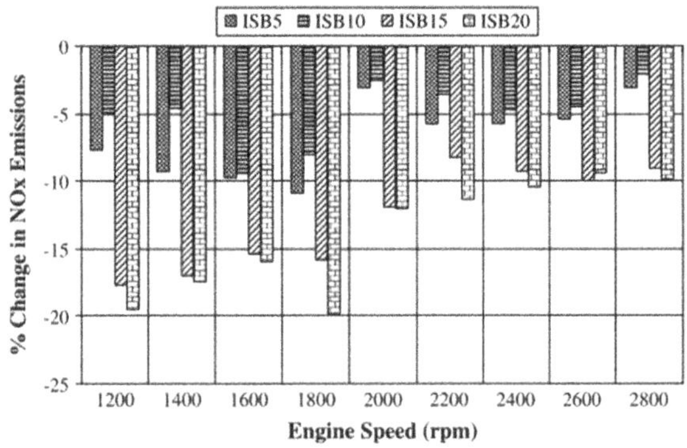

Fig. 7.25 Variation of NOx at different engine speeds (Al-Hasan and Al-Momany 2008)

significant cyclic cylinder pressure variation in CI engines. This may lead to degradation of performance and reliability of the engine, increase in engine noise, and exhaust gas temperature. In severe cases, incomplete burs or misfires may take place and can adversely affect the engine driveability and increase in engine pollution levels. As butanol has a lower CN number compared to diesel (although higher than lower alcohols like methanol and ethanol), but its addition to diesel fuel might lower the CN of diesel. This may result in cyclic inconsistency difficulties. Several researches (Wing 1975; Kouremenos et al. 1992; Sczomak and Henein 1979) have investigated these cyclic variability problems at different injection timing, under different loads, and pressure variation in diesel engines. An investigation of fuel line pressure by Rakopoulos et al. (2011b) revealed that n-butanol blends upto 24% v/v blends with diesel does not degrade the performance or emission of the engine. Yao et al. (2010) found that combustion of the engine at higher injection pressure, IT and CR, oxygenated fuel blends of 10% and 20% can be effectively be controlled by exhaust gas recirculation (EGR). Further, EGR was found to be more appropriate for enhancing engine performance and controlling engine emissions. Thus from the butanol application in CI engines, it can be summarized that peak gas pressure, heat release rate and ignition delay plays a significant role in the combustion process of the diesel engine with butanol blended fuels. Due to lower CN for both n-butanol and iso-butanol, the PCP, heat release rate and ignition delay increased with an increase in the percentage of butanol in diesel. In addition, increasing engine load increases the fuel consumption which in turn increases the volume of combustible fuel n the combustion chamber. Further, the use of butanol-diesel blends leads to improved thermal efficiency with a significant reduction of smoke and NOx emission in the CI engine. Table 7.4 Summarizes some of the investigations related to the performance, combustion and emission characteristics of alcohol use in diesel engines.

7.2.3 Emulsification in CI Engines

Emulsification is the process of mixing two liquids that are immiscible with each other (particularly water and oil), in which one liquid is dispersed as small droplets within the other (Jhalani et al. 2019). With the growing concerns of environmental degradation and stringent regulations on pollution control, emulsification process can be utilized in diesel engines for limiting emission levels. Particularly emulsification of water-diesel is reported to have shown better results in diesel engines compared to other techniques. Water-diesel emulsion has a high density and higher heat of vaporization rate, which in turn serves as primary cooling in diesel engines. As a result, the peak cylinder pressure reduces and consequently, the NOx and PM emissions are controlled. Several after-treatment technologies such as catalytic converters and diesel particulate filters also helps in the reduction of NOx and PM emission, but increases the fuel consumption rate and leads to a poorer economy of the engine. However, the water emulsification technique stands out in comparison to the after-treatment techniques (Gopidesi and Rajaram 2019). Water can be introduced in a diesel engine through three techniques viz. water injected directly into engine cylinder through the electronic unit, injection of water into the air through fumigation method, and emulsion of water with diesel (Jhalani et al. 2019). In direct water injection water can be mixed with fuel as per requirements by varying engine speed and load. However, the drawbacks of this technique are that the incompressible nature of water can cause hydrostatic lock condition and if the proportion of water exceeds certain limit can retard the speed of the engine. This condition can lead to physical damage to the piston and cylinder head of the engine. In addition, the water injection method is reported to be more corrosive compared to other methods (Zhang et al. 2017; Ma et al. 2014). Fumigation on the other hand is the process of mixing water with that of intake air, which results in a higher density of intake charge enhancing the volumetric efficiency of the engine. However, this technique causes corrosion problems in the engine and also engine starting problems in cold conditions (Khalife et al. 2017). Water emulsion gives better results of reduction of NOx and PM emission. One of the best features of this technique is that it can be applied without undergoing any engine modifications (Suresh and Amirthagadeswaran 2015). In this

Table 7.4 Summary of different alcohol use in CI engines

Investigator	Engine used	Fuels	Experimental condition	Research findings
Najafi and Yusuf (2009)	F4L 913 4-stroke diesel engine	Methanol and diesel	• Methanol blends of 10, 20 and 30%	• BTE and Torque increased • BSFC decreased
Lei et al. (Shen et al. 2011)	4-cylinder, naturally aspirated water-cooled 4-stroke DI diesel engine	Ethanol and diesel	• Methanol blends of 5, 10 and 15% • Stability additives of 0.8, 1 and 1.5%	• Increase in BSFC and BTE • Increase in HC emission • Decrease in NOx, CO and smoke emission
Canakci et al. (2009)	Single cylinder, 4- stroke naturally aspirated DI diesel engine	Methanol and Diesel	• Methanol blends of 5, 10 and 15% • Variation of injection pressure	• Increase in BSFC, BSEC and combustion efficiency • Decrease in BTE, heat release rate and PCP • Increase in NOx and CO_2 emission • Decrease in HC, CO and smoke emission
Singh et al. (2013)	KIRLOSKAR 1-cylinder, 4S, DI diesel engine	Propanol and diesel	• Propanol blends of 5, 10, 15 and 20% • Variation of engine load	• Increase in BTE at 15 and 20% propanol blends • Higher BSFC at low loads • NOx emission decreased upto 13.4 with 20% propanol blend • Smoke and CO emission decreased at high loads • Increase in HC emission

(continued)

Table 7.4 (continued)

Investigator	Engine used	Fuels	Experimental condition	Research findings
Dogan (2011)	1-cylinder, 4S, CR:18:1, AC, naturally aspirated, HS, DI diesel engine	Butanol and diesel	• Propanol blends of 5, 10, 15 and 20% • Variation of engine load	• Increase in BSFC with butanol blends • NOx, CO and smoke emission reduced • HC emission increased with butanol blends
Investigator	Engine used	Fuels	Experimental condition	Research findings
Banugopan et al. (2010)	Single cylinder 4-stroke water cooled diesel engine	Ethanol and diesel	• Ethanol blends of 10, 15, 20 and 25% • Variation of engine load	• BTE increased and BSFC decreased • Increased HC and CO emission
Ozsezen et al. (2011)	6-cylinder, 4S, RP:163KW, CR:16.4:1, WC, turbo-charged, HD, DI diesel engine	Iso-butanol and diesel	• 40% butanol-diesel and gasoline-diesel blends • Variation of engine speed • Variable BMEP	• BTE for 40% butanol-diesel blend higher than 40% gasoline-diesel blend • Smoke, NOx emission reduced for both blends • Increased HC and CO emission for both blends
Valentino et al. (2012)	4-cylinder, turbo- charged, CR:17.5:1, WC, CRDI diesel engine	Butanol, gasoline, diesel	• Methanol blends of 5, 10 and 15% • Variation of injection pressure	• Increase in BSFC, BSEC and combustion efficiency • Decrease in BTE, heat release rate and PCP • Increase in NOx and CO_2 emission • Decrease in HC, CO and smoke emission

(continued)

Table 7.4 (continued)

Investigator	Engine used	Fuels	Experimental condition	Research findings
Lu et al. (2005)	4-cylinder 4-stroke, naturally aspirated water- cooled, DI diesel engine	Ethanol and diesel	• Ethanol blends of 5, 10, 15 and 20% • CN improver	• Increase in BTE for all the ethanol blends • NOx emission reduced • CO and HC emission reduced with CN improver
Cheung et al. (2009)	4-cylinder, naturally aspirated DI diesel engine	Methanol and diesel	• Methanol blends of 10, 20 and 30% • Variation of engine load • Effect of DOC	• Reduction in BTE with methanol blends • NOx and smoke emission reduced • Addition of DOC reduced HC and CO emission
Balamurugan and Nalini (2014)	KIRSOLSKAR 1- cylinder, 4S, RP:5.2KW, CR:17.5:1, WC, DI diesel engine	Propanol and diesel	• Propanol blends of 4 and 8% • Variation of engine load	• Increase in BTE by 1.579% for 4% and 7.635% for 8% propanol blends • Decreased smoke, NOx and CO emission • Increased HC emission
Gu et al. (2012)	6-cylinder, 4S, CR:17.3:1, CRDI diesel engine with a WC EGR system	Iso-butanol, butanol and diesel	• 15 and 30% blends of butanol ad iso-butanol • Variable EGR rates • Advance and retarded IT upto 6°	• No change in BSFC with low EGR, but reduced BSFC at retarded IT • Reduced NOx emission at low EGR and retarded IT • Increased smoke emission high EGR while reduced at retarded IT • CO emission of Butanol > iso-butanol. Increased CO at high EGR and retarded IT

(continued)

Table 7.4 (continued)

Investigator	Engine used	Fuels	Experimental condition	Research findings
Kumar et al. (2007)	Direct injection single cylinder air-cooled naturally aspirated CI engine	Ethanol, diesel	• Ethanol blends of 13, 17 and 23% • Effect of emulsions 7, 13 and 17%	• Thermal efficiency and BSEC reduced • Smoke, NOx, HC and CO emission reduced at low loads • Increased HC and CO emission at high loads
Sayin et al. (2010)	Single cylinder, 4-stroke DI diesel engine	Methanol and diesel	• Ethanol blends of 5, 10, 15% • Variation of IT and pressure	• Increased BSFC and reduced BTE • Increased NOx emission • Reduced smoke HC and CO emission
Zheng et al. (2015)	1-cylinder, 4S, WC, CR:16:1, CRDI diesel engine with EGR system	Butanol, sec-butanol, iso-butanol, tert-butanol and diesel	• Alcohol blends of 20 and 40% with diesel • Effect of EGR	• Optimum results obtained for iso-butanol diesel blends • Enhanced BTE for all the blends • HC and smoke emission reduced with low EGR

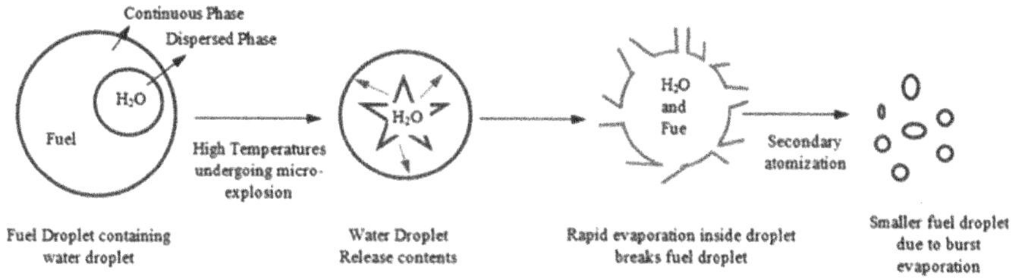

Fig. 7.26 Phenomenon of micro explosion (Gopidesi and Rajaram 2019)

method, two immiscible liquids, water, and diesel are mixed along with an added surfactant for increasing the stability of the emulsion. Water-diesel emulsification involves two essential characteristics; (i) the heat of vaporization of water reduces the peak combustion temperature which in turn reduce the NOx emissions, (ii) The phenomenon of micro-explosion of emulsified fuels improves combustion efficiency as it increases the volatility of diesel fuel (Lthnin et al. 2015; Ogunkoya et al. 2015). As illustrated in Fig. 7.26 micro-explosion is the phenomenon in which due to lower boiling temperature of water than oil, water droplets vaporize earlier on heating and burst the outside layer of oil which breaks the oil into small droplets (Gopidesi and Rajaram 2019). Another promising technology in the field of emulsification is the process of microchannel emulsification. It is the process of emulsion formation by injecting a dispersed phase through multiple micro fabricated microchannel arrays into continuous phase. These microfabrication technology is used to fabricate micro meter sized channels though microfluidic array plates. Further, the array plates results in formation of emulsion with sized controlled micro droplets with high efficiencies and high encapsulation. Through microchannel emulsification technology, variety of encapsulated products can be obtained ranging from single to multiple emulsions (Khalid et al. 2018).

7.2.3.1 Characterisation of Diesel-Water Emulsion

An emulsion is a high energy processing method that is achieved through mechanical homogenization, ultrasonication, and high-pressure mixing. Surfactants (emulsifiers) play a key role in stabilizing the emulsified fuels, although it cannot make them thermodynamically stable. In other words, over a period of time due to the phenomenon of flocculation, coalescence, and creaming, the emulsified fuels get separated into diesel and water. The function of surfactants is to lower the interfacial surface tension between the two liquids which helps to suspend the water particles in oil (Solans and Solé 2012). The preparation of a stable and homogenous emulsion greatly depends upon the type of surfactants used. Surfactants are mainly classified based on their hydrophile-lipophile balance (HLB) value. It is an empirical relation between water-loving (hydrophilic) and oil-loving (lipophilic) groups within the surfactant. HLB value ranges between 0 and 20, however, HLB value between the range of 4 and

8 generally produces water in oil (W/O) emulsions whereas HLB value above 10 produces better oil in water (O/W) emulsions (Al-Sabagh et al. 2013). Some of the surfactants commonly used along with their HLB values are shown in Table 7.5. The W/O emulsion indicates water is solute and oil is solvent and vice-versa in O/W emulsion. These type of emulsion is known as two-phase emulsions. Another type of emulsion is a three-phase emulsion also known as multiple emulsion, where inner and outer phases are detached by a dispersed phase. The schematic of the phase emulsion is shown in Figs. 7.27 and 7.28.

In most of the research work, it has been observed that a mixture of surfactants i.e. both W/O and O/W perform relatively better than pure surfactants with pure HLB value. The percentage of surfactants used significantly affects the stability of emulsified fuel. It has been found from different researches, amount of surfactants used varies in the range of 0.5–10% of total volume (Kumar and Kumar 2018). Based on polar head groups, surfactants are further classified as cationic, anionic, zwitter-ionic, and non-ionic. Among these groups, non-ionic surfactants have the desired properties of a sting with no soot and free nitrogen and sulphur that best suits

Table 7.5 HLB values of some commonly used surfactants (Jhalani et al. 2019)

Name of emulsifier	Emulsion type	HLB value	Name of emulsifier	Emulsion type	HLB value
Span 20	W/O	8.6	Tween 20	O/W	16.7
Span 40	W/O	6.7	Tween 40	O/W	15.6
Span 60	W/O	4.7	Tween 60	O/W	14.9
Span 80	W/O	4.3	Tween 61	O/W	9.6
Span 83	W/O	3.7	Tween 65	O/W	10.5
Span 120	W/O	4.7	Tween 80	O/W	15.0

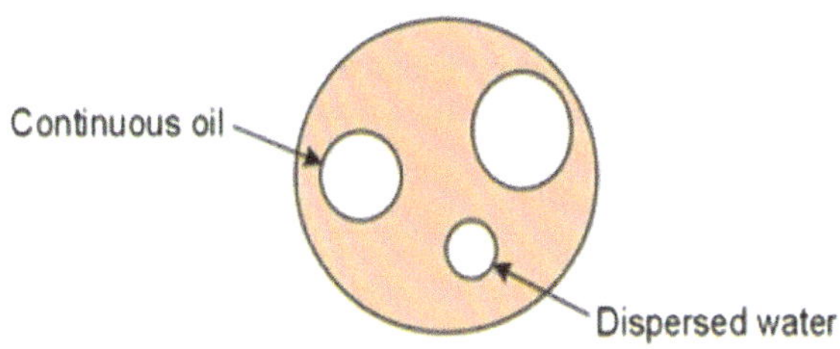

Fig. 7.27 Two phase emulsion

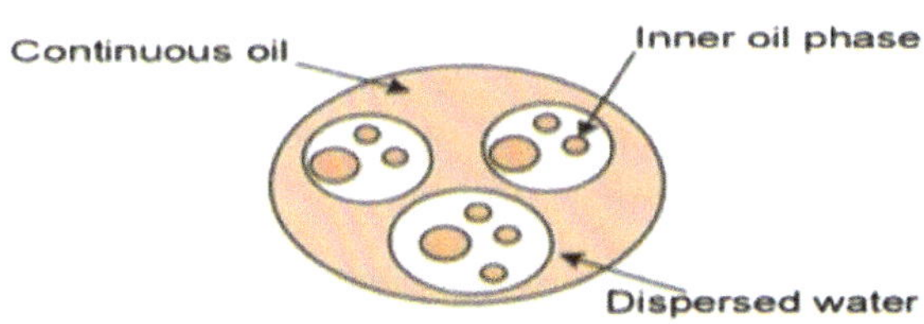

Fig. 7.28 Three phase emulsion

for water- diesel emulsions. Non-ionic surfactants have properties such as better lubrication, lesser corrosion, and anti-freezing properties that make them stand out among other polar head groups. There are various non-ionic surfactants used for making water-diesel emulsions viz. sorbitan mono-oleate (Span 80) and polyethylene glycol sorbitan mono-oleate (Tween 80), Triton X-100, sorbitol sesquioleate, TDS-30, sorbitan monolaurate, and various others including any suitable combination of these in varied proportions. Among these non-ionic surfactants, Span 80 and Tween 80 are the most commonly used (Vellaiyan and Amirthagadeswaran 2016; Al-Sabagh et al. 2012).

Three components viz. water, oil, and surfactants are mixed under high energy mechanical agitation or ultrasonification for preparation of emulsion. However, the stability and homogeneity of emulsions depend upon, the technique used for emulsification, amount of water added, duration of mixing, stirring speed, the concentration of surfactants, etc. Different types of equipments are used by various researchers for the preparation of emulsion. Emberson et al. (2016) prepared water- diesel emulsion using an ultrasonic generator with two different water proportions of 10 and 20% on a mass basis. Span 80 and Tween 80 with HLB value 6.4 were utilized as surfactants. It was reported that ultrasonicator produces emulsion with micro-size droplets within a short period. In another study (Hasannuddin et al. 2016), an electrical stirrer with a speed of 2500 rpm for stabilizing the emulsion using non-ionic surfactant span 80 by 1% ratio on volume basis. Lin and Chen (2008) compared the preparation of water-diesel emulsion using an ultrasonically vibrating method and mechanical homogenizer. The study revealed that the ultrasonically vibrating method provides effective preparation of water-diesel emulsion. Likewise, different equipment such as piezoelectric transducer, ultra-turrax machine, magnetic stirrer can also be used for the preparation of water-diesel blends using suitable surfactants.

The stability of emulsion plays a key role in preparing a suitable water-diesel emulsion. One of the major difficulties with an emulsion is its unstability because of liquid separation tendency, which results in reduction of its interfacial area with interfacial energy. The interfacial flim plays a significant role in the stability of emulsions. Emulsions are categorized based on kinetic stability, which is based on stability over period of time. Loose emulsion, in general, are less stable and separates within minutes whereas stable emulsion takes more days for separation. Some of the factors, which mainly affects the stability of the emulsion, are droplet and droplet size distribution, temperature, pH of brine, etc. Temperature affects the viscosity of emulsion which decreases with an increase in temperature. As a result, the water interfacial flims get destabilized. Similarly, the droplet size also affects the viscosity of the emulsion. With smaller droplet size, the viscosity of the emulsion is higher, and as such the stability of the emulsion is more which increases the separation time. Further, the pH of water also plays a significant role in the stability of emulsion. Generally, lower pH value is preferred for the preparation of water in oil emulsion (Gopidesi and Rajaram 2019).

Table 7.6 Comparison of properties of emulsified fuel with that of diesel (Mondal and Mandal 2019; Jhalani et al. 2019)

Properties	Diesel	Emulsion (5% v/v)	Emulsion (10% v/v)	Emulsion (15% v/v)
Density (kg/m^3)	835	839.8	850	852
Cetane number	52	–	44.5	45.5
Viscosity (cSt)	2.72	4.2	4.4	4.6
Lower heating value (MJ/kg)	42.49	42.9	42.10	39.51

7.2.3.2 Properties of Emulsified Fuel

The properties of emulsified fuel are greatly influenced by the density, dynamic viscosity, and kinematic viscosity of the fuel blends. Due to the higher viscosity and density of water, these properties increase with an increase in the percentage of water in emulsions. In addition, the lower heating value of water than diesel reduces the overall calorific value of the emulsions. The higher viscosity leads to higher consumption of fuel due to a reduction in backflow across the piston clearance of the injection pump (Jhalani et al. 2019). Moreover, as reported by Ertunc Tat et al. (2004) the higher viscosity of fuel also involves in fuel leakage past the plunger during compression. Noor et al. (El-Din et al. 2013) studied the behaviour of emulsion with different water proportions (5, 6, 7, 8 and 9% v/v) at varying temperature ranges of 10, 20, 30, 40 and 50 °C for durations of 0, 1 and 2 weeks respectively. The study revealed that all the samples under all the test conditions behaved like low viscosity Newtonian fluid upto 2 weeks' duration. Overall, the nano emulsion with 5% water content was observed to be transparent with 49.5 nm droplet size. Further, the variation of surfactants Tween and Span was also reported in this study. The results revealed that a substantial drop in droplet sized from 145.4 to 49.55 nm was found with increase in surfactant ratios of tween and span. In another study droplets size were found between 19.3 and 39 nm which depends on the amount of water and emulsifier used (Al-Sabagh et al. 2011). Mehta et al. (2015) investigated the variation of the size of droplets with four samples of 0.5, 1, 2.5 and 5% of water in diesel. The size of droplets increased with an increase in the percentage of water and also with time and as such the reduces the stability of the emulsion. In addition to this, the properties of emulsified fuel are also influenced by engine components like injector nozzle, fuel pump, and combustion period of the emulsified fuel. The properties of emulsified fuel with varying water content are illustrated in Table 7.6.

7.2.3.3 Application of Emulsified Fuel in CI Engines

Emulsified fuel use in diesel engine are increasing day by day due to suitable fuel enhancing characteristics. Two critical features of water-emulsified fuels are (i) the high evaporation rate of water reduces the peak combustion temperature thereby lowering the NOx emissions and (ii) the micro explosion phenomenon of emulsified fuel reduces the volatility of diesel fuel and improves the combustion efficiency. Various researches have reported the application of water emulsification in CI engines along with diesel as well as biodiesel blends. Kannan and Udaykumar (2009) explored the performance and emission characteristics of a diesel engine with 10% and 20% water emulsion. BTE was found to increase with increased water content while HC and CO emission decreased under all loading conditions. The presence of water results in increased expansion work and reduced compression work, thereby increasing work done during the cycle. In addition, the torque produced during the cycle increased with additional force of water vapour. This result in increase of BTE of emulsified fuel engine. Further, with reduction in adiabatic flame temperature of diesel due to vapourisation and sensible heat of water, 10% and 25% reduction in NOx emission was reported for 10 and 20% water content respectively. Maawa et al. (2020) investigated the effect of water emulsification on the performance, combustion, and emission of diesel- biodiesel blends at different proportions. The emulsification was performed at different water contents of 5, 20 and 30% with 20% biodiesel-diesel blend with 1% Span 80 and 1% Tween 80 as surfactants. The results revealed that the engine performance improved significantly with 5% water emulsification. Exhaust emission especially NOx emission reduced with each of the emulsified proportions, maximum reduction obtained with 30% water emulsion as shown in Fig. 7.29. The reliability of water emulsification uses in biodiesel-diesel fuel blend prolong usage in a diesel engine was analysed by Patidar and Raheman (2020). A comparative study of the engine performance and emission characteristics were performed between diesel, 20% biodiesel-diesel blend, and 10% water emulsifier biodiesel-diesel blend. The results revealed that BTE increased upto 5.14% at higher loading conditions for emulsified fuel compared to diesel. The emulsified fuel also resulted in upto 13.80 and 16.94% lower NOx and soot emission compared to conventional diesel. In another investigation, Roy et al. (2021) analysed the effect of emulsified fuel on dual biodiesel blend of Castor-Jatropha on the engine performance

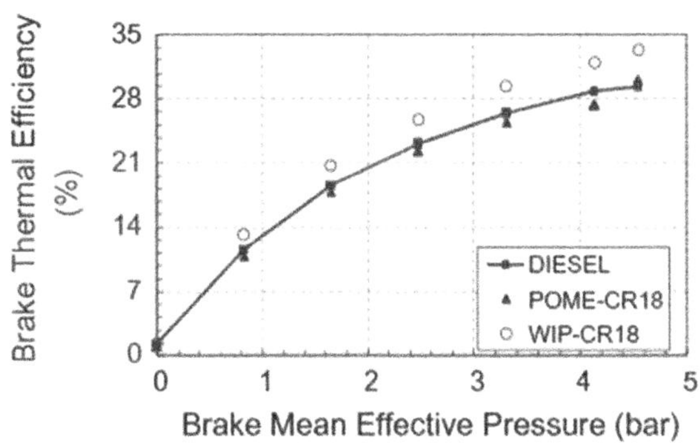

Fig. 7.29 Variation of BTE of emulsified biofuel compared to diesel (Maawa et al. 2020)

and emission characteristics. The emulsification was prepared with water concentration of (0, 1, 2, 3, 4, 5 v/v %) with biodiesel- diesel blend (10% jatropha, 10% castor oil and 80% diesel). The 5% emulsified fuel gave suitable engine performance with 14% higher BTE and approximately 60% lower NOx emission compared to other selected water concentration levels. Debnath et al. (2013) investigated the effect of operating parameters viz. CR and IT on the performance of emulsified palm oil methyl ester (POME) run diesel engine. The results revealed that at 18 CR and IT of 20° BTDC the emulsified fuel showed 11% higher BTE than diesel at 4.14 BMEP. However, at 4.55 bar BMEP higher BTE of 33.4, 30.2 and 29.4% were obtained for emulsified fuel, POME biodiesel and diesel respectively. Further, the emulsified fuel resulted in 43% and 20% CO and NOx emission reduction respectively compared to that of diesel. Annamalai et al. (2016) carried out an analysis on the performance, combustion and emission behaviour of a diesel engine powered by lemon grass oil (LGO) and ceria nanoparticle blended emulsified biofuel. The emulsification was prepared with 5% water, 93% lemon grass oil, 2% of span80 by volume basis, and ceria nanoparticle was dispersed with LGO is dosage of 30 ppm. The emulsified fuel resulted in improvement in BTE, BSFC along with a drastic reduction of CO, HC and NOx emission compared with neat LGO and diesel fuel. Srikanth et al. (2021) investigated the performance of a diesel engine using Niger seed oil biodiesel as an emulsifier for diesel-ethanol blend. The Niger seed oil biodiesel was used as additives from 5 v/v% to 20 v/v% in diesel-ethanol blends to enhance the solubility of ethanol in diesel fuel. Different blends were prepared with baseline fuel and diesohole biodiesel (DE5B10, DE10B10, DE15B10, DE5B20, DE10B20 and DE15B20) blends. The results indicated that maximum BTE was obtained for DE10B10 blends which were 11.19% higher compared to diesel fuel under full load conditions. Further DE5B20 blend resulted in 7.07% higher NOx emission than diesel under full load conditions. The variations are illustrated in Figs. 7.30, 7.31 and 7.32. Basha (Sadhik Basha 2018) analyzed the effect of carbon nanotubes (CNT) and diethyl ether (DEE)

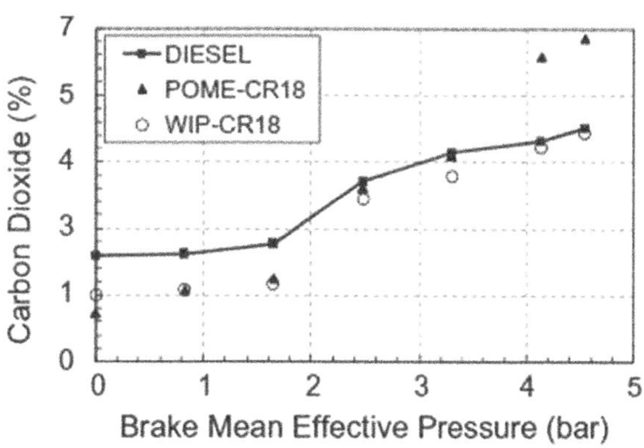

Fig. 7.30 Variation of CO_2 of emulsified biofuel compared to diesel (Srikanth et al. 2021)

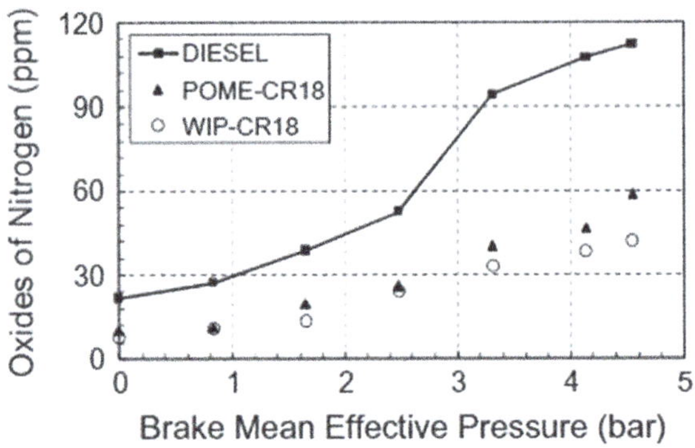

Fig. 7.31 Variation of NOx of emulsified biofuel compared to diesel (Srikanth et al. 2021)

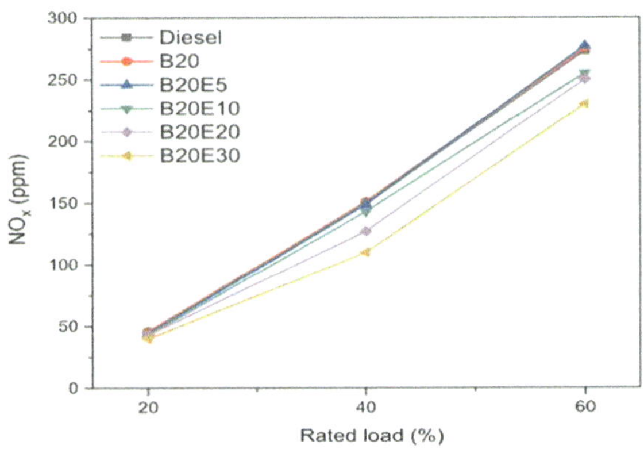

Fig. 7.32 Variation of NOx of different emulsified biofuel compared to diesel (Srikanth et al. 2021)

blends as additives to jatropha biodiesel emulsion in a diesel engine. The entire investigation was carried out in five stages viz. pure diesel and jatropha biodiesel, 91% of biodiesel, 5% of water and 4% of emulsifiers (by volume), CNT + DEE (50 ppm CNT and 50 ml DEE) as additives with emulsified fuel. The mixtures were then subjected to stability test followed by engine application. The results showed that the emulsification of CNT and DEE blended biodiesel showed improved results in terms of performance, emission and combustion compared to pure diesel and biodiesel. BTE, NO and smoke emission of CNT + DEE fuels was 28.8%, 895 ppm and 36%, while for pure diesel, it was 25.2%,1340 ppm and 71% respectively under full load conditions. The application of ethanol–diesel emulsification in diesel engine was reported by Shen and Chen (Lei et al. 2012). The investigation was carried out using ethanol blends of 10 and 15% along with emulsification of 1 and 1.5%. A novel mixed emulsifier of castor oil and biofuel resulted in suitable engine performance with 10% ethanol blends. BTE increased with the use of emulsifier and ethanol blends. On the other hand, the NOx emission reduced at low loads, along with lower smoke emission for all ethanol blends. Table 7.7 summarizes the use of emulsification on diesel engine performance, combustion and emission characteristics.

In addition to the above mentioned oxygenated fuel improvement techniques, some of the modern fuel additives such as 2,5 dimethyl furan, Di-tetra-butyl-peroxide, new metal-based additives TiO_2 nano-fluid, plays a significant role in improving the combustion and emission characteristics of CI engines. The combustion characteristics of 2,5 dimethyl furan shows significant effect on the laminar burning speed under fuel rich conditions. In addition, the change in ignition delay is significant at different temperature conditions. Ignition delay gradually decreased with increase in temperature due to higher reactivity of 2,5 methyl furan. In comparison to ethanol, 2,5 methyl furan is reported to be a better fuel when blended with iso-octane (Roy et al. 2019). In another study, while comparing the combustion and emission characteristics of 2,5 methyl furan-diesel (D30) and n butanol-diesel blends (B30), it is observed that D30 has longer ignition delay. This leads to faster burning rate due to lower CN. In addition, the use of D30 additives resulted in significant reduction in soot emission levels. This is due to the extended ignition delay and fuel oxygen of D30 blends (Chen et al. 2013). Di- tetra-butyl-peroxide is another promising fuel additives which improves the performance of CI engines. A 20% blend of this additive along with biodiesel resulted in 5.25, 7.4% and 11.12% reduction of CO, HC and NOx emission compared to raw biodiesel. Moreover, the smoke emission also reduced by 3.6% on application of di- tetra-butyl-peroxide with biodiesel (Rathinam et al. 2019). Fuel additives such as dimethyl carbonate and dimethoxy methane also contributes significantly in improving the BTE of engine along with reduction of emission levels. The addition of CN improver along with these further improves the CO an HC emission levels (Lü et al. 2005). TiO_2 additive is a nano fluid that provides high surface energy during the combustion stages and reduces the emission levels. The use of TiO_2 additive along with mustard biodiesel resulted in significant reduction of HC, NOx, CO and smoke emission. The suitable catalytic effects, better capability of oxidation and improved thermal conductivity contributes to superior properties of TiO_2 nanofluid (Kishore Pandian et al. 2017). In addition to above mentioned additives, certain metal based additives such as aluminium oxide also contributes towards improving the combustion and emission parameters of the CI engine (Jyothi et al. 2019).

7.3 Conclusion

With growing interest and usage of diesel engines both as a passenger vehicle as well as in heavy industries due to their higher efficiency, has led to faster diminution of fossil reservoirs globally. However, the higher emission levels of these engines along with increased emission regulations have led researchers to shift to alternative methods for improving the emissions drawbacks of diesel engines. Different engine side modification techniques resulted in significant improvement, however, limited for only recently developed engines. On the other hand, improvements in the fuel side

Table 7.7 Summary of emulsification of diesel/biodiesel blends in CI engine

Investigator	Engine used	Pilot fuel	Experimental condition	Research findings
Vellaiyan and Amirthagadeswaran (2020)	4-stroke, single cylinder, natural aspirated, variable compression ratio diesel engine	Diesel and lemon oil–water (LPO) emulsifier	• Water emulsification of upto 10% used (LPO10W) • Effect of different brake mean effective pressure (BMEP)	• LPO10W emulsifier resulted in improvements of LPO10W BSFC, BSEC and BTE by 5.6%, 10.4% and 11.6% compared to pure Lemon oil • LPO10W emulsion reduces HC, CO and smoke emissions by 18.7%, 33.3% and 11.9%. at peak BMEP • The water inclusion in LPO significantly reduces the NOx emission by 26% and increases the CO_2 emission by 3.9% compared to LPO
Abdollahi et al. (2020)	12hp Lombardini single cylinder air cooled diesel engine	Waste cooking oil, diesel, water emulsifier	• Nano emulsion of 5% waste cooking oil, 5% distilled water and 90% diesel • Effect of engine speed	• Lower power and torque for the nano emulsified fuel • Higher cylinder pressure • Lower CO, HC and NO emission compared to diesel

(continued)

Table 7.7 (continued)

Investigator	Engine used	Pilot fuel	Experimental condition	Research findings
Chen et al. (2010)	QC495, Anhui Quan-Chai Group Corp., China four cylinders; four strokes; direct injection; water-cooled	Diesel, palm oil biodiesel, bio-oil as emulsifier	• Effect of 16% emulsifier on different biodiesel blends • Effect of palm oil biodiesel-diesel blends	• E16P20 fuel (16 vol% bio-solution, 20 vol% palm-biodiesel, 64 vol% diesel, an additional 1 vol% surfactant) showed optimum results • Fuel saving of 12.4% and reduced PM by 90.1%
Ramalingam et al. (2018)	4-stroke, single cylinder direct injection diesel engine, water cooled	Diesel and Borassus Flabellifer Oil (BFO)	• Effect of water emulsification • Effect of additive 1-ascorbic acid	• Enhanced BTE and BSEC for additive based BFO emulsion • HC and CO reduced by 34.92 and 22.22% • NOx and smoke reduced by 20.08% and 20.83% respectively for additive based BFO emulsified fuel
Ithnin et al. (2015)	4-Stroke, single cylinder, air-cooled, direct injection diesel engine	Low grade diesel	• Use of four different grades of water emulsion (5, 10, 15 and 20%)	• Maximum cylinder pressure and higher HRR for 20% water emulsifier • NOx and PM reduced by 41% and 35% respectively • CO is higher and CO_2 is lower under low load conditions

(continued)

Table 7.7 (continued)

Investigator	Engine used	Pilot fuel	Experimental condition	Research findings
Elsanusi et al. (2017)	HATZ 2G40, Four-stroke, air-cooled, double cylinder diesel engine	Diesel, canola biodiesel	• Effect of different biodiesel blend • Effect of different water content	• EGT reduced with emulsified fuel • BTE increased by 7% for EB40W15 emulsified fuel • NOx and smoke emission reduced for EB40W15 blend • HC emission reduced by 70% approximately for EB40W15
Fahd et al. (2013)	Toyota 2KD-FTV, 4 Cylinder, 16 valves DOHC turbo	Diesel	• Effect of 10% water emulsification • Effect of loading conditions	• Emulsified fuel showed lower BTE and CO emission at low loading conditions • Higher load resulted comparable BTE compared to diesel • Higher BSEC and lower NOx under all loading conditions
Raheman and Kumari (2014)	10.3 kW, 4-stroke, single cylinder, water cooled, direct injection diesel engine	Diesel, Jatropha biodiesel	• Effect of 10% and 15% water emulsion • Effect of loading conditions	• BTE reduced by 1–2% at low loads while increased by 3–4% at higher load for emulsified fuel • Ignition delay 2–3° higher for emulsified fuel • NOx, CO and HC emission lower by 2–28%, 31.5–51.5% and 30–50% at higher loads

(continued)

Table 7.7 (continued)

Investigator	Engine used	Pilot fuel	Experimental condition	Research findings
Kannan and Gounder (2011)	Kirloskar Diesel engine of AV1 model, four stroke, DI, naturally aspirated, water cooled engine	Diesel and *Thevetia peruviana*	• Effect of 5, 10, 15 and 20% water emulsion • Effect of different loading conditions	• 15% emulsified fuel highest BTE increase of 6.87% • NOx emission reduced by 41% and 38% for 20% and 15% emulsion respectively at full load • HC emission and smoke opacity reduced by 1.94% and 7.22% respectively for 15% emulsified biodiesel fuel
Bora and Saha (2016)	Single cylinder, NA, four stroke, DI, water-cooled, 3.5 kW VCR diesel engine	Diesel, Rice bran biodiesel and biogas	• Effect of water based emulsion (5%) • Effect of IT and CR	• Optimum results obtained at 18 CR and 29°BTDC IT • Higher CR and IT, higher BTE and lower BSEC • NOx emission increased while HC and CO emission reduced with higher CR and IT

with minimum or no modification to the engine hardware can address entire range of CI engines that are essential to tackle the emission problems. Among different fuel improvement techniques, biodiesel, alcohols, emulsification of fuels are found to be effective for reducing the emission levels with no penalty on the engine performance. The characteristics performance of different biofuels in IC engine application is summarized below.

Biodiesel

- Biodiesel and its blend produces slightly lower brake thermal efficiency and higher brake specific fuel consumption because of its lower heating value
- Biodiesel and its blend with diesel has early start of combustion and shorter ignition delay because of its higher cetane number.
- Due to lower compressibility and higher viscosity biodiesel results in higher peak pressure and lower heat release rate
- Enriched oxygen content of biodiesel leads to improved combustion which results in formation of higher NOx and CO_2 than diesel, while emission of CO, HC and smoke decreases.
- A biodiesel blend of 20% with diesel is found to be suitable for reducing NOx and PM emissions in CI engines.
- Absence of Sulphur and aromatic compounds in biodiesel results in no emission of sulphurous component.

Alcohols

- Alcohols, categorized as lower (methanol, ethanol) and higher alcohols (propanol, butanol) are the most widely used oxygenated fuels in CI engines.
- Methanol and ethanol along with its blend with diesel significantly reduces CO, HC, smoke and BTE with higher NOx emission. This is primarily due to high latent heat of vaporization and longer ignition delay when used in diesel engines. However, methanol/ethanol fumigation provides a suitable technique for reduction of NOx emissions.
- Lower alcohols such as methanol and diesel has lower cetane number. The use of CN improver such as 2-methylhexyl nitrate (EHN) $C_{10}H_4C_{13}$–NO_2 in these fuels results in lower NOx emission.
- Propanol and Butanol due to their higher percentage of premixed combustion results in higher cylinder pressure and an increases in BTE.
- NOx and soot emission decreases with the use of propanol and butanol in diesel engines. However, CO and HC emission increased with higher alcohol fuels. EGR is most prominent method used in higher alcohols for reduction in NOx emission.
- Lower alcohols have limitations of lower heating value, poor miscibility while higher alcohol has a higher cost of production. Hence alcohol blends of upto 15% are found to be suitable for engine performance and emission characteristics.

Emulsification

- Emulsification of diesel in another process of reduction in emissions of CI engines which utilizes three components water, oil, and surfactants mixed with high energy mechanical homogenizer and ultrasonicator.
- Non- ionic surfactants such as Span 80 and Tween 80 are suitable for the stability enhancement of water–oil emulsion.
- The stability of emulsion plays a key role in preparing a suitable water-diesel emulsion. An emulsion is unstable due tendency of liquid separation and as a result reduces its interfacial area with interfacial energy.
- The droplet size affects the viscosity of the emulsion. With smaller droplet size, the viscosity of the emulsion is higher, and as such the stability of the emulsion is more which increases the separation time
- Emulsification of biodiesel is suitable method for NOx and smoke emission in diesel engine. 5%, 10% and 15% water emulsification of biodiesel resulted in 25–30% reduction of NOx emission. Further with higher engine load comparable performance of engine has been observed.

In addition to the above mentioned characteristics of biofuels, the applicability of different oxygenated fuel technologies in CI engines depends on several factors such as cost of production of fuels, availability of suitable feedstock, emission reduction without any significant compensation of the performance of the engine. Further, these new biofuel technologies are a boon towards future engine technologies moving a step closer towards clean and sustainable energy utilization in CI engines.

References

Aasberg PK (2004) Synthesis gas production for FT synthesis. Stud Surf Sci Catal 152

Abdollahi M, Ghobadian B, Najafi G, et al (2020) Impact of water—biodiesel—diesel nano-emulsion fuel on performance parameters and diesel engine emission. Fuel 280. https://doi.org/10.1016/j.fuel.2020.118576

Rao VT, Rao PG, Reddy KHC (2008) Experimental investigation of pongamia, jatropha and neem methylester as biodiesel on CI engine. Jordan J Mech Ind 2:117–122

Agarwal AK, Dhar A (2013) Experimental investigations of performance, emission and combustion characteristics of Karanja oil blends fuelled DICI engine. Renew Energy 52:283–291. https://doi.org/10.1016/j.renene.2012.10.015

Agarwal AK, Agarwal RA, Gupta T, Gurjar BR (2017) Biofuels: technology, challenges and prospects. Springer

Agarwal AK, Gautam A, Sharma N, Singh AP (2019) Methanol and the alternate fuel economy. Springer

Agarwal D, Agarwal AK (2007) Performance and emissions characteristics of Jatropha oil (preheated and blends) in a direct injection compression ignition engine. Appl Therm Eng 27. https://doi.org/10.1016/j.applthermaleng.2007.01.009

Ajav EA, Singh B, Bhattacharya TK (1999) Experimental study of some performance parameters of a constant speed stationary diesel engine using ethanol-diesel blends as fuel. Biomass Bioenerg 17:357–365. https://doi.org/10.1016/S0961-9534(99)00048-3

Al-Hasan MI, Al-Momany M (2008) The effect of iso-butanol-diesel blends on engine performance. Transport 23:306–310

Ali MH, Mashud M, Rubel MR, Ahmad RH (2013) Biodiesel from Neem oil as an alternative fuel for diesel engine. Procedia Eng 56:625–630. https://doi.org/10.1016/j.proeng.2013.03.169
Al-Sabagh AM, Emara MM, Noor El-Din MR, Aly WR (2011) Formation of water-in-diesel oil nano-emulsions using high energy method and studying some of their surface active properties. Egypt J Pet 20:17–23. https://doi.org/10.1016/j.ejpe.2011.06.005
Al-Sabagh AM, Emara MM, Noor El-Din MR, Aly WR (2012) Water-in-diesel fuel nanoemulsions prepared by high energy: emulsion drop size and stability, and emission characteristics. J Surfactants Deterg 15:139–145. https://doi.org/10.1007/s11743-011-1291-2
Al-Sabagh AM, Kandile NG, Nasser NM et al (2013) Novel surfactants incorporated with 1,3,5-triethanolhexahydro-1,3,5-triazine moiety as corrosion inhibitors for carbon steel in hydrochloric acid: electrochemical and quantum chemical investigations. Egypt J Pet 22:351–365. https://doi.org/10.1016/j.ejpe.2013.10.004
Ana GREE, Udofia BG (2015) Characterization of Oil and Biodiesel Produced from *Thevetia peruviana* (Yellow Oleander) Seeds. Int J Sustain Green Energy 4:150–158. https://doi.org/10.11648/j.ijrse.20150404.14
Annamalai M, Dhinesh B, Nanthagopal K et al (2016) An assessment on performance, combustion and emission behavior of a diesel engine powered by ceria nanoparticle blended emulsified biofuel. Energy Convers Manag 123:372–380. https://doi.org/10.1016/j.enconman.2016.06.062
Arunprasad S, Balusamy T (2018) Experimental investigation on the performance and emission characteristics of a diesel engine by varying the injection pressure and injection timing using mixed biodiesel. Int J Green Energy 15:376–384. https://doi.org/10.1080/15435075.2018.1464923
AsfarKR A-R (2003) Fuel blends in compression ignition engines. American Institute of Aeronautics and Astronautics (AIAA2003-6027). In: Proceedings of the1st international energy conversion engineering conference, 17–21 Aug
Atabani AE, Silitonga AS, Badruddin IA et al (2012) A comprehensive review on biodiesel as an alternative energy resource and its characteristics. Renew Sustain Energy Rev 16:2070–2093. https://doi.org/10.1016/j.rser.2012.01.003
Atabani AE, Silitonga AS, Ong HC et al (2013) Non-edible vegetable oils: a critical evaluation of oil extraction, fatty acid compositions, biodiesel production, characteristics, engine performance and emissions production. Renew Sustain Energy Rev 18:211–245. https://doi.org/10.1016/j.rser.2012.10.013
Azev VS, Gerasimova GN, Luneva VV (1986) Formulation of stable mixtures of diesel fuels with methanol. Chem Technol Fuels Oils (Engl Transl) (United States) 21
Balamurugan T, Nalini R (2014) Experimental investigation on performance, combustion and emission characteristics of four stroke diesel engine using diesel blended with alcohol as fuel. Energy 78:356–363. https://doi.org/10.1016/j.energy.2014.10.020
Baltz RA, Burcham AF, Sitton OC, Book NL (1982) The recycle of sulfuric acid and xylose in the prehydrolysis of corn stover. Energy 7:259–265. https://doi.org/10.1016/0360-5442(82)90075-5
Balusamy T, Marappan R (2009) A comparative analysis on the performance and emission characteristics of *Thevetia peruviana* seed oil (TPSO) with other non-edible oil in CI engine. Int Energy J 10:93–100
Banapurmath NR, Tewari PG, Hosmath RS (2008) Performance and emission characteristics of a DI compression ignition engine operated on Honge, Jatropha and sesame oil methyl esters. Renew Energy 33:1982–1988. https://doi.org/10.1016/j.renene.2007.11.012
Banugopan VN, Prabhakar S, Annamalai K et al (2010) Experimental investigation on D.I. diesel engine fuelled by ethanol diesel blend with varying inlet air temperature. Proc Int Conf Front Automob Mech Eng—2010. FAME-2010 128–134. https://doi.org/10.1109/FAME.2010.5714809
Bilgin A, Durgun O, Şahin Z (2002) The effects of diesel-ethanol blends on diesel engine performance. Energy Sour 24:431–440. https://doi.org/10.1080/00908310252889933
Bora BJ, Saha UK (2016) Estimating the theoretical performance limits of a biogas powered dual fuel diesel engine using emulsified rice bran biodiesel as pilot fuel. J Energy Resour Technol Trans ASME 138:1–10. https://doi.org/10.1115/1.4031836

Cai L, Vom Lehn F, Pitsch H (2021) Higher alcohol and ether biofuels for compression-ignition engine application: a review with emphasis on combustion kinetics. Energy Fuels 35:1890–1917. https://doi.org/10.1021/acs.energyfuels.0c03590

CAN Ö, ÇELİKTEN İ, USTA N (2005) Effects of ethanol blended diesel fuel on exhaust emissions from a diesel engine. Pamukkale Univ J Eng Sci 11:219–224

Canakci M, Sayin C, Ozsezen AN, Turkcan A (2009) Effect of injection pressure on the combustion, performance, and emission characteristics of a diesel engine fueled with methanol-blended diesel fuel. Energy Fuels 23:2908–2920. https://doi.org/10.1021/ef900060s

Chauhan BS, Kumar N, Cho HM (2012) A study on the performance and emission of a diesel engine fueled with Jatropha biodiesel oil and its blends. Energy 37:616–622. https://doi.org/10.1016/j.energy.2011.10.043

Chauhan BS, Kumar N, Cho HM, Lim HC (2013) A study on the performance and emission of a diesel engine fueled with Karanja biodiesel and its blends. Energy 56:1–7. https://doi.org/10.1016/j.energy.2013.03.083

Chen KS, Lin YC, Te HL et al (2010) Saving energy and reducing pollution by use of emulsified palm-biodiesel blends with bio-solution additive. Energy 35:2043–2048. https://doi.org/10.1016/j.energy.2010.01.021

Chen G, Shen Y, Zhang Q et al (2013) Experimental study on combustion and emission characteristics of a diesel engine fueled with 2,5-dimethylfuran-diesel, n-butanol-diesel and gasoline-diesel blends. Energy 54:333–342. https://doi.org/10.1016/j.energy.2013.02.069

Cheung CS, Zhang ZH, Chan TL, Yao C (2009) Investigation on the effect of port-injected methanol on the performance and emissions of a diesel engine at different engine speeds. Energy Fuels 23:5684–5694. https://doi.org/10.1021/ef9005516

Cornils B (2004) Handbook of commercial catalysts. Heterogeneous catalysts. Von Howard F. Rase.

Debnath BK, Sahoo N, Saha UK (2013) Adjusting the operating characteristics to improve the performance of an emulsified palm oil methyl ester run diesel engine. Energy Convers Manag 69:191–198. https://doi.org/10.1016/j.enconman.2013.01.031

Debnath BK, Saha UK, Sahoo N (2015) A comprehensive review on the application of emulsions as an alternative fuel for diesel engines. Renew Sustain Energy Rev 42:196–211. https://doi.org/10.1016/j.rser.2014.10.023

Demirbas A (2008) Relationships derived from physical properties of vegetable oil and biodiesel fuels. Fuel 87:1743–1748. https://doi.org/10.1016/j.fuel.2007.08.007

Devan PK, Mahalakshmi NV (2009) A study of the performance, emission and combustion characteristics of a compression ignition engine using methyl ester of paradise oil-eucalyptus oil blends. Appl Energy 86:675–680. https://doi.org/10.1016/j.apenergy.2008.07.008

Doğan O (2011) The influence of n-butanol/diesel fuel blends utilization on a small diesel engine performance and emissions. Fuel 90:2467–2472. https://doi.org/10.1016/j.fuel.2011.02.033

Dorado MP, Ballesteros E, Arnal JM et al (2003) Exhaust emissions from a diesel engine fueled with transesterified waste olive oil. Fuel 82:1311–1315. https://doi.org/10.1016/S0016-2361(03)00034-6

Dürre P (2007) Biobutanol: an attractive biofuel. Biotechnol J Healthc Nutr Technol 2:1525–1534

Dürre P (2008) Fermentative butanol production: bulk chemical and biofuel. Ann N Y Acad Sci 1125:353–362

El-Din MRN, El-Hamouly SH MH et al (2013) Water-in diesel fuel nano emulsions: preparation, stability and physical properties. Egypt J Pet 22:517–530. https://doi.org/10.1016/j.ejpe.2013.11.006

El-Kasaby M, Nemit-Allah MA (2013) Experimental investigations of ignition delay period and performance of a diesel engine operated with Jatropha oil biodiesel. Alexandria Eng J 52:141–149. https://doi.org/10.1016/j.aej.2012.12.006

Elsanusi OA, Roy MM, Sidhu MS (2017) Experimental investigation on a diesel engine fueled by diesel-biodiesel blends and their emulsions at various engine operating conditions. Appl Energy 203:582–593. https://doi.org/10.1016/j.apenergy.2017.06.052

Emberson DR, Ihracska B, Imran S, Diez A (2016) Optical characterization of diesel and water emulsion fuel injection sprays using shadowgraphy. Fuel 172:253–262. https://doi.org/10.1016/j.fuel.2016.01.015

Tat ME, Van Gerpen JH, Wang PS (2004) Fuel property effects on injection timing, ignition timing and oxides of nitrogen emissions from biodiesel-fueled engines

Fahd MEA, Wenming Y, Lee PS et al (2013) Experimental investigation of the performance and emission characteristics of direct injection diesel engine by water emulsion diesel under varying engine load condition. Appl Energy 102:1042–1049. https://doi.org/10.1016/j.apenergy.2012.06.041

Ganesan S, Sivasubramaninan R, Sajin JB et al (2019) Performance and emission study on the effect of oxygenated additive in neat biodiesel fueled diesel engine. Energy Sour Part A Recover Util Environ Eff 41:2017–2027. https://doi.org/10.1080/15567036.2018.1549148

Giuffrè AM, Capocasale M, Zappia C, Poiana M (2017) Biodiesel from tomato seed oil: transesterification and characterisation of chemical-physical properties. Agron Res 15:133–143

Godiganur S, Suryanarayana Murthy CH, Reddy RP (2009) 6BTA 5.9 G2–1 Cummins engine performance and emission tests using methyl ester mahua (*Madhuca indica*) oil/diesel blends. Renew Energy 34:2172–2177. https://doi.org/10.1016/j.renene.2008.12.035

Gogoi TK, Baruah DC (2011) The use of Koroch seed oil methyl ester blends as fuel in a diesel engine. Appl Energy 88:2713–2725. https://doi.org/10.1016/j.apenergy.2011.02.023

Gopidesi RK, Rajaram PS (2019) A review on emulsified fuels and their application in diesel engine. Int J Ambient Energy 0:1–9. https://doi.org/10.1080/01430750.2019.1667435

Gu X, Li G, Jiang X et al (2012) Experimental study on the performance of and emissions from a low-speed light-duty diesel engine fueled with n-butanol–diesel and isobutanol–diesel blends. Proc Inst Mech Eng Part D J Automob Eng 227:261–271. https://doi.org/10.1177/0954407012453231

Hagos FY, Aziz ARA, Tan IM (2011) Water-in-diesel emulsion and its micro-explosion phenomenon-review. 2011 IEEE 3rd Int Conf Commun Softw Netw ICCSN 2011:314–318. https://doi.org/10.1109/ICCSN.2011.6014903

Hagos FY, Ali OM, Mamat R, Abdullah AA (2017) Effect of emulsification and blending on the oxygenation and substitution of diesel fuel for compression ignition engine. Renew Sustain Energy Rev 75:1281–1294. https://doi.org/10.1016/j.rser.2016.11.113

Haiter L, Ravi R, Arumugham S, Thyagarajan K (2012) Performance, emission and combustion evaluation of diesel engine using methyl esters of Mahua oil. Int J Environ Sci 3:639–649. https://doi.org/10.6088/ijes.2012030131062

Hansen AC, Zhang Q, Lyne PWL (2005) Ethanol-diesel fuel blends—a review. Bioresour Technol 96:277–285. https://doi.org/10.1016/j.biortech.2004.04.007

Harvey BG, Meylemans HA (2011) The role of butanol in the development of sustainable fuel technologies. J Chem Technol Biotechnol 86:2–9

Hasannuddin AK, Wira JY, Sarah S et al (2016) Performance, emissions and lubricant oil analysis of diesel engine running on emulsion fuel. Energy Convers Manag 117:548–557. https://doi.org/10.1016/j.enconman.2016.03.057

He BQ (2016) Advances in emission characteristics of diesel engines using different biodiesel fuels. Renew Sustain Energy Rev 60:570–586. https://doi.org/10.1016/j.rser.2016.01.093

Huang ZH, Lu HB, Jiang DM et al (2004) Engine performance and emissions of a compression ignition engine operating on the diesel-methanol blends. Proc Inst Mech Eng Part D J Automob Eng 218:435–447

Huang J, Wang Y, Li S et al (2009) Experimental investigation on the performance and emissions of a diesel engine fuelled with ethanol-diesel blends. Appl Therm Eng 29:2484–2490. https://doi.org/10.1016/j.applthermaleng.2008.12.016

Ithnin AM, Ahmad MA, Bakar MAA et al (2015) Combustion performance and emission analysis of diesel engine fuelled with water-in-diesel emulsion fuel made from low-grade diesel fuel. Energy Convers Manag 90:375–382. https://doi.org/10.1016/j.enconman.2014.11.025

Janousek GS (2010) Evaluation of ethanol and water introduction via fumigation on efficiency and emissions of a compression ignition engine using an atomization technique. 124

Jhalani A, Sharma D, Soni SL et al (2019) A comprehensive review on water-emulsified diesel fuel: chemistry, engine performance and exhaust emissions. Environ Sci Pollut Res 26:4570–4587. https://doi.org/10.1007/s11356-018-3958-y

Jindal S, Nandwana BP, Rathore NS, Vashistha V (2010) Experimental investigation of the effect of compression ratio and injection pressure in a direct injection diesel engine running on Jatropha methyl ester. Appl Therm Eng 30:442–448. https://doi.org/10.1016/j.applthermaleng.2009.10.004

Jindal S (2011) Experimental investigation of the effect of compression ratio and injection pressure in a direct injection diesel engine running on Karanj methyl ester. Int J Sustain Energy 30. https://doi.org/10.1080/14786451.2011.598936

Jyothi US, Gopinath L, Shankarachary B (2019) Effect of oxygenated and metal-based additives on performance, combustion and emission characteristics of Mahua bio-diesel at optimal blend. AIP Conf Proc 2200. https://doi.org/10.1063/1.5141199

Kannan U, Udayakumar M (2009) NOx and HC emission control using water emulsified diesel in single cylinder diesel engine. J Eng Appl Sci 4:59–62

Kannan TK, Rakkiyanna Gounder M (2011) The *vetia peruviana* biodiesel emulsion used as a fuel in a single cylinder diesel engine reduces NOx and smoke. Therm Sci 15:1185–1191. https://doi.org/10.2298/TSCI100810045K

Karabektas M, Hosoz M (2009) Performance and emission characteristics of a diesel engine using isobutanol-diesel fuel blends. Renew Energy 34:1554–1559. https://doi.org/10.1016/j.renene.2008.11.003

Khalid N, Kobayashi I, Neves MA et al (2018) Microchannel emulsification: a promising technique towards encapsulation of functional compounds. Crit Rev Food Sci Nutr 58:2364–2385. https://doi.org/10.1080/10408398.2017.1323724

Khalife E, Tabatabaei M, Demirbas A, Aghbashlo M (2017) Impacts of additives on performance and emission characteristics of diesel engines during steady state operation. Prog Energy Combust Sci 59:32–78. https://doi.org/10.1016/j.pecs.2016.10.001

Kim HN, Choi BC (2008) Effect of ethanol-diesel blend fuels on emission and particle size distribution in a common-rail direct injection diesel engine with warm-up catalytic converter. Renew Energy 33:2222–2228. https://doi.org/10.1016/j.renene.2008.01.002

Kim S, Dale BE (2005) Life cycle assessment of various cropping systems utilized for producing biofuels: bioethanol and biodiesel. Biomass Bioenerg 29:426–439. https://doi.org/10.1016/j.biombioe.2005.06.004

Kishore Pandian A, Munuswamy DB, Radhakrishana S et al (2017) Influence of an oxygenated additive on emission of an engine fueled with neat biodiesel. Pet Sci 14:791–797. https://doi.org/10.1007/s12182-017-0186-x

Kouremenos DA, Rakopoulos CD, Kotsos KG (1992) A stochastic-experimental investigation of the cyclic pressure variation in a di single-cylinder diesel engine. Int J Energy Res 16:865–877

Kullolli S, Sakthivel G, Ilangkumaran M (2016) A neural network model for the prediction of compression ignition engine performance at different injection timings. Int J Ambient Energy 37:227–236. https://doi.org/10.1080/01430750.2014.931297

Kumar H, Kumar V (2018) Ultrasonication assisted formation and stability of water-in-oil nanoemulsions: optimization and ternary diagram analysis. Ultrason Sonochem 49:79–88. https://doi.org/10.1016/j.ultsonch.2018.07.022

Kumar S, Cho JH, Park J, Moon I (2013) Advances in diesel-alcohol blends and their effects on the performance and emissions of diesel engines. Renew Sustain Energy Rev 22:46–72. https://doi.org/10.1016/j.rser.2013.01.017

Kumar C, Athawe M, Aghav YV et al (2007) Effects of ethanol addition on performance, emission and combustion of DI diesel engine running at different injection pressures. SAE Techn Paper

Kumar Patidar S, Raheman H (2020) Performance and durability analysis of a single-cylinder direct injection diesel engine operated with water emulsified biodiesel-diesel fuel blend. Fuel 273:117779. https://doi.org/10.1016/j.fuel.2020.117779

Laza T, Bereczky Á (2009) Influence of higher alcohols on the combustion pressure of diesel engine operated with rape seed oil. Acta Mech Slovaca 13:54–61

Lei J, Shen L, Bi Y, Chen H (2012) A novel emulsifier for ethanol-diesel blends and its effect on performance and emissions of diesel engine. Fuel 93:305–311. https://doi.org/10.1016/j.fuel.2011.06.013

Lin CY, Chen LW (2008) Comparison of fuel properties and emission characteristics of two- and three-phase emulsions prepared by ultrasonically vibrating and mechanically homogenizing emulsification methods. Fuel 87:2154–2161. https://doi.org/10.1016/j.fuel.2007.12.017

Liu S, Zhu Z, Zhang Z et al (2010) Effect of a cetane number (CN) improver on combustion and emission characteristics of a compression-ignition (CI) engine fueled with an ethanol-diesel blend. Energy Fuels 24:2449–2454. https://doi.org/10.1021/ef901543m

Liu K, Atiyeh HK, Stevenson BS et al (2014) Continuous syngas fermentation for the production of ethanol, n-propanol and n-butanol. Bioresour Technol 151:69–77. https://doi.org/10.1016/j.biortech.2013.10.059

Lthnin AM, Ahmad MA, Bakar MAA (2015) Combustion performance and emission analysis of diesel engine fuelled with water-in-diesel emulsion fuels made from low grade diesel fuel. Energ Convers Manag 90:375–382

Lü XC, Yang JG, Zhang WG, Huang Z (2005) Improving the combustion and emissions of direct injection compression ignition engines using oxygenated fuel additives combined with a cetane number improver. Energy Fuels 19:1879–1888. https://doi.org/10.1021/ef0500179

Ma X, Zhang F, Han K et al (2014) Effects of intake manifold water injection on combustion and emissions of diesel engine. Energy Procedia 61:777–781. https://doi.org/10.1016/j.egypro.2014.11.963

Maawa WN, Mamat R, Najafi G, De Goey LPH (2020) Performance, combustion, and emission characteristics of a CI engine fueled with emulsified diesel-biodiesel blends at different water contents. Fuel 267. https://doi.org/10.1016/j.fuel.2020.117265

Mattiello M, Cosmai L, Pistone L et al (1992) Experimental evidence for microexplosions in water/fuel oil emulsion flames inferred by laser light scattering. Symp Combust 24:1573–1578. https://doi.org/10.1016/S0082-0784(06)80183-9

Mehta RN, More U, Malek N et al (2015) Study of stability and thermodynamic properties of water-in-diesel nano emulsion fuels with nano-Al additive. Appl Nanosci 5:891–900. https://doi.org/10.1007/s13204-014-0385-3

Mofijur M, Masjuki HH, Kalam MA, Atabani AE (2013) Evaluation of biodiesel blending, engine performance and emissions characteristics of Jatropha curcas methyl ester: Malaysian perspective. Energy 55:879–887. https://doi.org/10.1016/j.energy.2013.02.059

Mondal PK, Mandal BK (2019) A comprehensive review on the feasibility of using water emulsified diesel as a CI engine fuel. Fuel 237:937–960. https://doi.org/10.1016/j.fuel.2018.10.076

Moses CA, Ryan TW, Likos WE (1980) Experiments with alcohol/diesel fuel blends in compression-ignition engines. In: VI international symposium on alcohol fuels technology, Guaruja, Brazil

Murugesan A, Umarani C, Subramanian R, Nedunchezhian N (2009) Bio-diesel as an alternative fuel for diesel engines—a review. Renew Sustain Energy Rev 13:653–662. https://doi.org/10.1016/j.rser.2007.10.007

Najafi G, Yusaf TF (2009) Experimental investigation of using methanol-diesel blended fuels in diesel engine. Proc Fourth Int Conf Therm Eng Theory Appl 1–5

Nalgundwar A, Paul B, Sharma SK (2016) Comparison of performance and emissions characteristics of di CI engine fueled with dual biodiesel blends of palm and Jatropha. Fuel 173:172–179. https://doi.org/10.1016/j.fuel.2016.01.022

No SY (2011) Inedible vegetable oils and their derivatives for alternative diesel fuels in CI engines: a review. Renew Sustain Energy Rev 15:131–149. https://doi.org/10.1016/j.rser.2010.08.012

Noh HK, No SY (2017) Effect of bioethanol on combustion and emissions in advanced CI engines: HCCI, PPC and GCI mode—a review. Appl Energy 208:782–802. https://doi.org/10.1016/j.apenergy.2017.09.071

Ogunkoya D, Li S, Rojas OJ, Fang T (2015) Performance, combustion, and emissions in a diesel engine operated with fuel-in-water emulsions based on lignin. Appl Energy 154:851–861. https://doi.org/10.1016/j.apenergy.2015.05.036

Onoji SE, Iyuke SE, Igbafe AI, Nkazi DB (2016) Rubber seed oil: a potential renewable source of biodiesel for sustainable development in sub-Saharan Africa. Energy Convers Manag 110:125–134. https://doi.org/10.1016/j.enconman.2015.12.002

Ozsezen A, Turkcan A, Sayin C, Canakci M (2011) Comparison of performance and combustion parameters in a heavy-duty diesel engine fueled with iso-butanol/diesel fuel blends. Energy Explor Exploit 29:525–541. https://doi.org/10.1260/0144-5987.29.5.525

Patil PD, Deng S (2009) Optimization of biodiesel production from edible and non-edible vegetable oils. Fuel 88:1302–1306. https://doi.org/10.1016/j.fuel.2009.01.016

Paul G, Datta A, Mandal BK (2014) An experimental and numerical investigation of the performance, combustion and emission characteristics of a diesel engine fueled with Jatropha biodiesel. Energy Procedia 54:455–467. https://doi.org/10.1016/j.egypro.2014.07.288

Puhan S, Vedaraman N, Rambrahamam BV, Nagarajan G (2005a) Mahua (*Madhuca indica*) seed oil: a source of renewable energy in India. J Sci Ind Res (india) 64:890–896

Puhan S, Vedaraman N, Sankaranarayanan G, Ram BVB (2005b) Performance and emission study of Mahua oil (madhuca indica oil) ethyl ester in a 4-stroke natural aspirated direct injection diesel engine. Renew Energy 30:1269–1278. https://doi.org/10.1016/j.renene.2004.09.010

Pulkrabek WW (2003) Engineering fundamentals of the internal combustion engine, 2nd edn. Pearson

Qureshi N, Saha BC, Dien B et al (2010) Production of butanol (a biofuel) from agricultural residues: Part I—use of barley straw hydrolysate. Biomass Bioenerg 34:559–565. https://doi.org/10.1016/j.biombioe.2009.12.024

Raheman H, Ghadge SV (2008) Performance of diesel engine with biodiesel at varying compression ratio and ignition timing. Fuel 87:2659–2666. https://doi.org/10.1016/j.fuel.2008.03.006

Raheman H, Kumari S (2014) Combustion characteristics and emissions of a compression ignition engine using emulsified Jatropha biodiesel blend. Biosyst Eng 123:29–39. https://doi.org/10.1016/j.biosystemseng.2014.05.001

Rahman KM, Mashud M, Roknuzzaman M, Al GA (2010) Biodiesel from Jatropha oil as an alternative fuel for diesel engine. Int J Mech Mech Eng 10:1–6

Rajan K, Senthil Kumar KR (2020) Experimental study on diesel engine working characteristics using yellow oleander biodiesel with the effect of different injection timings. Energy Sour Part A Recover Util Environ Eff 00:1–14. https://doi.org/10.1080/15567036.2020.1722295

Rajesh Kumar B, Saravanan S (2016) Use of higher alcohol biofuels in diesel engines: a review. Renew Sustain Energy Rev 60:84–115. https://doi.org/10.1016/j.rser.2016.01.085

Rakopoulos CD, Antonopoulos KA, Rakopoulos DC (2007) Experimental heat release analysis and emissions of a HSDI diesel engine fueled with ethanol-diesel fuel blends. Energy 32:1791–1808. https://doi.org/10.1016/j.energy.2007.03.005

Rakopoulos DC, Rakopoulos CD, Giakoumis EG et al (2010) Effects of butanol-diesel fuel blends on the performance and emissions of a high-speed di diesel engine. Energy Convers Manag 51:1989–1997. https://doi.org/10.1016/j.enconman.2010.02.032

Rakopoulos DC, Rakopoulos CD, Papagiannakis RG, Kyritsis DC (2011a) Combustion heat release analysis of ethanol or n-butanol diesel fuel blends in heavy-duty di diesel engine. Fuel 90:1855–1867. https://doi.org/10.1016/j.fuel.2010.12.003

Rakopoulos CD, Rakopoulos DC, Giakoumis EG, Kyritsis DC (2011b) The combustion of n-butanol/diesel fuel blends and its cyclic variability in a direct injection diesel engine. Proc Inst Mech Eng Part A J Power Energy 225:289–308

Ramalingam S, Rajendran S, Ganesan P, Govindasamy M (2018) Effect of operating parameters and antioxidant additives with biodiesels to improve the performance and reducing the emissions

in a compression ignition engine—a review. Renew Sustain Energy Rev 81:775–788. https://doi.org/10.1016/j.rser.2017.08.026
Roy S, Zare S, Askari O (2019) Understanding the effect of oxygenated additives on combustion characteristics of gasoline. J Energy Resour Technol Trans ASME 141:1–10. https://doi.org/10.1115/1.4041316
Roy A, Dabhi Y, Brahmbhatt H, Chourasia SK (2021) Effect of emulsified fuel based on dual blend of Castor-Jatropha biodiesel on CI engine performance and emissions. Alexandria Eng J 60:1981–1990. https://doi.org/10.1016/j.aej.2020.12.003
Sadhik Basha J (2018) Impact of carbon nanotubes and di-ethyl ether as additives with biodiesel emulsion fuels in a diesel engine—an experimental investigation. J Energy Inst 91:289–303. https://doi.org/10.1016/j.joei.2016.11.006
Sakthivel R, Ramesh K, Purnachandran R, Mohamed Shameer P (2018) A review on the properties, performance and emission aspects of the third generation biodiesels. Renew Sustain Energy Rev 82:2970–2992. https://doi.org/10.1016/j.rser.2017.10.037
Saravanan N, Nagarajan G, Puhan S (2010) Experimental investigation on a DI diesel engine fuelled with Madhuca Indica ester and diesel blend. Biomass Bioenerg 34:838–843. https://doi.org/10.1016/j.biombioe.2010.01.028
Sayin C (2010) Engine performance and exhaust gas emissions of methanol and ethanol-diesel blends. Fuel 89:3410–3415. https://doi.org/10.1016/j.fuel.2010.02.017
Sayin C, Ilhan M, Canakci M, Gumus M (2009) Effect of injection timing on the exhaust emissions of a diesel engine using diesel-methanol blends. Renew Energy 34:1261–1269. https://doi.org/10.1016/j.renene.2008.10.010
Sayin C, Ozsezen AN, Canakci M (2010) The influence of operating parameters on the performance and emissions of a DI diesel engine using methanol-blended-diesel fuel. Fuel 89:1407–1414. https://doi.org/10.1016/j.fuel.2009.10.035
Sczomak DP, Henein NA (1979) Cycle-to-cycle variation with low ignition quality fuels in a CFR diesel engine. SAE Trans 3124–3144
Shen L, Lei J, Bi Y (2011) Performance and emission characteristics of diesel engine fueled with ethanol-diesel blends in different altitude regions. J Biomed Biotechnol 2011. https://doi.org/10.1155/2011/417421
Shen CR, Liao JC (2011) A synthetic iterative pathway for keto acid elongation. Methods Enzymol 497:469–481
Singh AK, Mishra C, Vibhanshu V, Kumar N (2013) Performance evaluation and emission studies of a single cylinder diesel engine fuelled with isopropyl alcohol and diesel. SAE Techn Paper
Solans C, Solé I (2012) Nano-emulsions: Formation by low-energy methods. Curr Opin Colloid Interface Sci 17:246–254. https://doi.org/10.1016/j.cocis.2012.07.003
Song R, Liu J, Wang L, Liu S (2008) Performance and emissions of a diesel engine fueled with methanol. Energy Fuels 22:3883–3888
Srikanth HV, Godiganur S, Manne B, Bharat Kumar (2021) Niger seed oil biodiesel as an emulsifier in diesel-ethanol blends for compression ignition engine. Renew Energy 163:1467–1478. https://doi.org/10.1016/j.renene.2020.07.010
Srirangan K, Liu X, Westbrook A et al (2014) Biochemical, genetic, and metabolic engineering strategies to enhance coproduction of 1-propanol and ethanol in engineered *Escherichia coli*. Appl Microbiol Biotechnol 98:9499–9515
Srivastava DK, Agarwal AK, Datta A, Maurya RK (2018) Advances in internal combustion engine research. Springer
Stępień Z, Żak G, Markowski J, Wojtasik M (2021) Investigation into the impact of the composition of ethanol fuel deposit control additives on their effectiveness. Energies 14. https://doi.org/10.3390/en14030604
Suresh V, Amirthagadeswaran KS (2015) Combustion and performance characteristics of water-in-diesel emulsion fuel. Energy Sour Part A-Recovery Util Environ Eff 37:2020–2028. https://doi.org/10.1080/15567036.2015.1072605

Sureshkumar K, Velraj R, Ganesan R (2008) Performance and exhaust emission characteristics of a CI engine fueled with Pongamia pinnata methyl ester (PPME) and its blends with diesel. Renew Energy 33:2294–2302. https://doi.org/10.1016/j.renene.2008.01.011
Tamilselvan P, Nallusamy N, Rajkumar S (2017) A comprehensive review on performance, combustion and emission characteristics of biodiesel fuelled diesel engines. Renew Sustain Energy Rev 79:1134–1159. https://doi.org/10.1016/j.rser.2017.05.176
Tsang KS, Zhang ZH, Cheung CS, Chan TL (2010) Reducing emissions of a diesel engine using fumigation ethanol and a diesel oxidation catalyst. Energy Fuels 24:6156–6165. https://doi.org/10.1021/ef100899z
Tziourtzioumis DN, Stamatelos AM (2017) Experimental investigation of the effect of biodiesel blends on a DI diesel engine's injection and combustion. Energies 10. https://doi.org/10.3390/en10070970
Valentino G, Corcione FE, Iannuzzi SE (2012) Effects of gasoline–diesel and n-butanol–diesel blends on performance and emissions of an automotive direct-injection diesel engine. Int J Engine Res 13:199–215. https://doi.org/10.1177/1468087412441879
Vallinayagam R, Vedharaj S, Yang WM et al (2014) Pine oil-biodiesel blends: A double biofuel strategy to completely eliminate the use of diesel in a diesel engine. Appl Energy 130:466–473. https://doi.org/10.1016/j.apenergy.2013.11.025
Vasudevan P, Sharma SK (2005) Liquid fuel from biomass: an overview. J Sci Indus Res 64:822–83
Vellaiyan S, Amirthagadeswaran KS (2016) The role of water-in-diesel emulsion and its additives on diesel engine performance and emission levels: a retrospective review. Alexandria Eng J 55:2463–2472. https://doi.org/10.1016/j.aej.2016.07.021
Vellaiyan S, Amirthagadeswaran KS (2020) Compatibility test in a CI engine using lemon peel oil and water emulsion as fuel. Fuel 279:118520. https://doi.org/10.1016/j.fuel.2020.118520
Verma G, Nigam P, Singh D, Chaudhary K (2000) Bioconversion of starch to ethanol in a single-step process by coculture of amylolytic yeasts and *Saccharomyces cerevisiae* 21. Bioresour Technol 72:261–266. https://doi.org/10.1016/S0960-8524(99)00117-0
Watanabe H, Okazaki K (2013) Visualization of secondary atomization in emulsified-fuel spray flow by shadow imaging. Proc Combust Inst 34:1651–1658. https://doi.org/10.1016/j.proci.2012.07.005
Wing RD (1975) The rotary fuel-injection pump as a source of cyclic variation in diesel engines, and its effect on nitric oxide emissions. Proc Inst Mech Eng 189:497–505
Yadav AK, Khan ME, Pal A, Dubey AM (2018) Performance, emission and combustion characteristics of an Indica diesel engine operated with Yellow Oleander (*Thevetia peruviana*) oil biodiesel produced through hydrodynamic cavitation method. Int J Ambient Energy 39:365–371. https://doi.org/10.1080/01430750.2017.1303631
Yahaya Khan M, Abdul Karim ZA, Hagos FY et al (2014) Current trends in water-in-diesel emulsion as a fuel. Sci World J 2014. https://doi.org/10.1155/2014/527472
Yao M, Zhang Q, Liu H, et al (2010) Diesel engine combustion control: medium or heavy EGR? SAE Techn Paper
Zhang Z, Kang Z, Jiang L et al (2017) Effect of direct water injection during compression stroke on thermal efficiency optimization of common rail diesel engine. Energy Procedia 142:1251–1258. https://doi.org/10.1016/j.egypro.2017.12.514
Zheng Z, Li C, Liu H et al (2015) Experimental study on diesel conventional and low temperature combustion by fueling four isomers of butanol. Fuel 141:109–119. https://doi.org/10.1016/j.fuel.2014.10.053

Part IV
Carbon Neutrality

Chapter 8
Future Sustainable Transport Fuels for Indian Heavy Duty Vehicles

Subhanker Dev

Abstract Heavy duty vehicles fitted with internal combustion engines have been instrumental in helping human for meeting their economic and social goals. In the pursuit we have put enormous burden on our environment. And now we are facing the challenges to limit and reduce the burden on environment over a period of time. Presently, there are plethora of technology concepts which are worldwide discussed and projected as a sustainable solution for future heavy duty transportation. However, there are number of considerations for a nation like India that needs careful analysis before we adopt a particular transport solution. There is no denying the fact that eventually we have to make heavy duty vehicles fossil and emission free for reducing their impact on environment to minimum. But this would happen over a period of time. Now given the large number of heavy duty vehicles on road and new vehicles that are forecasted to be produced, combustion engines will be around for another two decades. Meeting the goal of net zero GHG emission by 2050 requires number of parallel approaches. Hence, we have to work on various fronts, like related to combustion engines we need to put effort on improving newly developed heavy duty engine efficiencies, improving existing fleet emission, introducing low carbon fuels like natural gas and carbon neutral bio fuels and drop-in fuels. Simultaneously we also have to progress on the long term new developments like electrification of urban transport and hydrogen fuel cell truck for long haul operations. In this regard, we will briefly discuss here the several ways to make transportation in India more environment benign, role and potential of several prospective bio-fuels and various base and advanced engine technologies to improve the carbon footprint of existing and new developed fleet of heavy duty vehicles.

Keywords Heavy duty vehicles · Internal combustion engines · Low carbon fuels · Methanol · Ethanol · DME · HVO · Synthetic fuels (GTL/BTL) · Retro-fitment · Direct injection technologies · Advanced combustion and engine technologies

---

S. Dev (✉)
The Automotive Research Association of India (ARAI), Pune, India
e-mail: dev.edl@araiindia.com

A. V. Agarwal and H. Valera (eds.), *Potential and Challenges of Low Carbon Fuels for Sustainable Transport*, Energy, Environment, and Sustainability,
https://doi.org/10.1007/978-981-16-8414-2_8

8.1 Introduction

8.1.1 *Internal Combustion Engines*

Heavy duty vehicle (HDV) fitted with internal combustion engines have played a key role in our shaping our civilization in the way we wanted (Agarwal et al. 2021). The need for transportation, power generation and other industrial activity has emitted enormous amount of greenhouse gas emissions (GHG) and air pollutants in our planet. Considering the present problems and challenges discussed below, there are number of technology approaches country should consider implementing for heavy duty transportation. Further talking about the regulations there is a need to implement fuel efficiency regulations for heavy duty vehicles, as per a release (Air Pollution Press Release 2021) it is expected by Jan 2022.

8.1.2 *Present Problems and Challenges*

- *Global warming* and its adverse effects are already being witnessed in many part of world and its future potential threat is now becoming a serious concern for everyone. As described in Beneath the waves "Norway and CO2 Emission", according to IPCC we can only emit approximately 330 Gtons of CO_2 for limiting the rise in global temperature to stay below 1.5 °C. Now considering the present annual GHG emission of 42 Gtons, we may exhaust our budgetary CO_2 by 2028, if we continue at same rate. However, if we refer the trend for the annual increase in CO_2 emission each year for the past 50 years (https://ourworldindata.org/co2-emissions) then considering the same business as usual case we will definitely cross the limit of 330 Gtons even earlier than 2028. Hence, worldwide many agreements and targets are being signed to not only avoid such a situation but also to achieve ambitious target of net zero CO_2 emissions by 2050.
- *Air quality* another serious issue in India. As per a report from Ponskshe (2020), 17 Indian cities ranked among world most polluted cities and significant proportion of air pollution is caused by vehicles especially in big cities. Almost 30–40% of the total human made sources of $PM_{2.5}$ is attributed to vehicular emission for big cities.
- *Energy security*, as per a press release (Air Pollution Press Release 2021) the energy demand for road transport is projected to double over next two decades for India and also oil demand is expected to increase by almost four million barrels per day in 2040—largest increase for any country.

8.1.3 Different Approaches Toward CO_2 and Air Pollution Mitigation

Given the pace at which the world economy is growing we need to work on multiple approaches to not only mitigate the CO_2 and air pollution problems but also embrace a sustainable transport solution over a time. There are number of studies (Beneath the waves "Norway and CO_2 Emission"; Decarbonising the Indian Transport Sector Pathways and Policies 2020; Fast Tracking Freight in India 2021) and forecast which proposed that we have to have multiple strategies to achieve the target of net zero CO_2 footprints by 2050. Some of these are discussed here:

- *Urban mobility*, urbanization is a leading trend which many developing countries have been seeing. Though for country like India presently major part of population lives in rural, will be seeing a trend toward shifting to cities. This will only strain the existing infrastructures and resources of these cities. Intensive urban planning would be required in the direction of waste management, for urban mobility making public transport more approachable, efficient and cleaner, developing urban infrastructure, moving to connected and shared mobility etc. would be a wiser step.
- *Transport modal shift* will be required particularly for freight transportation. Presently road transport is responsible for major freight transportation in India. As per Improving fuel efficiency for heavy duty vehicles (2019), Karali and Gopal (2017), India's contribution to total world GHG emissions are around 7% i.e. ~3 Gtons, of all the sectors transport contribution is ~20% i.e. 600 million tons, road transport contributes to ~90% of transport related CO_2 emission i.e. 540 million tons. The HDV accounts for ~15% i.e. 80 million tons of CO_2 emission and HDV above 12 ton accounts for 60% of total fuel usage and 60% of CO_2 emissions of total HDV fleet i.e. 50 million tons of CO_2 emissions. As mentioned by Fast Tracking Freight in India (2021), increase of rail mode and intermodal transport including waterway etc. would be needed in coming future as one of the opportunity to reduced carbon emissions, reduced logistic cost and reduced air pollution.
- *Freight efficiency* typically defined as payload multiplied by fuel economy and is expressed as ton-mpg or ton-kmpl (Stanton et al. 2013). Though as mentioned by Stanton et al. (2013), freight efficiency has significant relationship with engine brake thermal efficiency (Stanton et al. 2013), improving freight carrying capacity (Fast Tracking Freight in India 2021), improving daily utilization (Fast Tracking Freight in India 2021), reducing empty running (Fast Tracking Freight in India 2021, optimizing load factors (Fast Tracking Freight in India 2021) etc. can also lead to improving the freight efficiency. As per Fast Tracking Freight in India (2021), older trucks which predominates in India use outdated technologies and have less efficient drive trains are one of the major reasons for lower inefficiencies of vehicles. Talking about engine efficiency, improving the engine brake thermal efficiency is one of the core of this chapter and hence is discussed later in the chapter.

- *Electrification* more importantly renewable based electricity use for transportation is the ultimate sustainable solution. However, at present with major chunk of electricity generated in India from coal, issues of limited vehicle range, limited charging infrastructure, longer charging time and high fixed cost electrification can be a solution adopted for urban transport with government participation and measures.
- *Cleaner fuels* like natural gas (compressed-CNG/liquefied-LNG), bio fuels like (ethanol and methanol), DME (Di-Methyl Ether) and drop-in fuels (synthetic diesel and HVO) can be adopted long way to replace fossil fuels and help mitigate the CO_2 footprints and air pollution. Though stringent emission regulations like BS VI have controlled the new vehicular emissions but there is still enormous amount of work needed to be done for the existing and aged fleet of HDV in India. According to a report (Centre for Science and Environment (CSE) Report 2020), total number of end-of-life commercial vehicles (both goods and passenger) would be 13 lakh by 2025. In India total number of heavy duty vehicles with GVW 3.5–12 tonne are 48 lakh and out of that vehicles above 12 tonne are 28 lakh for year 2014 (Improving fuel efficiency for heavy duty vehicles of 3.5–12 tonne in India 2019; Karali and Gopal 2017). Hence the number of end-of-life vehicles for commercial use is significant. There are also other predictions (Joshi 2021), which says that since vehicle lifetime is over a decade even with full electrification by 2040 there will be significant number of vehicles powered by an internal combustion engine beyond 2050, next report by The Energy Transformation (www.shell.com) says oil to remain as largest source of energy in transport beyond 2050. Hence it would be effective on large scale if vehicles approaching end-of-life or for that matter even existing vehicles can be made to use the synthetic fuels or drop-in fuels and can contribute significantly toward reducing carbon footprints.

8.2 Prospective Transport Solution for India

In the previous sections we have discussed the critical issues like global warming and air quality for transport sector and some of the important approaches needed for the transport sector to address these issues in true sense. Elaborating all the approaches would be out of scope of this chapter hence we will discuss here the scope of improving freight efficiency by optimizing base engine technologies, by adopting cleaner low carbon fuels like natural gas or carbon neutral bio-fuels or drop-in fuels, by retro-fitting older and aged vehicles and advanced engine combustion technologies for new heavy duty vehicles:

- Existing heavy duty vehicles
  - Natural gas

    Retro-fitment technology (dual fuel combustion)

  - Drop-in fuels

    Synthetic-Diesel
    HVO (hydrotreated vegetable oil)

- New heavy duty vehicles
  - Engine technologies
  - Natural gas with advanced combustion

    Dual fuel combustion
    Direct injection—(CNG/LNG)

  - Bio-fuels with advanced combustion

    Methanol
    Ethanol.

Additionally very importantly we need to progress on the quick implementation of fuel economy and CO_2 standards as mentioned in Sect. 8.1.1. Though in India we have taken the big step in form of BS VI to curb the vehicular emissions however India still has to witness the measures to improve fuel efficiency and CO_2 emissions of heavy duty vehicles (Banerjee 2021).

8.2.1 Natural Gas

Natural gas is one of the most abundant fossil fuel produced in association with crude oil. Methane being one of its main constituent is typically produced from

- organic matter decomposition under heat
- organic matter converted under action of micro-organism
- coal releases methane over time
- methane also found in large quantity deep within crust.

Some of the favorable properties and challenges associated with use of natural gas as transport fuel are as mentioned below:

- CO_2 advantage: For getting 1 kWh work from engine, burning NG would produce 20% less CO_2 than diesel engine, and burning ethanol would produce 5% more CO_2 than diesel, assuming same BTE (45%) for all 3 engines.
- H/C ratio is higher at 4 and which has advantage of cleaner combustion and contributes to lower GHG emissions.
- Combustion of natural gas is almost free of particulate and with addition of suitable catalyst is possible to achieve lower emission levels of NO_x, CO and NMHC.
- Lower running cost.

However, there are also certain challenges as mentioned below:

- Being a gas storing adequate quantities storage on board the vehicle is challenging. Storage as high pressure compressed gas 350 MPa is expensive and requires substantial volume even for limited vehicle range.
- Adverse GHG impact from methane leaks in well to tank life cycle along with lower efficiency of natural gas engines could remove the well to wheel GHG advantages of their lower carbon content fuel relative to diesel and could result in higher GHG emissions in some scenarios (Cohn and Bromberg 2018).
- According to https://dieselnet.com/tech/engine_natural-gas_heavy-duty.php, it was mentioned that main sources of methane losses to exhaust are methane blow-by-piston crevice and bulk combustion losses.
- For methane to be taken care in TWC, the minimum temperature required is ~600 °C.
- As mentioned (Boretti et al. 2013), lower flame speed combustion characteristic of natural gas also contributes to power drop. Also due to lower burning speed of natural gas, the cylinder pressure develops slow and results in lower maximum cylinder pressure and cycle work however this has positive effect of more ignition timing advance possible.
- As per Neame et al. (1995) it was mentioned that methane because of lower laminar flame speed has lower tolerance for EGR and require higher spark advance. The term EGR tolerance means the ability of an engine to avoid cyclic variation with EGR.
- Lower specific power output, poor transient response and lower thermal efficiency (Figer 2014).
- Natural gas engine runs hotter than diesel engine and can accelerate lubricant degradation. Natural gas engine can suffer (MD/HD Natural Gas engine Efficiency Research Needs 2017):
  - Accelerated oxidation and nitration
  - Earlier onset of HTHS degradation
  - Faster loss of oil remaining useful life
  - More rapid TBN depletion and acidic corrosion.

Natural gas derived from fossil sources can produce GHG emissions on order of 20–30% lower than gasoline and diesel fuel whereas biogas has potential to effectively reduce GHG emissions to near zero depending on feedstock and production process (Besch et al. 2015). Hence natural gas has good potential to replace fossil fuel in transport especially if it is bio-gas. Moreover, use of natural gas in form of CNG (Compressed Natural Gas) and LNG (Liquefied Natural Gas) together can play a key role in addressing the transport related issues in India.

In India presently there are ~3000 CNG fueling stations and government has already announced in 2018 to set-up 10,000 CNG fueling stations by next 10 years (https://energy.economictimes.indiatimes.com/news/oil-and-gas/india-to-have-10000-cng-stations-in-next-10-years-on-track-to-adapt-cleaner-fuels-dharmendra-pradhan/65700258). Talking about bio-gas plants we have 4 major Indian states with higher number of bio-gas plant and Maharashtra being the state with over 9 lakh biogas plant (Number of biogas plants across India 2020 by state). LNG is another area were government is set to put 10,000 crore of investment for setting up 1000 LNG stations by 2022 (https://www.business-standard.com/article/economy-policy/india-set-to-attract-rs-10-000-cr-for-1-000-lng-stations-in-three-years-120111900653_1.html).

For use in heavy duty vehicles natural gas can be combusted inside cylinder in following ways (Wallner et al. 2017):

1. Stoichiometric spark ignition
2. Lean premixed diesel pilot
3. Lean premixed spark ignition
4. Direct injection diesel pilot.

Out of the above 4 technologies, the lean premixed diesel pilot and direct injection diesel pilot are more promising for heavy duty vehicles due to their higher engine efficiency potential. Both these engine technologies are also sometime referred to as dual fuel combustion (Natural Gas + Diesel). These dual fuel combustion technologies can be further classified as below:

1. Retro-fitment technologies for existing fleet of heavy duty vehicles
   - Dual fuel kit using fumigated natural gas
   - Dual fuel kit using port injection of natural gas
2. Direct injection technologies for new fleet of heavy duty vehicles
   - Direct injection of natural gas and diesel fuel.

The above dual fuel combustion technologies are discussed in sections ahead.

8.2.2 Methanol and Ethanol

As discussed in Agarwal et al. (2021), methanol can be produced from number of ways like coal, natural gas and biomass yielding synthetic gas. In India methanol is largely produced from imported natural gas, hence it would be better economical option for India to meet 90% of its requirement by importing methanol from Saudi Arabia and Iran as there it is abundant and cheaper. On the other side not only large reserves of coal of India can be used for its production but also biomass can be utilized to produce methanol. For India two suitable alternate would be high ash coal and CO_2 emissions from plants and factories.

About ethanol production in India, as already mentioned in Agarwal et al. (2021) that India is the second largest producer of sugarcane and even if entire crop is used for production of ethanol the yield would be 360 crore liter. For India to achieve 20% blending target for 2/3/4 wheelers program, then it would require ~900 crore liter by 2030. There are other challenges too like land availability, ground water consumption etc. However, if it is second generation based ethanol production then the scenario is as shown in Fig. 8.1.

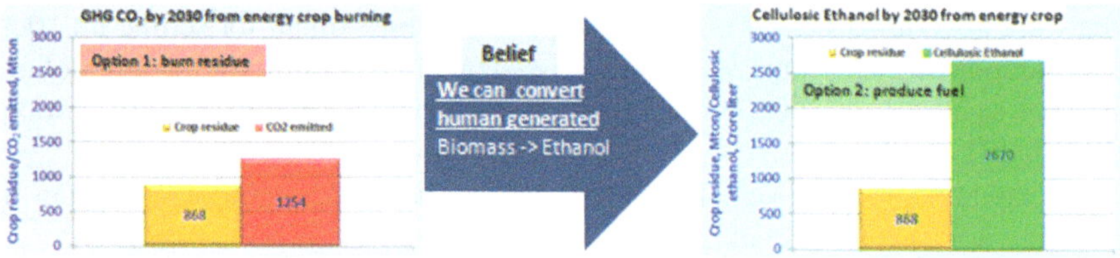

Fig. 8.1 Crop burn emitting CO_2 and agricultural residue conversion to ethanol

The above estimate based on a report (Pavlenko et al. 2019) shows how two ironical options are available to us and how using option 2 we can tackle the energy crisis and environmental concern together. There are even 3rd generation bio-fuels (algae based) which are not with advantage of high energy density yield in terms of land usage i.e. they are not as demanding on land area for production of same energy as like bio-fuels produced from 1st and 2nd generation methodology and further they can be grown even without ground water like wastewater.

Alcohol fuels including ethanol and methanol have excellent combustion properties. According to Hardenberg and Schaefer (1981), below are some of the interesting aspects of a ethanol as fuel when used in CI engines:

- It was mentioned that combustion of ethanol in CI engine reveals some attractive properties of this fuel like, e.g. low black smoke, NO_x and THC emissions. Vehicle tests over more than 1 million kilometers showed other favorable results like good fuel economy and low engine wear but yes ethanol fuel caused corrosion problem.
- He mentioned that for DI diesel engine maximum power output is limited to lambda 1.35 i.e. 35% excess air, while same engine when operating with ethanol can run with only 1.1 times the stoichiometric AFR with correspondingly higher power output. In other words, ethanol proves to be an extremely low smoke fuel in diesel engine provided that problem of its insufficient ignition quality can be overcome.
- Because ethanol engines operate with higher amount of H_2O than diesel engine, the specific heat ratios are higher for ethanol engines causing lower combustion temperatures and higher polytropic exponent for expansion. Hence ethanol engine would operate with higher thermodynamic efficiency.
- He also mentioned about the impact of ethanol as fuel on engine oil. From engine durability test low contamination of engine oil and better engine wear characteristics were noted when using ethanol in comparison to diesel fuel. In case of diesel vehicle, the TBN (total base number) loss is much higher than ethanol engine. This benefit in TBN with ethanol engine could be primarily due to lack of sulphur in ethanol fuel and consequently lower quantity of acidic products of combustion. This finding would not be much exciting as present diesel fuels available in market already have drastically reduced sulphur and now the sulphur content in commercial diesel is close to 10 ppm.
- Higher flame speed, ~40% higher than gasoline, which is beneficial for efficiency (more isochoric combustion).

- Flammability limit of ethanol and methanol is better than gasoline, hence combustion stability at lower load is better than gasoline.
- Has high enthalpy of vaporization that can increase the charge density.

However, there are certain concerns which are mentioned here and should be considered before adopting such alternate fuels:

- According to Hardenberg and Schaefer (1981), because of insufficient ignition quality pure ethanol is unsuitable for combustion in CI engine even when reasonably higher compression ratios and other conventional means of reducing the ignition delay are applied.
- Another important consideration when using ethanol in diesel engine is that the volume of fuel injected is 70% higher than diesel for getting same engine output. This would necessitate careful selection of nozzle.
- According to Hardenberg and Schaefer (1981), one disappointing damage of vehicle test conducted with ethanol is corrosion damage. Material corrosion like tank coating or in injection system and deposits of corrosion products from various sources of both vehicles and filing stations (e.g. blocked fuel filters or nozzle failure due to deposits on injector needle causing it stick in guide were found at various points in fuel system). From this point quality of ethanol itself appears to be of considerable significance. Large fluctuation in ethanol composition (water content, acid content) cause different corrosion characteristic which are difficult to counteract by desired minimal admixture of inhibitors. There is a need to regulate the content of fuels like acids, aldehydes, ester, copper and ash are also important.
- The low lubricity of alcohols requires additive in fuel to avoid problems in with diesel type of fuel pumps or injectors.

There was another study by Caterpillar on Richards et al. (1990), that gave worth mentioning findings that though components like piston ring, liner and bearing after long trials were investigated for wear found no degradation, iron concentration were less for methanol, average concentration of copper and lead were same as diesel, engine oil viscosity showed no appreciable increase but life of some of the diesel engine components like exhaust valve and seat, fuel injector nozzle were lower than diesel. Hence, such components need careful consideration or may be with time has evolved for such adoption to alternate fuels.

8.2.3 DME

DME (Di-Methyl Ether) is a synthetic fuel produced from natural gas, coal and biomass and relatively low cost manufacturing technology has been developed in recent years. Since DME is relatively easy to liquefy DME can be carried on vehicle as a liquid (Tsuchiya et al. 2006).

As per a report Ghosh (2016), in India the production capacity for DME is not very encouraging however there are several emerging technologies in the area of methanol/DME production. And the most promising is the use of plasma arc for gasification and CO_2 hydrogenation. These technologies have great promise for waste disposal and carbon dioxide fixing. Plasma based technology can handle wide variety of feed materials with production of harmless slag.

Some of the combustion properties making it another potential fuel are as mentioned below:

- High cetane number with soot free combustion.
- Better NO_x emission control high NO_x reduction possible with high EGR.
- Higher latent of vaporization so lower charge temperature in spite of 50% more fueling.

However, some of the major consideration when using DME are as below:

- Because of lower LCV and density than diesel, about twice the DME fuel must be injected compared to diesel fuel in order to ensure same power output as diesel fuel. Fuel tank size needs to be doubled. Also each injection event will have 50% more fuel than diesel so NO_x will on higher side.
- Compressibility of DME is significantly higher even in liquid state and also has lower viscosity and lubricity. Therefore, it is not possible to use conventional unmodified diesel fuel system for DME. To enhance lubricity additive required to be added to DME.
- According to Arcoumanis et al. (2007), because of low boiling point of DME (−25 °C), it is a gas at standard atmospheric conditions. Hence fuel system must be closed pressurized including the storage tank and handling should be like LPG (liquefied petroleum gas). Vapor pressure of DME similar to LPG, storage and handling should be like LPG.
- Cavitation in FIE (fuel injection equipment) is high due to high vapor pressure of DME. To avoid cavitation, the feed supply pressure from storage tank to fuel pump should be at 12–30 bar.
- Low injection pressure in range of 200–300 bar sufficient due to low boiling point of DME which results in immediate vapor formation of DME after liquid injection.
- Due to low viscosity leakage is a problem specially under higher pressure. Leakage problem reported for heavy duty engines using rotary type fuel system between plunger and barrel.
- Low lubricity can cause wear problem. This can be enhanced with addition of Lubrizol (1000 ppm), Hitec (100 ppm) and Infineum (500 ppm).
- Due to lower LHV and density, almost 1.8 times the volume of diesel fuel to be injected for getting same energy which necessitates longer injection durations.
- Due to higher compressibility of DME than diesel (4–6 times higher than diesel) requires more almost 10% higher compression work.
- According to Kajitani et al. (2001), needle opening pressure needs to set lower than in case of diesel. For diesel it is 200 bar and for DME it is 88 bar.
- According to Kato et al. (2014), it is said that since years DME is known as good diesel type fuel however due to low viscosity, high vapor pressure, low lubricity

and lower bulk modulus has restricted the injection pressure available for DME to 500 bar. However, they adapted a common rail system to deliver maximum injection pressure of up to 1000 bar. It is said that combination of low injection pressure and large injection nozzle flow area for DME resulted in fuel spray which is relatively bushier and has lower penetration than equivalent diesel. Denso has developed fuel injection system up to 1000 bar.

- In another study by Kato et al. (2015), spray was studied for chamber condition of 800 K and 45 bar. For DME the injection pressure was 600 bar and diesel it was 2000 bar. The nozzle hole diameter for diesel was 0.173 mm and for DME it was 0.35 mm. It was found that diesel took more time to evaporate than DME. Also the liquid portion for DME was shorter than diesel. Also for diesel the liquid core extends to certain length up to certain time then liquid core stabilizes, however for DME the core is shorter. And lastly more importantly both diesel and DME spray length penetration are almost similar under same chamber conditions.

8.2.4 HVO

In Europe HVO have been used as transport fuel on commercial basis. These fuels are actually manufactured from variety of feedstock including used cooking oil, vegetable oil, residue from oil industry, animal fats, fatty acids etc. Instead of using esterification to form FAME (Fatty Acid Methyl Ester), vegetable oil and animal fat can be hydrotreated to form HVO. The hydrotreatment uses hydrogen to remove the double bond and oxygen that leads to stability issues in FAME and thus gives HVO improved storage properties as well as reduced risk for oil contamination. HVO has much lower aromatic content. The first commercial production was commissioned at Porvoo refinery (Neste Oil) with capacity of 170,000 tons per year (Shukla et al. 2018).

- According to Aatola et al. (2008), hydrotreated vegetable oils do not have detrimental effect of ester type of bio-diesel fuels like increased NO_x emissions, deposit formation (see biodiesel barriers) storage stability, more rapid aging of oil and poor cold properties. HVO are straight chain paraffinic hydrocarbon that are free of aromatics, oxygen and sulphur and have high cetane numbers. As HVO fuels are hydrocarbons they meet conventional diesel fuel requirements like EN 590 and the FAME ester specification do not apply for HVO. HVO fuels has calorific value higher than ethanol. Further the quality of FAME depends on type of feedstock. For example, biodiesel derived from rapeseed and canola are more susceptible to oxidation issues. HVO can be produced from many kind of vegetable oil without compromising on fuel quality.

8.2.5 Synthetic Fuel

Synthetic diesel produced from sustainable feedstock can be of another key step to reduce transport related CO_2 emissions. As per https://www.neste.com/products/all-products/renewable-road-transport/neste-my-renewable-diesel#248febd3, renewable diesel made from 100% renewable raw materials can reduce 90% less greenhouse gases over fuel life cycle when compared to fossil diesel. Even they are working on diversified range of renewable feedstock like present renewable oil and fats to future options like recyclable waste plastic, ligno-cellulosic, agricultural residue & forestry, municipal solid waste, algae, CO_2-Power-to-X etc.

The renewable diesel can be used as drop-in fuels for transport purpose in both older and existing vehicles in India considering the range of sustainable feedstock available in India. As per a report Pavlenko et al. (2019), India has the potential to produce ~1300 crore liter of biofuel (GTL/FT-Diesel) by 2030 from municipal solid waste alone.

According to Gill et al. (2010), Fischer Tropsch diesel fuels can be derived from raw material like natural gas (GTL), Biomass (BTL) or Coal (CTL). Studies related to both BTL and CTL are low and GTL has become more attractive.

According to Uchida et al. (2014), FT diesel have favorable properties as mentioned below:

- High cetane to shorten ignition delay, reduce combustion noise and suppress HC for low load
- Very low aromatic to suppress soot formation
- Lower T90 so better volatility, better mixture formation and better fuel economy
- Low density so poor volumetric fuel consumption, hence increased injection duration which has penalty on thermal efficiency.

As per a report from https://www.norsk-e-fuel.com/en, Norsk e-fuels have released production plant which produces e-fuels. The e-fuels have been produced by capturing CO_2 from air and then syngas can be produced from CO_2 and water using 100% renewable energy. Each plant has a capacity to produce 1 crore liter of fuel per year by 2023 and this they have planned to increase to 10 crore liter by 2026.

Though at present these approaches are energy and cost intensive, with more research and commercialization such issue may be addressed. As mentioned for e-fuels by dieselnet.com, almost 6.25 MJ of electricity would be needed to supply 1 MJ of energy to wheels. And this would directly demand for higher power generation and cost of production.

8.3 Review of Various Technologies and Suitable Alternate Fuels to Improve Older, Present Modern and Future Heavy Duty Diesel Engines

For older vehicles retro-fitment and drop-in fuels could be one possible option to make them more tolerant to environment. Under Sect. 8.3.1 we have discussed these technologies in detail.

Today's modern heavy duty diesel engines are result of a number of state of art innovations and developments, which has made them more efficient and cleaner than ever. It is well known that a major chunk of fuel heat energy (close to 55–60%) for the internal combustion engines accounts for losses including heat transfer losses, exhaust losses, pumping losses, friction losses and emission losses. And though there is a tremendous conscious effort from the engine research and development team across the world to reduce these losses and maximize the amount of useful work available, certain engine irreversibilities in summation limits the amount of useful work that can be obtained from such internal combustion engines. According to Stanton et al. (2013), the combined effect of such irreversibilities restricts the internal combustion engines utilizing the slider crank mechanism to future engine having brake thermal efficiency of approximately 60%. Presently in the market we have heavy duty diesel engines operating with 45–46% brake efficiency. The awareness on practical limit on brake efficiency of such engines has however on the other hand given a ray of hope to the engine enthusiast and hence many engine researcher and laboratories have already set targets to develop heavy duty diesel engines with brake efficiency of 50% and >55% with technologies like waste heat recovery. Reaching such targets of engine brake efficiency and being still cleaner from the present modern heavy duty diesel engines would require understanding the present levels of losses and scope of improvements.

For India the modern heavy duty diesel engines are typically at brake thermal efficiency of ~40%. The base heavy duty diesel engines technologies should be optimized better to make our engines 10–12% improved. Under Sect. 8.3.2 we will discuss the baseline improvement strategy. Under this section we will also discuss how using various suitable low carbon and alternate fuels we can achieve baseline improvement i.e. target brake thermal efficiency of ~45%.

Further in Sects. 8.3.3 and 8.3.4 we will discuss the future engine technologies and to make engines 50% brake thermal efficient and role of alternate fuels in making heavy duty engines sustainable.

8.3.1 Older/Aged Fleet of Heavy Duty Vehicles

As already discussed in Sects. 8.1.3 and 8.2 about the existing fleet of old and aged vehicles which can be improved for their carbon footprint by adopting retro-fitment with natural gas dual fuel technologies or by using carbon–neutral drop-in fuels like synthetic diesel or HVO.

Retrofitment technology is an attractive option to both government and heavy duty fleet owners. For government it is one of the way to reduce the older or aged heavy duty vehicle's carbon footprints. For fleet owners retrofitment can be adopted for older or aged vehicles to avail the benefit of reduced operating cost. Retro-fitment generally done from the point of view of reducing the burden on environment created by older or aged fleet of heavy duty vehicles. Retro-fitment could mean altering the bulk fuel that is being combusted thereby altering the fuel related emissions and carbon footprints or by changing the emission control devices of the existing heavy duty diesel engine to reduce the present level of tailpipe emission. Here we are talking about the former i.e. using dual-fuel (natural gas + diesel) retrofitment where depending upon the engine operating point the majority of diesel fuel can be replaced by natural gas.

Retro-fitment options though require huge investment and considerations dependent on the level of vehicle engineering and number of older or aged heavy duty vehicles available in market but such retro-fitment programs can get return which can be almost 20 times the invested amount and these return are result of air pollution related health benefits or number of premature deaths which could be avoided due to reduction in pollution associated (Kubsh 2017).

The existing heavy duty vehicles are operating on diesel engines which are known for their high engine efficiency. With retro-fitment we can use the same heavy duty diesel engine compression ratio and can achieve same or even higher engine efficiency, moreover these retro-fitted engines will be cleaner and environmentally friendly. The simplification with retro-fitment of adopting the same heavy duty diesel engine hardwares like compression ratio, cylinder head makes the option further interesting. Here diesel injection is the same conventional direct injection however the natural gas injection can adopt any of the following techniques:

- natural gas fumigation (pre-compressor)
- port injection (sequential)
- manifold injection
- post intercooler injection.

The control strategy adopted here is to substitute the diesel quantity with natural gas charge and maintain a critical diesel quantity for maintaining the original diesel power output and emission levels. Basically for any operating point of the engine we try to introduce the natural gas and for any increase in engine torque we reduce the diesel quantity to maintain the same engine output. The input for the natural gas substitution typically depends on the driver accelerator position signal and engine speed signal. And the external control unit is calibrated for the entire engine operating map for relatively lower emissions and same or better efficiency and maintaining the same diesel engine power output.

Table 8.1 shows the results obtained for dual fuel based retro-fitment as compare to diesel operation.

Table 8.1 Percentage values when compared to diesel fuel operation over FTP cycle (Besch et al. 2015)

Parameters	Values
Diesel substitution	~20–45%
CO_2 reduction	~3–8%
NO_x reduction	~25–40%
Methane	~7–35 g/kWh

The reduction in CO_2 emissions could be attributed to higher H/C ratio of natural gas as fuel. The NO_x emissions reduction for dual fuel combustion are generally attributed to the change in in-cylinder charge properties with introduction of natural gas. With introduction of natural gas the specific heat property increases and this leads to lower manifold temperatures and temperatures at top dead center, lower premixed combustion and lower combustion rates (Besch et al. 2015). These all factors may lead to lower NO_x emission. However, with relatively simplified retro-fitment as compared to modern heavy duty diesel engines the chances of hydrocarbon emission could be very high especially at lower load conditions. It is known that at lower equivalence ratio or low natural gas substitution typical with lower load conditions the combustion rates of the premixed charge are mostly dependent on spray entrainment and mixture stratification of diesel fuel with dominant portion of combustion occurring as diffusion combustion i.e. no bulk ignition or flame propagation. This poses a significant challenge for dual fuel combustion and can lead to very high hydrocarbon emissions including methane. This high hydrocarbon emissions can also seen as decreased fuel economy or deteriorated brake thermal efficiency. Whereas for higher equivalence ratio or higher natural gas substitution typical with higher load conditions the dominant combustion is through flame propagation which results in high temperature combustion reactions throughout the chamber provided the premixed charge has exceeded certain threshold value (Besch et al. 2015).

Summary

- Retro-fitment using dual fuel technology can be a potential way to reduce the older/aged vehicles carbon footprints and vehicular emissions. However, there are certain points that should be considered while executing such programs:
  - The system should not be too simplified so that the problem of hydrocarbon emissions including methane are escaping in-cylinder combustion and are becoming burden for the existing diesel oxidation catalyst. If that is the case then options for raising the catalytic efficiency or light-off adequate for methane oxidation (~500 °C) should be considered so that any CO_2 advantage obtained should not get offset with methane emissions due to their higher global warming potential then CO_2.
  - NO_x aftertreatment systems should be carefully re-selected so that any excess methane emissions do not lead to HC poisoning effect of such NO_x control systems.

- Depending upon the duty cycle of the selected vehicle, the low load operating regions should be carefully optimized so that the effect of lower charge temperatures inherent with introduction of natural gas do not lead to any excess loss of volumetric efficiency or loss of manifold temperatures leading to higher hydrocarbon emission or lower operating efficiency of vehicles under such operating regions.

8.3.2 Modern Heavy Duty Engines

Heavy duty engine manufacturers and country should see the road map of these prime movers as shown in Fig. 8.2.

Modern heavy duty diesel engines are developed with strong emphasis on high brake efficiency and low emission. The need for 25–30% reduction in CO_2 in next 7–9 years and 90% further reduction in NO_x emissions are driving the need for making diesel engines more efficient and cleaner.

8.3.2.1 Engine Technologies

The engine technologies on the frontline for making this happen are higher compression ratio, better matched combustion bowl-swirl-cone angle, advanced turbocharging, EGR system and advanced injection systems. Daimler (Gruden et al.) developed engines with peak brake efficiency of >46% and >43% at maximum power

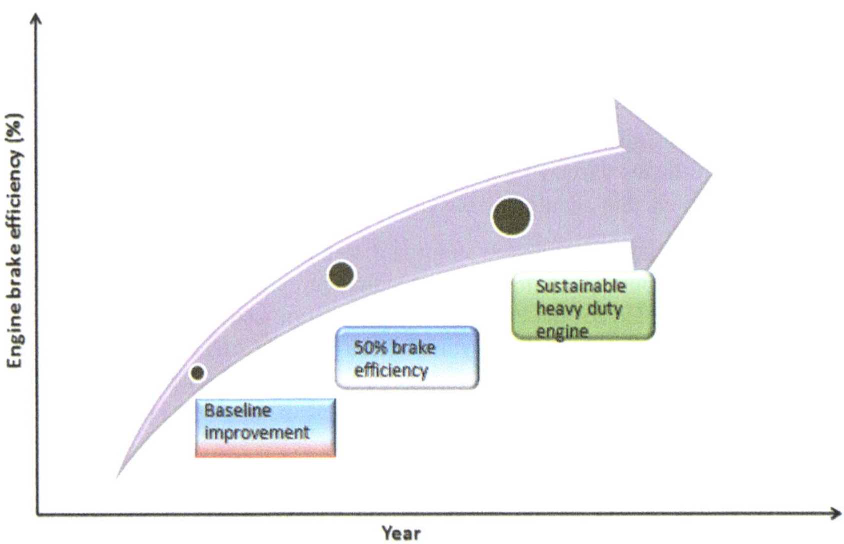

Fig. 8.2 Road map for heavy duty diesel engine

Table 8.2 Engine configuration and efficiencies (Gruden et al.)

Displacement, l	14.8
CR	18.0
Turbocharger	Asymmetric
Maximum injection pressure, bar	2500
Efficiency at rated power, %	>43
Peak efficiency, %	>46

point. Table 8.2 details some of the advanced engine technologies adapted for such engines.

Achieving high engine brake efficiencies requires careful considerations on the areas were engine losses are still substantial and has enormous potential for improvement. Presently, in India the highest heavy duty diesel efficiencies are lower than the initial baseline efficiencies in Europe or US. Hence, we would require to improve our new heavy duty vehicles efficiency by at least 10–12% through baseline improvement from present Indian heavy duty diesel engines (as shown in Fig. 8.2).

Table 8.3 shows the target for peak engine brake efficiency and engine out emissions for meeting present (BS VI) and upcoming norms (ULN-ultra low NO_x) toward following the road map shown in Fig. 8.2.

Generally, manufacturers tend to achieve engine out efficiency improvements by calibrating engine out NO_x for higher levels (>5 g/kWh) and utilizing marginal or no EGR strategy. There are several interesting points that we should consider when going for such high engine out NO_x calibrations:

- Such strategies not only limit the engine combustion capabilities but also put tremendous demand on after treatment systems especially under critical real driving conditions were exhaust temperatures are not suitable or typical to specific duty cycle. Presently in BS VI regulations, the cold start emissions are not given much importance and weightage for such emissions are 14%. As per a report

Table 8.3 Target for engine brake efficiency and emission

Parameters	Baseline improvement	50% brake efficiency	Sustainable heavy duty engine	
BTE, %	45	50	>50	45–50
Fuel used	Diesel	Diesel/natural gas/gasoline	Biofuels/synthetic fuels/HVO/DME/natural gas	Biofuels/natural gas
NO_x emission norms, g/kWh	BS VI (0.4 g/kWh)	ULN (~0.027)		
Engine out NO_x, g/kWh	2.5	1.5–2.5	0.25–2.5	2.5
NO_x conversion efficiency, %	85	98–99	90	99

Sathiamoorthy et al. (2020), it was mentioned that it is likely that in several start-and-go application in winter the engine will operate more than 14% of the total operational time. Policymakers may consider increasing the weightage for cold start emissions.

- Duty cycle dependent real world emissions of heavy duty vehicles. In a report (Badshah et al. 2019), it was found that during low vehicle speed operation, particular to urban operation, a disproportionate amount of NO_x emitted from HDV. And these emissions are found to be 5 times higher than certification limits for the average heavy duty vehicle in the study.
- OBD requirements add another level of complexity for such strategies.
- Moreover, such strategies would require further use of more advanced NO_x control systems for meeting, if not immediately the upcoming future ultra-low NO_x (ULN) norms (Lynch et al. 2020).

Hence for achieving the target diesel engine efficiencies of 45% as intended for baseline improvement we need to adopt different approach toward development of heavy duty diesel combustion by embracing key engine technologies like exhaust gas recirculation, high adequate boost, increased peak firing pressures and higher injection pressures in coming future and also understand the scope of improvement and losses associated with present heavy duty diesel engines.

Figure 8.3 shows the energy audit for a typical heavy duty diesel engine in India.

We can further explain the energy audit in detail with below equations, in terms of mean effective pressure:

$$\eta_{\text{combustion}} = \text{Qhrmep}/\text{Fuelmep} \tag{8.1}$$

$$\eta_{\text{thermodynamic}} = \text{Qhrmep}/\text{IMEPg} \tag{8.2}$$

$$\eta_{\text{gasexchange}} = \text{IMEPg}/\text{IMEPn} \tag{8.3}$$

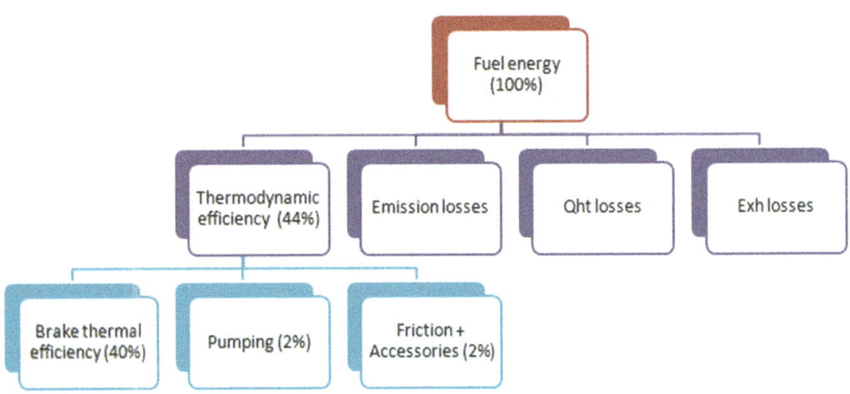

Fig. 8.3 Energy audit for typical heavy duty diesel engine

$$\eta_{\text{mechanical}} = \text{IMEPn}/\text{BMEP} \tag{8.4}$$

Multiplying above all Eqs. 8.1–8.4, we get as below:

$$\eta_{\text{brakethermal efficiency}} = \eta_{\text{combustion}} \times \eta_{\text{thermodynamic}} \times \eta_{\text{gasexchange}} \times \eta_{\text{mechanical}} \tag{8.5}$$

$$= (\text{BMEP}/\text{Fuelmep}) \times 100 \tag{8.6}$$

So we see that improving brake efficiency means maximizing efficiencies or minimizing losses at each of them through 1–4. In general combustion efficiency are no worries for heavy duty diesel engines as because of lean combustion and high load factors. For baseline improvements of about 10–12% can approached by following areas:

- Increasing thermodynamic efficiency of base engine ($\eta_{\text{thermodynamic}} \geq 50\%$)
- Increasing or atleast maintaining the same gas exchange efficiency of base engine ($\eta_{\text{gas exchange efficiency}} \geq 96.5\%$ or PMEP $\leq 2\%$ Fuelmep)
- Increasing or atleast maintaining the same mechanical efficiency ($\eta_{\text{mechanical efficiency}} \geq 92.5\%$ or FMEP $\leq 3\%$ of Fuelmep).

Figure 8.4 shows the new proposed heavy duty diesel engine with improved baseline performance.

Let us now consider a hypothetical case (Table 8.4) to understand analytically the importance of various parameters of engine thermodynamic process and how these parameters influence the engine efficiency and emissions. This hypothetical case closely represents correct correlation of various engine specific parameters, efficiencies and emissions.

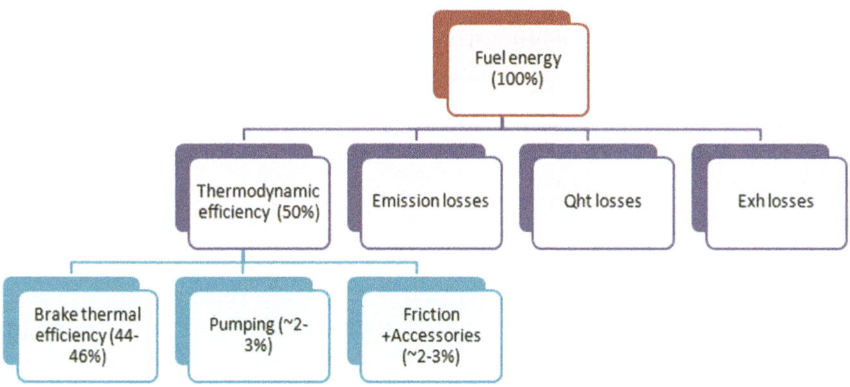

Fig. 8.4 Energy audit for proposed heavy duty diesel engine for baseline improvement

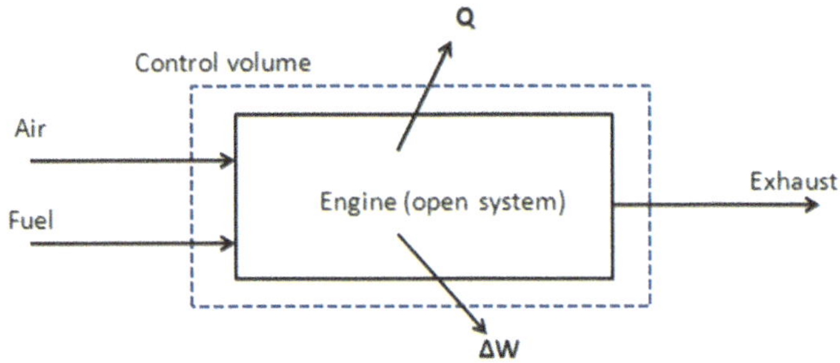

Fig. 8.5 Engine as open system

Thermodynamic efficiency, is basically how effectively we convert the Qhrmep (heat released from fuel chemical energy) to gross indicated work. For understanding more on gross indicated work we consider the engine as open system as shown in Fig. 8.5.

Figure 8.5 shows the engine as open system were fuel and air are at the inlet and exhaust gas at the outlet. The Q is the heat loss from the cylinder charge to walls and ΔW is the useful work output from the engine or the system. Applying thermodynamic 1st law to the system and neglecting combustion losses and blow by we have,

$$-\mathrm{Q} = (\mathrm{Hp} - \mathrm{Hr}) + \Delta\mathrm{W} \tag{8.7}$$

re-arranging the terms of Eq. 8.7 we get,

$$\mathrm{Hr} = \Delta\mathrm{W} + \mathrm{Q} + \mathrm{Hp} \tag{8.8}$$

Now here Hr, enthalpy of reactant can be written as below:

$$\mathrm{Hr} = \mathrm{Hinlet} + \mathrm{Qfuel}$$

$$\mathrm{Hinlet} = \text{enthalpy of inlet air mass} = \mathrm{mr} \times \mathrm{C} \times (\Delta\mathrm{T}) \tag{8.9}$$

mr mass flow rate of air.
Qfuel fuel energy.

$$\mathrm{Q} = \mathrm{ha} \times \mathrm{Awall} \times (\mathrm{Tcyl} - \mathrm{Twall}) \tag{8.10}$$

Now from Eqs. 8.8 and 8.9 it can be seen that for given enthalpy of reactant i.e. given mass flow rate of air if we reduce the Q and Hp we can increase the useful work. In other words if we increase inlet air mass then for given Q we can increase useful work. And with higher charge and lower Q, we can maximize the ΔW i.e. the gross work output from the system. Engine boosting is needed for getting higher gross output but as explained by Johansson et al. (2010), Q has also to be reduced by decreasing the Tcyl which otherwise would increase and cause the Q or heat transfer losses to increase. Hence with use of EGR we can reduce the Tcyl and with adequate lambda we can maintain the mass flow and also control the emissions. Thus maximizing thermodynamic efficiency basically involves the optimization of the heat release rate and centroid of combustion so that we have minimum heat transfer and maximum effective expansion work. Better optimization of heat release rates by placing the centroid of combustion (CA50) at optimum crank angle location and reducing the combustion duration we can maximize the thermodynamic efficiency. So maximizing thermodynamic efficiency is all about optimizing the placement of heat release and reducing the combustion duration. Typically, we try to achieve CA50 at around close to 5–15° atdc depending on load and combustion duration between 10 and 30 crank angle deg for getting maximum effective expansion stroke thereby maximizing thermodynamic efficiency (Johansson et al. 2009). Typically, for some of the heavy duty engine in Indian market we largely reply on SCR systems to control majority of engine out NO_x (with SCR eff. requirements close to 95–96%) and marginally using EGR. This is done considering the advantages on fuel economy, component durability and transient behavior. Hence, with low dilution and with limited advanced timings we do not have optimized CA50 and combustion rates resulting in thermodynamic efficiency close to 43–44%. We would now discuss the below engine out targets and combustion system requirements required for seeing baseline improvement (as shown in Fig. 8.6):

Engine out targets: As per Johansson et al. (2010), Dreisbach et al. (2007) we get to know the general trend for EGR% requirements to achieve reducing levels of engine out NO_x. We see that around 20% EGR at peak torque would give us an engine out NO_x of 2.5 g/kWh. For diesel like fuel which are highly prone to soot formation due to their alignment to requisite conditions for soot formation as mentioned by Heywood (2018), i.e. carbon atoms in range of 12–22 and H/C ratio above 2 when under in-cylinder condition of diesel combustion leads to formation of $C_{2n}H_2$ and PAH (polycyclic aromatic hydrocarbon), meet the smoke limiting targets sooner. Hence, higher injection pressures and adequate lambda has to be maintained to limit the soot emissions. As per Zhang et al. (2018a), Moser et al. (2004), with injections pressures as high as 2000 bar would be sufficient enough to full load smoke emissions to ~0.6 FSN (or ~0.03 g/kWh) and lambda close to 1.5. Though with adequate lambda, higher injection pressures and low sulphur fuel, the fuel derived PM can be reduced however the oil derived PM contribution can increase. Hence, keeping the heavy duty engine oil consumption at minimum would also be required toward achieving total low PM emissions from engine. Typically, the modern heavy duty engines can target oil consumption rates in range of 6–35 g/h (Zhang et al. 2018b).

Turbocharging and EGR system requirements, is one of the essential change that would be required to be investigated for such baseline improvement and hence

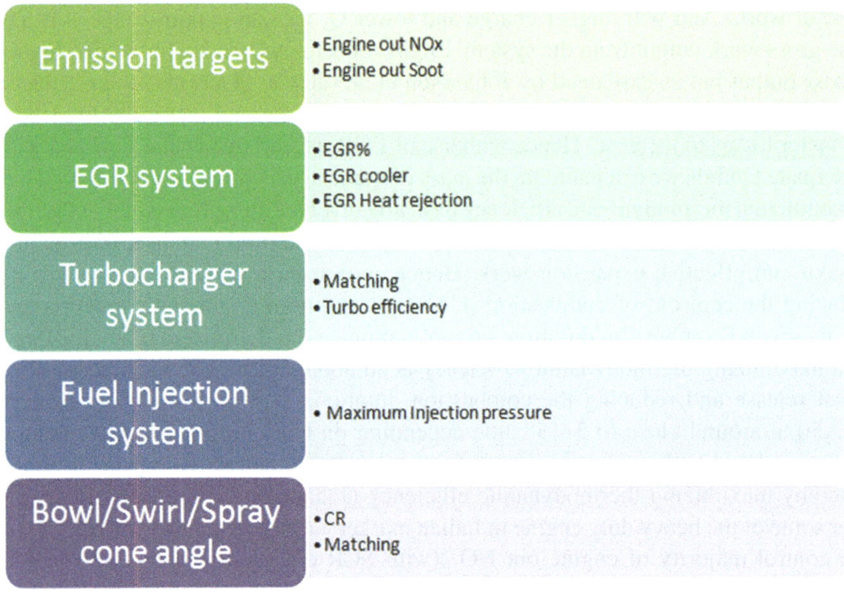

Fig. 8.6 Engine combustion system considerations

would therefore need detail consideration. We have to target gas exchange efficiency of at least 96.5% for baseline improvement. Typically, when using low dilution approach, the requirements on gas exchange efficiency is not much demanding and we could achieve ≥97% or PMEP ≤ 1.1% of total fuelmep. But we have to strive hard to maintain almost same existing gas exchange efficiency levels when looking for baseline improvements due to higher EGR requirements. As per Table 8.4 data, with 20% EGR the gas exchange efficiency is 94.2% with PMEP as 1.2 bar, which is 2.75% of total fuelmep. This PMEP here is higher than the required as mentioned above. Hence, with 20% EGR target at full load we will need to select higher overall turbocharger efficiency to maintain the same gas exchange efficiency and PMEP levels. Generally, with dilution the mass flow across turbocharger tends to increase and it becomes tough to match the turbocharger with such range of flow over the engine speed range. Further difficulty adds with following below considerations:

- What is the highest overall turbocharger efficiency available? And what are the compressor outlet temperatures available?
- What are the pressure requirements needed to maintain lambda with the available over all turbocharger efficiency? How well is the engine volumetric efficiency?
- What are the exhaust backpressure created to maintain the required intake manifold pressures? And what are the associated pumping loss (PMEP)?
- What are EGR cooler out temperatures available? And what are related intake manifold temperatures available?

Table 8.4 Engine operating conditions

BMEP, bar	17.1
EGR%	20
Lambda	1.76
Int man pressure, bar-abs	3.5
Exh man pressure, bar-abs	4.2
Fuelmep, bar	43.1
IMEPg, bar	20.7
IMEPn, bar	19.5
PMEP, bar	1.2
FMEP, bar	2.4
Peak firing pressure, bar	152
Main timing, ° btdc	2
Injection pressure, bar	1800
Exh flow, kg/h	807.7
Exh man temperature, °C	540.8
NO_x, g/kWh	2.7
Smoke, FSN	1.0
BIC temperature, °C	190
IMT, °C	70
EGR cooler out temp—EGR cooler, °C	210
Exhaust temperature after low pressure turbine, °C	340
Thermodynamic efficiency, %	48
Gas exchange efficiency, %	94
Mechanical efficiency, %	88
Engine brake efficiency, %	~40

- What is the maximum heat rejection happening at maximum power and peak torque points?
- Whether adequate turbine inlet temperatures necessary for aftertreatment available with determined dilution?

With 20% EGR requirement for meeting the target engine out NO_x. The design of bowl-swirl-cone angle matching would play an important role in achieving the engine out soot emissions so that optimum lambda can be maintained with required EGR level. This is because the lambda requirements would determine the boost levels or the compressor out pressure levels and which in turn will depend on the compressor and turbine work balance.

From Table 8.4, we will be calculating the overall turbocharger efficiency for the given case. We know the equation for turbocharger work balance involving the overall turbocharger efficiency as mentioned below:

$$\text{Compressor work} = \eta\text{Turbine work} \tag{8.11}$$

where,

$$\eta_{OTC} = \text{overall turbocharger efficiency} = \eta c \times \eta t \times \eta m$$

$$\text{Compressor work} = (\text{mc} \times \text{Cpc} \times \text{Tamb}) \left[(\text{Pint}/\text{Pamb})^{\gamma c - 1/\gamma c} - 1 \right] \times (1/\eta c) \tag{8.12}$$

$$\text{Turbine work} = \eta m \times (\text{mt} \times \text{Cpt} \times \text{Texh}) \left[1 - (\text{Pamb}/\text{Pexh})^{\gamma t - 1/\gamma t} - 1 \right] \tag{8.13}$$

m	mass flow rate, kg/h.
C	specific heat of gas, kJ/kg-h.
Tamb	ambient temperature, °K.
Pint	intake manifold temperature, mbar-abs.
Pamb	ambient pressure, mbar.
Pexh	exhaust manifold pressure, mbar-abs.
ηc	compressor isentropic efficiency.
ηt	turbine isentropic efficiency.
ηm	mechanical efficiency.
γ	polytropic exponent.

Using the Eqs. 8.11, 8.12 and 8.13, we get η_{OTC} for Table 8.4 data we get the overall turbocharger efficiency as below:

$$\eta_{OTC} = 46\%$$

For the present case (Table 8.4 data) the overall turbocharger efficiency comes to be 46% and the associated gas exchange efficiency for 20% EGR is 94%. The PMEP is 2.78% of total fuelmep. However, for baseline improvement the targets as mentioned above are written below:

- Gas exchange efficiency = $\geq$96.5%
- PMEP $\leq$ 2% Fuelmep.

For improving the gas exchange efficiency we need to work on below following aspects:

- reduce the pumping work by reducing the pressure losses in intake ports, intake valves, intake manifolds i.e. improving the volumetric efficiency
- reduce the pressure losses across EGR valves and cooler
- improve the turbocharger work balance.

The turbocharger work balance is the work required by the turbocharger to operate compressor to meet both the desired fresh air flow and EGR (Stanton et al. 2013). From the Eq. 8.11 and expanded Eqs. 8.12 and 8.13 we find following aspects:

- Any increase in mass flow rate, increases required compressor work
- Any increase in compressor inlet temperature, increases required compressor work
- Any increase in intake manifold pressure demand, increases the required compressor work
- Any drop in compressor inlet pressure, increases the required compressor work
- Any increase in compressor work would put more demand on turbine work
- Any drop in turbine efficiency, would demand higher turbine inlet pressure
- Any drop in turbine inlet mass flow, would demand higher turbine inlet pressure
- Any drop in turbine inlet temperature, would demand higher turbine inlet pressure to meet the compressor work demand
- Any rise in turbine inlet pressure, would increase the engine backpressure
- Any increase in engine backpressure would increase the fuel penalty via negative pumping work.

Hence, from the above points it is clear that to improve the gas exchange efficiency for Table 8.4 data, we need to improve on above all factors including turbocharger efficiency. If we analyze any compressor map, we find that compressor has well defined central eye region were the compressor is operating at its peak efficiency. Any increase or decrease in reduced mass flow parameter to compressor together with increase in pressure ratio demand may shift the compressor operating zone to either choke or surge region i.e. far away from the eye. Generally with higher engine mass flow range matching then becomes difficult. And such situations arises either when we are working with higher bmep engines, higher dilution as it is the case with advanced combustion concepts. In such cases matching the maximum torque and power points of engine operation with turbocharger eye region becomes difficult. And if turbocharger operating efficiencies are poor it will not help to achieve the gas exchange and engine efficiency targets.

The present lower overall turbocharger efficiency seems lower. We can get more direction to the required target overall turbocharger efficiency from a study done by Chadwell et al. were it is observed that we would require ≥50% overall turbocharger efficiency for meeting the engine brake efficiency target close to 45% at full load with 20% EGR requirement. As per another article from Stanton et al. (2013), Watson and Janota (1982) turbocharger efficiency has significant influence over engine fuel efficiency and he cited an example that improving 50–57% turbocharger efficiency has 1–2% improvement in engine brake thermal efficiency. The full load lambda here is also higher than actually required. Typically, the lambda required for full load should be sufficient at 1.4 (Chadwell et al. 2011). The demand for lambda would depend on the engine smoke level which in turn depends on bowl-spray-swirl interaction and bowl's air utilization. For any excess lambda it increases the demand of pressure ratio delivery at compressor outlet. And this would further increase the compressor work demand from turbine incurring loss on engine brake efficiency due to rise in ΔP (Exh Pr − Int Pr). The ΔP value here is 700 bar which seems to be quite high and PMEP (pumping mean effective pressure) is 1.2 bar which in terms of percentage of fuelmep (fuel equivalent mean effective pressure, which is here 43.1 bar) is almost 3% as discussed above. This PMEP has to be limited to

2% for baseline improvement. With higher overall higher turbocharger efficiency we can improve the turbocharger work balance and this will reduce the ΔP required to get required boost and this will further reduce the PMEP. For 10%-point increase in overall turbocharger efficiency would result in almost 0.3 bar reduction (Lavoie et al. 2012) in PMEP i.e. almost 2%-point increase in gas exchange efficiency. Also volumetric efficiency plays another key role. Higher the efficiency lower the pressure requirements. Hence when trying to improve the baseline efficiency there are two basic consideration required from lambda and turbocharger selection to ensure that PMEP lies in range of 2–3% of total fuelmepas mentioned below:

- Improve the selection of turbocharger so that overall turbocharger efficiency selected for the improved base heavy duty diesel engine is ≥55%
- Optimize lambda so that pressure ratio requirements are reduced and there is reduction in engine back pressure requirements.

Hence, with above requirements when using 20% EGR and close to 1.5 lambda we would likely see following improvement in gas exchange capabilities:

- With selection of overall turbocharger efficiency of 55% that is almost 10%-point increase from existing 46% we can improve the gas exchange efficiency by 2%-point increase i.e. from 94 to 96% (i.e. almost 0.3–0.4 bar drop in absolute PMEP) for the Table 8.4 data, improve the PMEP% from 2.75% to ~1.8% (i.e. almost 1%-point improvement in PMEP% of total Fuelmep) for the same Table 8.4 data.

Now next step is we need to investigate the engine boost requirements for achieving EGR as 20% and full load lambda as 1.4. Now for the full load point we will estimate the engine boost as below:

Now in Table 8.5 we have the bsfc value after considering the target peak brake thermal efficiency as 45% and target volumetric efficiency considered here is 93%. Also now here we have estimated the intake manifold temperature as 66 °C by considering the EGR gas out temperature to be as 170 °C and charge air cooler out temperature as 40 °C. The estimation is explained below:

Intake manifold temperature can be estimated using below empirical equation:

Table 8.5 Data for estimation of engine boost at full load

Parameters	Values
Speed	1400 rpm (as per Table 8.4 data)
Load	17.1 bar
EGR	20%
Lambda	1.4
BSFC (target BTE = 45%)	183.9 g/kWh
Volumetric efficiency	93%
R	287 J/g mol-K
Tman (intake manifold temperature)	66 °C

$$\text{Tman} = (0.20) \times (170) + (0.80) \times (40) = 66\,^{\circ}\text{C}$$

- Theoretical vol air flow, m^3/h = (engine size, l) × (3600) × (N/(2 × 120)) × 10^{-3}
- Actual mass air flow, kg/h = (lambda) × ((bsfc × P)/(1000)) × (14.6)
- Density at manifold condition, kg/m^3 = (actual mass air flow)/(vol eff × theoretical vol air flow)
- Intake manifold pressure = (density) × (RT).

Now here using above formula we can calculate the engine boost requirement as = 1.85 bar abs. Typically this engine boost level is relatively lower than usual boost level. And this is because of higher target volumetric efficiency and lower intake manifold mixing temperature. Also the EGR gas out temperature has an important role to play.

Now with respect to EGR system and charge air cooler we need to ensure that the EGR cooler out temperatures and charge air cooler out temperatures are optimum so that the final charge temperature after EGR mixing in the intake manifold are reasonable, in our case it may be 66 °C.

Below equation determines the charge air cooler effectiveness:

$$\varepsilon = (\text{actual heat transfer})/(\text{maximum possible heat transfer}) \tag{8.14}$$

$$\varepsilon = (T_2 - T_3)/(T_2 - T_w) \tag{8.15}$$

T_3 intake manifold temperature.
T_w cooling medium inlet temperature.

Next we also need to check the compressor outlet temperature for the overall turbocharger efficiency as 55%.

Consider the following ideal gas equation for approximately knowing the relation between the charge temperature, mass flow and pressure requirements:

$$PV = (m/M)RT \tag{8.16}$$

From Eq. 8.16 we see that with increase in final charge temperature we need to raise the charge pressure so that same fresh air mass flow is maintained for meeting the desired lambda. Hence, for any increase in final manifold charge temperature would put more demand on the compressor outlet delivery pressure.

Here we will now estimate the compressor outlet temperature for the full load point using the below equation after assuming gamma coefficients suitably:

This can be verified using the below Eq. 8.17 for Table 8.6 data. We will now consider two set of data as below:

$$\Delta T = T_{inlet}[(P_{in}/P_{amb})(\gamma_c - 1)/(\gamma_c) - 1]/(\eta_{comp}) = T_2 - T_{inlet} \tag{8.17}$$

T_{inlet} compressor inlet temperature.

Table 8.6 Data for estimation of compressor outlet temperature at full load

Parameters	Value
Overall turbo efficiency	55%
Compressor efficiency	75%
Turbine efficiency	75%
Mechanical efficiency	98%
EGR%	20%
IMT	66 °C
Intake pressure	1.85 bar abs

P_{in} compressor outlet pressure.
η_{comp} compressor isentropic efficiency.

Using the Eq. 8.5 we get the compressor outlet temperature = 109 °C. This is quite low but actually the compressor outlet temperature would be more useful to know at engine rated operating point because of following aspects:

- Lower volumetric efficiency at rated power
- Higher pressure drops so higher compressor delivery pressure requirements.

For Table 8.4 data case with 20% EGR we get EGR mass flow of 194 kg/h. We will calculate the heat rejection for this mass flow of EGR as mentioned below:

$$\text{Heat rejection} = m \times c \times (\Delta T)$$
$$= 17\ \text{kW}$$

$$\text{Total heat input} = (\text{mep}) \times (N/(2 \times 60)) \times (7.2) \quad (8.18)$$

The total heat input here is 362 kW. Hence, for the required EGR mass flow the total heat rejection is almost 5% of total heat input. The heat balance sheet must be evaluated when using EGR. As per Teng et al. (2009), especially the EGR cooler out temperature and temperature of exhaust leaving the turbine or the low pressure turbine must be compared. For our case Table 8.7 shows the estimate.

For the baseline case with better optimized lambda we can reduce the actual air flow and thus the EGR mass flow and this in turn will reduce the amount of heat rejection related to EGR.

Till now we have seen the role of different parameters and their contribution in achieving target engine brake thermal efficiency and emissions. The above analysis was mostly focused on full load. However, we also need to investigate the engine maximum power point because of following aspects:

- What fraction of engine duty cycle does it operates at peak power? This will decide the engine efficiency requirements.

Table 8.7 Heat balance for the considered engine operating point

Parameters	Values	% of total heat input
Heat rejection after turbine, kW	72	~20
Heat rejection EGR cooler, kW	17	~5
Heat rejection charge air cooler, kW	35	~10
Heat rejection coolant, kW	94	~25
Useful work, kW	144	~40

- What are the EGR target level at peak power? What engine out emissions are we targeting at peak power? This will decide the engine lambda and EGR requirements.
- What is the volumetric efficiency available at peak power? This will decide to an extent the engine boost requirements.
- How is the EGR gas out temperature at engine peak power? This will decide the engine boost requirements for target lambda.
- How is the turbocharger work balance at peak power while meeting the lambda and EGR requirements? This will decide the gas exchange efficiency of the engine at peak power.
- What are EGR heat rejection at peak power?

Turbocharger hardware requirement i.e. whether single stage or 2 stage will be decided based on the engine mass flow range from peak torque to peak power and at what efficiency level the turbocharger is operating at peak power. There are many other factors that needs to be considered apart from the ones listed above. Next we will discuss about the combustion system.

Combustion system requirements for targeting baseline efficiency improvements would require optimal matching of bowl-swirl-spray cone angle for getting higher fuel–air mixing resulting in fast combustion and lower soot emissions. Such faster heat release rate would require advancing the injection timings and utilization of higher injection pressures. With retarded injection timings as still used for engine out NO_x control we would face limitation on CA50 optimization due to retarded combustion thereby limiting thermodynamic efficiencies. Combustion bowl design largely depend on the size of engine. The quiescent type bowl with lower swirl levels are normally used for large bore size, for heavy duty engines with medium bore size typically we go for swirl supported bowl systems like stepped bowl with cone angles in range of 140–145° and swirl in range of 1.4. Such bowl are used for better air utilization especially when using EGR and they have better S/V ratios than re-entrant bowls and also have better tolerance for advance timings due to their wider bowl diameters implementing lower compression ratios. These bowls are also benefiting for lower soot in oil. However, simulation results would finally help toward the selection of bowl-swirl-spray cone angle with optimal matching resulting in improvement in indicated efficiency and soot emissions.

Multiple injection strategies and EGR dilution has shown benefits of improving engine efficiencies by reducing the wall heat transfer losses (Uchida et al. 2013).

Regarding the compression ratio selection though higher compression ratio increases the theoretical thermal efficiency of engine, followings aspects has also to be considered:

- not high enough to reach the base engine peak firing pressure limit
- not low enough to affect the cold startability requirement of engine.

Engine compression ratio selection is limited by the engine mechanical load limits i.e. engine's peak firing pressure limits. Generally we are using CR of 16, however from cold startability point of view as per Dreisbach et al. (2007), a CR of 16.5 is good enough for cold startability. In Europe and US the CR used are typically in range of 16–18 (Vojtech 2018). However, these engine would be operating at peak firing pressures of 200–230 bar, which is quite high compare to our heavy duty diesel engines. It is well known that increasing compression ratio, reducing cut-off ratio and increasing γ (gamma) all increases the theoretical thermal efficiency. However, to get proportional rise in engine brake thermal efficiency would require reduction of losses like frictional losses and heat transfer losses. With increase in compression ratio there are effects of rising peak firing pressures, residing pressures between the piston ring and cylinder wall, thinning of lube oil film, more chances of contact between rubbing parts of connecting rod and crankshaft bearing etc. and these all factors can contribute in raising the engine FMEP (friction mean effective pressures). As a result the benefit obtained from increase in theoretical efficiency and thermodynamic efficiency would not reflect in engine brake efficiency proportionally. Similarly the heat transfer losses may increase due to use of higher compression ratio causing the compression end gases to be at higher temperatures. This would further offset the gain in theoretical efficiency of engine. However, such increased losses has to be dealt with intensive optimization of not only the combustion system of the base engine but also the engine sub system design, so that the target engine brake thermal efficiency of the engine is achieved. Engine operating at higher boost experiences high peak firing pressures.

From a study by Funayama et al. (2016), it was observed that almost 70 bar increase in PFP happened resulting in 1.5% point increase in FMEP due to rise in CR by 9 points or for 2 point increase in CR would lead to 15 bar rise in PFP, 0.3% point increase in FMEP or 0.5% drop in mechanical efficiency and 0.2 bar rise in FMEP value. It is known that with rise in PFP generally the FMEP increases. As per one of the study by Kouremenos et al. (2001), the FMEP rise from increasing the PFP can be predicted to an extent using the below expression:

$$\mathrm{FMEP} = c \times 0.0104 \times \mathrm{IMEP} + 0.15 \times \mathrm{Vp} + 1.0955 \times 10^{-3} \times \mathrm{Pmax} \quad (8.19)$$

Equation 8.19 shows FMEP has a linear relation with Pmax. Here c is a co-efficient which has to be adjusted to match the with experimental value of FMEP, here the value is 2 for 100% load. Vp is the mean piston speed, Pmax maximum combustion pressure and IMEP is indicated mean effective pressure. Before using this equation the coefficient needs to be determined with existing data and later used to determine the FMEP. Once the peak firing pressure estimate for required compression ratio and engine boost is done, the FMEP can be estimated using the Eq. 8.9.

As per Kouremenos et al. (2001), no serious decrease of mechanical efficiency or increase of FMEP is noticed for PFP in range of 180–190 bar. In certain cases if the desired mechanical efficiency or FMEP level are not reached we may have to adopt to certain advanced sub-system engine technologies to achieve the target engine FMEP levels. As per Horvath et al. (2020) for their heavy duty engine project with PFP target of atleast 280 bar from existing base engine PFP of 230 bar was to be achieved with an additional requirement of maintaining the same existing base engine FMEP at 0.61 bar. And they include use of technologies like variable oil pump, low viscosity lube oil, friction optimized seal, map controlled water pump and optimized belt drive. Apart from this there were suitable design changes related to cranktrain, piston ring tension etc. which are beyond the scope of this chapter to maintain same existing base engine FMEP despite of increase in PFP.

8.3.2.2 Alternate Fuel

Till now we have discussed the base engine technologies to achieve the target of 45% brake efficiency, moving forward we will discuss the role of various low carbon fuels and carbon neutral fuels in conjunction with latest technologies to achieve 45% efficiency and how further these engines fitted with latest technologies can be can forward to achieve 50% brake thermal efficiency.

Natural gas in the form of LNG and *direct injection technology* can be a promising latest technology in making the new fleet of heavy duty vehicles more efficient and cleaner. When using natural gas in form of LNG offers the range advantage which is a limitation for CNG vehicles especially for long haul application.

As per report (https://evreporter.com/lng-for-heavy-commercial-vehicles/) India plans to increase the share of natural gas from present 6.2–15% in its primary energy mix by 2030. For achieving this efforts are underway to increase domestic production of gas and enhance LNG import capacity. Presently there are ~6 LNG fuel stations in India but government plans to set-up 1000 LNG stations as mentioned in Sect. 8.2.1. As per another report (https://www.forbesindia.com/blog/economy-policy/why-trucks-should-consider-switching-to-liquefied-natural-gas-lng/) the overall cost of ownership for LNG trucks are 32% lower than diesel trucks. Also they can lower the greenhouse gases by ~30% from conventional fuels (https://auto.economictimes.indiatimes.com/news/commercial-vehicle/mhcv/tata-motors-completes-delivery-of-indias-first-lng-bus-order/74439051). Hence they are seen as an suitable option for long haul applications. Some of the options to make use of LNG in long haul trucks are as mentioned below:

- LNG system adopted in buses and trucks as retro-fitment options discussed in Sect. 8.3.1

- New LNG buses and trucks manufactured by original equipment manufacturer like TATA Motors recently launched Starbus running on LNG
- New advanced LNG trucks and buses using direct injection technology as mentioned above.

We will here discuss the last technology. Direct injection of natural gas following the 5–10% (McTaggart-Cowan et al. 2017) of diesel direct injection can achieve similar or even higher than modern heavy duty diesel engine efficiency, better combustion efficiency and better transient response than the fumigated natural gas technologies. Typically here the combustion is unlike other dual-fuel combustion approaches where we have premixed charge and flame propagation. Here like conventional diesel combustion we have mixing controlled combustion. The better part of such combustion is here we have combustion similar to conventional diesel combustion but the bulk of fuel is not a high carbon fossil fuel like diesel instead we have low carbon fuel like natural gas.

There are different approaches to introduce both natural gas and diesel directly into the cylinder. There can be two separate injectors for both the fuels or it can be HPDI (High Pressure Direct Injection) (McTaggart-Cowan et al. 2015).

HPDI combustion has the fresh charge inducted externally mixed with EGR the charge is then compressed and near the TDC the diesel pilot is injected following that happens the natural gas direct injection. The diesel pilot acts as the multiple source of ignition that ignites the natural gas jets (McTaggart-Cowan et al. 2015). Some of the advantages of using HPDI are as mentioned below:

- Because the load control is via natural gas control so throttling or part-throttling related losses
- End-gas knock not a concern so we can retain the diesel like compression ratio and diesel or even higher brake thermal efficiency
- Even a low load unlike fumigated dual fuel combustion techniques or port injected techniques lower methane emissions due to avoidance of over-mixing of natural gas due to injection adjustment, also no fuel observed in crevice region or end-gas region.

There are lot of research presently going on to optimize the diesel pilot and natural gas injection for better emission control thereby retaining the same engine efficiency levels achieved by such direct injection technologies. A study by McTaggart-Cowan et al. (2017), reported interesting results on such optimization. By allowing slightly more premixing of natural gas before start of natural gas ignition there were report on significant reduction (90%) in particulate levels with almost same NO_x and methane emissions and slight (2%) improvement in efficiency. However, optimization of injection process plays a crucial role to avoid methane emissions as though with slightly more premixed charge heat release increases but further high increase in premixing can result in positive ignition dwell thereby significantly affecting methane emission. Also issue of high pressure rates and cycle to cycle variation are also reported with this approach. Below Table shows the result for slightly premixed combustion (SPC) using HPDI (Table 8.8).

Table 8.8 Performance and emission using (SPC) in HPDI technology (McTaggart-Cowan et al. 2017)

S. No.	Parameters	Values
1	Fuel used	• Natural gas (Direct injection) • Diesel (Direct injection)
2	Test conditions	• Speed and load: 1500 RPM and 16.5 bar IMEPg
3	Performance and emissions	• Brake thermal efficiency: ~45% • NO_x: 1.3 g/kWh • PM: <0.002 g/kWh • MHC: 0.6 g/kWh • PRR: 8 bar/° CA

Table 8.9 Performance and emission using synthetic diesel (Uchida et al. 2014)

S. No.	Parameters	Values
1	Fuel used	• Synthetic diesel (Gas to liquid)
2	Test conditions	• Speed and load: 1620 RPM and 13.0 bar BMEP
3	Performance and emissions	• Brake specific energy consumption: ~same as base diesel • NO_x: ~3.8 g/kWh (same as diesel) • Soot: 0.02 FSN (almost 4 times lower than diesel)

Synthetic fuel which are also seen as drop-in fuels for older or new diesel vehicles can be another option utilized for reducing CO_2 emissions and even fuel efficiency improvements.

Table 8.9 shows the results of emission and performance obtained by using Fischer-Tropsch diesel in a work by Uchida et al. (2014).

In the same study the emission over transient with aftertreatment were reported to be as NO_x less than 0.7 g/kWh and PM close to 0.01 g/kWh. It is observed that such smoke advantage for synthetic fuels over conventional diesel fuels could be due to lower aromatic fractions and also such fuels can better tolerate EGR for NO_x reductions.

Hydrotreated vegetable oil (HVO) is another renewable bio-based diesel fuel, also called as renewable diesel fuel. These fuels can be better than bio-diesels for the reasons already discussed in Sect. 8.2.4. They are another type of drop-in fuels that can be used for existing and new fleet of heavy duty vehicles due to their properties similar to fossil diesel.

Table 8.10 highlight one of the interesting emission and performance result obtained by using HVO.

The HVO fuels which is a paraffinic fuel has high H/C ratio, with no aromatics and sulphur reduces the soot formation. Whereas the effect of HVO as a fuel for NO_x formation has dual theories, one says that due to absence of aromatics and oxygen as fuel component the adiabatic flame temperatures are low however other theory say that due to high cetane number of fuel the ignition delay is lower and the start of

Table 8.10 Performance and emission using HVO (Aatola et al. 2008)

S. No.	Parameters	Values
1	Fuel used	• HVO
2	Test conditions	• Speed and load: 1500 RPM and 100% load
3	Performance and emissions	• Brake specific energy consumption: ~same as base diesel • NO_X: ~4.0 g/kWh • Soot: 20–40% lower than diesel • SFC: 4–7% lower than diesel

combustion is close to TDC leading to high temperature and pressures causing high NO_x formations. Hence a clear trend is often not understood. Though in the result discussed here the trend observed was reducing NO_x emissions with HVO as fuel.

Di-methyl Ether *(DME)* is another clean alternate for heavy duty transportation. However due to large differences between the conventional diesel fuel and DME with regard to its physical and chemical properties a significant amount of changes are required for engine and vehicle system as described under Sect. 8.2.3. Hence adopting these fuels directly on existing or older fleet of vehicles would not be possible and hence require new development.

Table 8.11 demonstrate engine emission and performance obtained from using DME as fuel in diesel engine:

- Till now we have discussed the base engine technologies that can be adopted to make heavy duty engines 45% brake efficient and cleaner. Moving ahead we will discuss the latest technologies and better alternate fuels to achieve 50% brake efficiency.
- Though not covered in this chapter thermal management would be another area which has to be more understood and implemented especially when we are using EGR for better effectiveness of the aftertreatment system.
- Moreover, though not discussed there are even other areas like aftertreatment optimization to reduce the fuel penalty related to DPF active regeneration, correct prediction of ash accumulation, urea consumption optimization which can also reduce the heavy duty total ownership cost.

Table 8.11 Performance and emission using DME (Kato et al. 2015)

S. No.	Parameters	Values
1	Fuel used	• DME
2	Test conditions	• Speed and load: 1300 RPM and 12 bar BMEP
3	Performance and emissions	• ISFC: 209 g/kWh • NO_X: ~0.8 g/kWh • Soot: Very low

8.3.3 Modern Heavy Duty Engines on Path of Achieving 50% Brake Thermal Efficiency

In the Sect. 8.3.1 above we have discussed how we can aim for baseline improvement on engine efficiency of 10–12% from existing heavy duty diesel engines in Indian market and still be cleaner. This baseline improvement would see peak brake thermal efficiency of 45%. However, these improved new heavy duty diesel engines would need to achieve the target brake thermal efficiency of 50% thereby demonstrating their capability to contribute significantly (by atleast 20% reduction as compare to base engines) in meeting the global green house gas reduction targets of 30% by 2030. To achieve another milestone as predicted by Fig. 8.2, Table 8.12 shows the target engine efficiencies and some of engine technologies that can be adopted.

Improving the thermodynamic efficiency would require maximizing the effective expansion work, reducing the combustion duration, optimizing the CA50 and reducing the heat transfer and exhaust losses. Also when moving toward achieving the engine brake efficiency of 50%, higher level of compression ratio are needed. Geometric compression ratio higher than >19.0 are required, presently in market the trend is to use CR in range of 16–18 as mentioned in Sect. 8.3.2. With higher CR the

Table 8.12 Target on engine efficiencies and technologies for achieving 50% BTE for heavy duty diesel engines

Engine parameters	40% BTE	45% BTE	50% BTE
Engine efficiencies			
$\eta_{thermodynamic}$	~48	~50	~53
$\eta_{gas\ exchange}$	~94	~96	~97
$\eta_{mechanical}$	~88	~93	~97
$\eta_{brake\ efficiency}$	~40	~45	~50
PMEP	~2.7%/1.2 bar	~2%/0.9 bar	~1.6%/0.7 bar
FMEP	~5.5%/2.4 bar	~3%/1.5 bar	~1.5%/0.7 bar
Thermodynamic improvement		• CR increase + EGR • PFP increase • Bowl opt	• CR increase + Miller + EGR • PFP increase • Bowl opt • Flexible rate shaping • Thermal coating
PMEP reducing area		• Turbo eff	• Turbo eff • 2-stage/EGR scroll/E-Turbo • LP EGR/Split EGR cooling
FMEP reducing area	• Low friction ring, improved bearing, piston cooling, variable oil pump, map controlled water pump, low viscous oil • Downspeeding, optimized manifolds etc.		

limit for peak firing pressures also needs to be raised. Further with increase in bmep levels there are higher need for boost which further increases the peak firing pressure levels. The rise in boost levels comes from different reasons like higher boost demand due to use of miller cycle, higher dilution, higher bmep levels etc. High CR and miller cycle ensure great benefit by improving the engine brake efficiency at high load and also offer exhaust temperature benefits for aftertreatment system due to more optimized lambda at low load. This is achieved by reducing the effective compression ratio at high load and maintaining the same effective expansion ratio with increased geometric compression ratio. Hence miller cycle when combined with higher compression ratio is beneficial for achieving higher engine brake efficiency while still being within the limits of engine peak firing pressure limits. The use of miller cycle also lowers the NO_x emissions. There are number of ways to use miller cycle. A study performed by Garcia et al. (2020) following approached were analyzed when using miller cycle:

- maintaining constant intake pressure
- maintaining constant lambda
- maintaining constant lambda + improved turbocharger efficiency.

The first approach though yield benefit on reducing effective compression, reducing EGR requirements for same NO_x levels but has high deterioration on combustion which ultimately penalize fuel consumption severely. However the second approach also reduces effective compression ratio, maintains the engine peak firing pressures within limits, and minimizes the penalty on fuel consumption and smoke levels. However, when maintaining lambda with improved turbocharger efficiency can even achieve no penalty on fuel consumption and smoke levels.

Talking about the injection rate modulation, these are effective way to control emissions particularly NO_x emission. Here a number of injection rate shapes are adopted across various regions of engine operating map. As per a study by Benajes et al. (2006) by adopting boot shape we reduce the mass flow of fuel at start of injection. This reduces the amount of fuel mixed with air during early phase of combustion leading to lower heat release rate and lower in-cylinder local temperatures resulting in lower NO_x emissions. Further using such injection rate shape we have later peaks of heat release and by that time we already have reduced local O_2 which also contributes to lower NO_x emission.

Now discussing about the pumping and friction reduction area the target for PMEP and FMEP are almost as 1.5% of total fuel related mean effective pressures. This requires the gas exchange efficiency and mechanical efficiency to be in the range of 97%. There are study which however shows that for targeting 50% BTE the reasonable gas exchange and mechanical efficiencies could be as high as 100% and ~94.3% for closed cycle efficiency of 53% (Mohr et al. 2019). The challenges to achieve such high values of gas exchange efficiencies becomes tougher as:

- there is need to introduce more EGR% (>25%) at higher bmep to simultaneously reduce engine out emissions and achieve higher engine efficiencies

- there is need to operate both maximum torque and maximum power points at best compressor efficiency zone.

The overall turbocharger efficiency required can be ≥60%. There are certain turbocharging technologies and EGR systems like 2-stage turbocharging, twin scroll turbocharger, use of electric turbocharger and use of low pressure EGR that can be very effective in achieving above targets for gas exchange efficiencies.

2-stage turbocharger is an efficient way to meet the engine boost requirement for higher EGR level/higher BMEP levels. With rise in EGR% levels to lower engine out NO_x emissions the need for higher intake manifold pressure also rises to maintain same lambda and with this the limit of single stage compressor reaches. Also the need to limit the intake manifold temperature further restricts the compressor out temperature. With higher EGR levels the pressure requirements are much higher than 2.5 bar abs (Watson and Janota 1982). Such pressure levels marks the operation limit for single stage turbochargers due to restriction on impeller speed of compressor. Moving to 2 stage compressor allows more flexibility in this case. The high pressure turbocharger can be matched to maximum torque and low pressure turbocharger can be matched to maximum power and this combination together with intercooler can achieve not only the higher pressure requirements but can also limit the pumping work and include more number for engine operating points to operate at higher compressor efficiency regions. However, such 2 stage turbocharger are not without complexity. They have complicated piping, larger inertia and sophisticated control.

Twin-scroll asymmetric turbochargers are another suitable option when there is demand for higher EGR% and higher overall turbocharger efficiency. Here we have two scrolls at the inlet of turbine. One scroll is sized to meet the EGR% requirement for NO_x control and another one is for lambda. These two scrolls are connected to separate banks of engine cylinder. When the EGR valve is open the one set of cylinders works as EGR pump. The remaining exhaust flow of these banks of cylinder flows through the EGR scroll of the turbine. The other bank of cylinders are not connected to this EGR scroll. Using this architecture we always able to transport higher EGR% and still able to have positive charging cycle i.e. the difference between the intake manifold pressure and average turbine inlet pressure of both cylinder group is >0 (Chebli et al. 2013).

Low pressure and high pressure EGR system is another approach to attain higher overall turbocharger efficiency when EGR% requirements are high. Using both LP and HP EGR system we can regulate and optimize the flow through turbine during the engine operating range. At lower speed where the engine flow is less the LP EGR is operated and at higher engine speed and load where the flow is high the HP EGR is operated. Using such a strategy we can effectively reduce the mass flow range through which the engine has the turbocharger has to work and thus resulted in overall higher turbocharger efficiency (Svensson et al. 2017).

Future scope of investigation could be as follow:

- How we can further increase the compression ratio (~20) using miller cycle to get higher effective expansion ratio and still maintain FMEP at ~2%?Also how with optimized bowl we can reduce cooling heat losses?
- How with efficient turbocharger ($\eta = 65$–70%) and EGR system (EGR = 30%) we can achieve engine out $No_x = 1.5$ g/kWh and still maintain PMEP at 2%?
- How LP + HP EGR can be used to significantly reduce PMEP?
- How multiple injection + quality dilution + optimized bowl system (cone angle + swirl + bowl) can be used to reduce wall heat losses and get benefit on engine efficiency?
- How downsizing + efficient turbocharging can benefit us in both friction reduction and efficiency improvement by 5–7%?
- How use of turbocompound especially at rated speed can yield benefit of 6–7% on BTE by producing net + MEP offsetting marginal PMEP penalty after considering minimum PMEP for driving EGR?

So till now we have seen how some of the latest technologies can be adopted to make heavy duty engines reaching 50% brake efficient and cleaner. Moving onward we shall discuss some of the ways to make these engines sustainable.

There are other approaches too that can be adopted for making these engines 50% brake efficient like engine downsizing, turbo-compounding, electrically assisted compressor, cylinder de-activation and even more efficient with technologies like waste heat recovery and hybridization.

8.3.4 Future Sustainable Heavy Duty Engines on Path of Achieving ≥ 50% Brake Thermal Efficiency

We have already discussed diesel fuel for 45–50% BTE and low carbon and carbon neutral fuels role for achieving 45% BTE under Sects. 8.3.2 and 8.3.3. Now we will discuss under this section how we can proceed toward the vision of seeing heavy duty diesel engines achieving 50% BTE with sustainability using low carbon and carbon neutral fuel.

8.3.4.1 Ethanol

This advanced combustion potential has already been discussed in detail under Sect. 4.4.1 of publication (Agarwal et al. 2021). Already there are research work which has shown that using such a advanced combustion concept how engine brake efficiency of 48.5% was achieved and engine out emission including NO_x (0.25 g/kWh) and soot (0.06 FSN) were at its lowest levels. Even engine advanced combustion concepts like RCCI (Reactivity Controlled Compression Ignition) can be used to achieve such high engine efficiencies when using both diesel and ethanol as fuel.

Table 8.13 Performance and emission using dual fuel-fumigated technology (Nieman et al. 2019)

S. No.	Parameters	Values
1	Fuel used	• Natural gas (Fumigated) • Diesel (Direct injection)
2	Test conditions	• Speed and load: 1000 RPM and 15 bar bmep
3	Performance and emissions	• Brake thermal efficiency: 47.2% • NO_x: 5–6 g/kWh • Smoke: ~0.08FSN • HC: 2–3 g/kWh (mostly methane) • PRR: 10 bar/° CA

8.3.4.2 Natural Gas

Natural gas is one such fuel which has been researched with a wide range of conventional and advanced combustion approaches like HPDI (High Pressure Direct Injection), Stoichiometric combustion and Dual Fuel Combustion and surprisingly all have their own potential.

As per https://dieselnet.com/tech/engine_natural-gas_heavy-duty.php, HPDI 2.0 which addressed several challenges of HPDI 1.0 such as incompatibility with modern heavy duty engines that use smaller sized injectors are anticipated to achieve BTE as high as 48%.

Another engine technology with natural gas—fumigated dual fuel combustion has achieved BTE of 47.2% as shown in Table 8.13

Future scope of investigation could be as follow:

- How NO_x-smoke trade off can be further improved?
- How diesel injection timing can be optimized to reduce the PRR without sacrificing on brake engine efficiency?

8.4 Summary

The world is changing fast in terms of 3 E's increasing economy size, energy and environment crisis. There are several possible areas including urban mobility, transport modal transition, electrification, freight efficiency and cleaner fuels to alleviate the situation. In this chapter we have discussed the last two options to an extent.

- Stricter emission norms like BS VI has reduced the heavy duty vehicular emission to large extent. However, there is a large fleet of older and aged vehicles which needs to be made cleaner using retro-fitment options running on low carbon fuels like natural gas. Drop-in fuels with fuel characteristics similar to existing diesel fuels (HVO and synthetic diesel) with certain emission reducing properties can make relatively these vehicles with lesser carbon footprint.

- Using base engine technology like EGR, turbocharger, injection systems etc. we can upgrade our heavy duty engine efficiencies to atleast 45%.
- Using latest engine technologies like direct injection (HPDI), efficient turbocharger, increased firing pressures, optimized combustion system, efficient EGR coolers and other friction reduction technologies we can target 50% BTE.
- Moving toward making these heavy duty engines sustainable we can adopt advanced bio-fuels like ethanol, methanol and DME.

References

Aatola H, Mikkonen S et al (2008) Hydrotreated vegetable oil (HVO) as a renewable diesel fuel: trade-off between NO_x, particulate emission, and fuel consumption of a heavy duty engine. SAE technical paper 2008-01-2500

Agarwal AK et al (2021) Methanol—a sustainable transport fuel for CI engines, Part II and Chap 4. Springer, Berlin

Air pollution press release (2021). Centre for Science and Environment

Arcoumanis C, Bae C et al (2007) The potential of di-methyl ether (DME) as an alternative fuel for compression-ignition engines: a review. ScienceDirect

Badshah H et al (2019) Current state of NO_x emissions from in-use heavy duty diesel vehicles in the United States. ICCT, Nov 2019

Banerjee S (2021) India cannot afford to delay and weaken fuel efficiency regulations for vehicles, says new CSE analysis. Centre for Science and Environment. Press release Aug 2021

Benajes J et al (2006) Influence of injection rate shaping on combustion and emissions for a medium duty diesel engine. J Mech Sci Technol (KSME Int. J.) 20(9):1436–1448

Beneath the waves "Norway and CO_2 Emission". Documentary film

Besch MC et al (2015) Emission characterization from different technology heavy-duty engines retrofitted for CNG/diesel dual-fuel operation. SAE technical paper 2015-01-1085

Boretti A et al. (2013) CNG fueling strategies for commercial vehicles engines-a literature review. SAE Technical Paper 2013-01-2812

Centre for Science and Environment (CSE) report—India will have over 2 crore end-of-life vehicles by 2025, Sept 2020

Chadwell C et al (2011) Boosting simulation of high efficiency alternative combustion mode engines. SAE technical paper 2011-01-0358

Chebli E et al (2013) Development of an Exhaust-gas turbocharger for HD Daimler CV engines. Autotech Review

Cohn D, Bromberg L (2018) Dual fuel gasoline alcohol engines for heavy duty trucks: lower emissions, flexible—fuel alternative to diesel engines. SAE technical paper 2018-01-0888

Conference report: SAE WCX 2021 digital summit. dieselnet.com

Decarbonising the Indian transport sector pathways and policies. Climate Action Tracker, Dec 2020

Dreisbach R et al (2007) HD base engine development to meet future emission and power density challenges of a DDI™ engine. SAE technical paper 2007-01-4225

Fast tracking freight in India, NITI Aayog, June 2021

Figer G (2014) Commercial vehicles natural gas engines with diesel like efficiency. MTZ

Funayama Y et al (2016) A study on the effects of a higher compression ratio in the combustion chamber on diesel engine performance. SAE technical paper 2016-01-0722

Garcia E et al (2020) Impact of Miller cycle strategies on combustion characteristics, emissions and efficiency in heavy-duty diesel engine. SAE technical paper 2020-01-1127

Ghosh S (2016) Production of methanol and DME. TIFAC, DST

Gill SS et al (2010) Combustion characteristics and emissions of Fischer-Tropsch diesel fuels in IC engines. Prog Energy Combust Sci. https://doi.org/10.1016/j.pecs.2010.09.001
Global CO_2 emission trend. Our world in data. https://ourworldindata.org/co2-emissions
Gruden I et al (2013) New 14.8-L HD truck engine from Daimler for NAFTA, MTZ Worldwide November 2013
Hardenberg HO, Schaefer SJ (1981) The use of ethanol as a fuel for compression ignition engine. SAE technical paper 811211
Heywood JB (2018) Internal combustion engine fundamental, 2nd edn.
Horvath A et al (2020) The future heavy-duty engine—basic engine concept for maximum CO_2 reduction. AVL, MTZ
https://auto.economictimes.indiatimes.com/news/commercial-vehicle/mhcv/tata-motors-completes-delivery-of-indias-first-lng-bus-order/74439051
https://dieselnet.com/tech/engine_natural-gas_heavy-duty.php
https://energy.economictimes.indiatimes.com/news/oil-and-gas/india-to-have-10000-cng-stations-in-next-10-years-on-track-to-adapt-cleaner-fuels-dharmendra-pradhan/65700258
https://evreporter.com/lng-for-heavy-commercial-vehicles/
https://www.business-standard.com/article/economy-policy/india-set-to-attract-rs-10-000-cr-for-1-000-lng-stations-in-three-years-120111900653_1.html
https://www.forbesindia.com/blog/economy-policy/why-trucks-should-consider-switching-to-liquefied-natural-gas-lng/
https://www.norsk-e-fuel.com/en/
Improving fuel efficiency for heavy duty vehicles of 3.5–12 tonne in India: benefits, cost and environmental impact. ICCT, Feb 2019
Johansson B et al (2009) Effects of different type of gasoline fuels on heavy duty partially premixed combustion. SAE technical paper 2009-01-2668
Johansson B et al (2010) Influence of inlet pressure, EGR, combustion phasing, speed and pilot ratio on high load gasoline partially premixed combustion. SAE technical paper 2010-01-1471
Joshi A (2021) Review of vehicle engine efficiency and emissions. SAE technical paper 2021-01-0575
Kajitani S et al (2001) A study of low compression ratio dimethyl ether diesel engine. Ibaraki University, Japan
Karali N, Gopal AR (2017) Improved heavy duty vehicle fuel efficiency in India. ICCT, April 2017
Kato M, Gill DW et al (2014) An improvement into the effect of fuel injection system improvement on the injection and combustion of dimethyl ether in a diesel cycle engine. SAE technical paper 2014-01-2658
Kato M et al (2015) An experimental study of injection and combustion with dimethyl ether. SAE technical paper 2015-01-0932
Kouremenos et al (2001) Development of a detailed friction model to predict mechanical losses at elevated maximum combustion pressures. SAE technical paper 2001-01-0333
Kubsh J (2017) Diesel retrofit technologies and experience for on-road and off-road vehicles
Lavoie GA et al (2012) Thermodynamic sweet spot for high efficiency, dilute, boosted gasoline engines. Int J Engine Res. https://doi.org/10.1177/1468087412455372
Lynch LA et al (2020) On-road heavy-duty low NO_x technology cost study. National Renewable Energy Laboratory
McTaggart-Cowan G et al (2015) Direct injection of natural gas at up to 600 bar in a pilot ignited heavy duty engine. SAE technical paper 2015-01-0865
McTaggart-Cowan G et al (2017) Effect of injection strategies on emissions from a pilot-ignited direct-injection natural-gas engine—part II: slightly premixed combustion. SAE technical paper 2017-01-0763
MD/HD natural gas engine efficiency research needs (2017). Argonne, Sandia, Oak Ridge and NREL
Mohr D et al (2019) The thermodynamic design, analysis and test of Cummins Supertruck 2 50% brake thermal efficiency engine system. SAE technical paper 2019-01-0247

Moser FX et al (2004) Lowest engine out emissions as the key to the future of the heavy duty diesel engine—new development results. AVL
Neame GR et al (1995) Improving the fuel economy of stoichiometric fuelled SI engine by means of EGR and enhanced ignition—a comparison of gasoline, methanol and natural gas. SAE technical paper 952376
NESTE MY renewable diesel—low carbon biodiesel. https://www.neste.com/products/all-products/renewable-road-transport/neste-my-renewable-diesel#248febd3
Nieman DE et al (2019) Utilizing multiple combustion modes to increase efficiency and achieve full load dual-fuel operation in a heavy duty engine. SAE technical paper 2019-01-1157
Number of biogas plants across India 2020 by state. Statista
Pavlenko N et al (2019) The potential for advanced biofuels in India: assessing the availability of feedstocks and deployable technologies. ICCT
Ponskshe A (2020) Policymaking towards green mobility in India. Observer Research Foundation
Richards BG et al (1990) Methanol-fueled caterpillar 3406 engine experience in on-highway trucks. SAE technical paper 902160
Sathiamoorthy B et al (2020) Testing the pollutant emissions and fuel efficiency of a commercial bus in India. ICCT, June 2020
Shukla PC et al (2018) Investigation of particle number emission characteristics in a heavy duty compression ignition engine fueled with hydrotreated vegetable oil (HVO). SAE technical paper 2018-01-0909
Stanton DW et al (2013) Systematic development of highly efficient and clean engines to meet future commercial vehicle green house gas regulations. L. Ray Buckendale lecture. SAE technical paper 2013-01-2421
Svensson E et al (2017) Combined low and high pressure EGR for higher brake efficiency with partially premixed combustion. SAE technical paper 2017-01-2267
Teng H et al (2009) Improving fuel economy for HD diesel engines with WHR Rankine cycle driven by EGR cooler heat rejection. SAE technical paper 2009-01-2913
The energy transformation. SHELL scenario sky 1.5. www.shell.com
Tsuchiya T et al (2006) Development of DME engine for heavy duty truck. SAE technical paper 2006-01-0052. http://doi.org/10.4271/2006-01-0052
Uchida N et al (2013) Reexamination of multiple fuel injections for improving the thermal efficiency of a heavy-duty diesel engine. SAE technical paper 2013-01-0909
Uchida et al (2014) Simultaneous improvement in both exhaust emissions and fuel consumption by means of Fischer–Tropsch diesel fuels. Int J Engine Res. https://doi.org/10.1177/1468087412456528
Vojtech R (2018) Advanced combustion for improved thermal efficiency in an advanced on-road heavy duty diesel engine. SAE technical paper 2018-01-0237
Wallner T, Musculus M, Curan S, Zigler B (2017) MD/HD natural gas engine efficiency research needs. Argonne National Lab, Sandia National Lab, Oak Ridge National Lab and NREL
Watson N, Janota MS (1982) *Turbocharging the internal combustion engine*, The Macmillan Press Ltd., ISBN 0 333 24290 4
Zhang J et al (2018a) Quantitative estimation of the impact of ash accumulation on diesel particulate filter related fuel penalty for a typical modern on-road heavy duty diesel engine. Appl Energy Sci. https://doi.org/10.1016/j.apenergy.2018.08.071
Zhang Y et al (2018b) An experimental and computational investigation of gasoline compression ignition using conventional and higher reactivity gasolines in a multi-cylinder heavy-duty diesel engine. SAE technical paper 2018-01-0226

Chapter 9
Potential and Challenges of Using Biodiesel in a Compression Ignition Engine

Akshay Garg, Balendra V. S. Chauhan, Ajitanshu Vedrantam, Siddharth Jain, and Sawan Bharti

Abstract Alternative green fuels play a vital role in meeting the energy demand. Biodiesel has been recognized as a sustainable low-carbon alternative fuel for diesel engines. This chapter reviews the potential and various challenges associated with the use of biodiesel as a fuel in a compression ignition engine. The chapter further reviews various thermo-physical–chemical properties of biodiesel and their significant effects on combustion characteristics like cylinder pressure, rate of pressure rise, mean gas temperature, and engine performance parameters including Brake specific fuel consumption (BSFC), Brake Thermal Efficiency (BTE), and Brake power (BP). Moreover, engine emissions and various emission control techniques i.e. blending of fuels, hardware modifications in engines are thoroughly reviewed. The literature reports the use of different refined oils and feedstocks to produce biodiesel utilizing different methods e.g., pyrolysis, wet washing, blending or direct use, transesterification, and microemulsions. But still, considerable efforts are needed for the commercialization of biodiesels due to their properties of higher viscosity, corrosivity, and less stable nature. So, the present work concludes with an optimized approach to consider the potentiate biodiesel as a fuel in CI engines.

Keywords Biodiesel · Potential · Challenges · Engine · Modifications

A. Garg · S. Jain
Department of Mechanical Engineering, College of Engineering Roorkee, Roorkee, India

B. V. S. Chauhan
Automotive Fuels and Lubricants Application Division, CSIR-Indian Institute of Petroleum, Dehradun, Uttarakhand, India

A. Vedrantam (✉)
Universidad de Alcalá, Alcalá de Henares, Spain
e-mail: ajitanshu.m@invertis.org

S. Bharti
Invertis University, Bareilly, India

A. V. Agarwal and H. Valera (eds.), *Potential and Challenges of Low Carbon Fuels for Sustainable Transport*, Energy, Environment, and Sustainability,
https://doi.org/10.1007/978-981-16-8414-2_9

Nomenclature

BP	Brake power
BMEP	Brake mean effective pressure
BSFC	Brake specific fuel consumption
BTE	Brake thermal efficiency
Bx	Biodiesel proportion in blend
CI	Compression ignition
CO	Carbon monoxide
CO_2	Carbon dioxide
CR	Compression ratio
DI	Direct injection
EGR	Exhaust gas recirculation
EGT	Exhaust gas temperature
HRR	Heat release rate
ID	Ignition delay
IMEP	Indicated mean effective pressure
IT	Injection timing
IP	Injection pressure
NO_x	Oxides of nitrogen
SCR	Selective catalytic reduction
UHC	Unburnt hydrocarbons

9.1 Introduction

The growth of our world and its well-being is directly based on energy growth. Rapid urbanization and industrial evolution have increased the high demand for energy overall in the world. Hence in this time of comprehensive exchange precariousness, energy plays an important role to support our economic and social growth (Majeed et al. 2021). But allocating this energy around the globe is quite a daunting task as it must be fulfilled with greater responsibility. Thus, it has become a serious concern to develop various sources of energy to meet the high demand for energy in the world. Our over-dependency on fossil fuels is causing their depletion rapidly. This depletion can lead us to fail to meet the energy demand in the upcoming years (Raheem et al. 2020). Besides energy security, the burning of fossil fuels also causes high harmful gaseous emissions which are affecting the environment as well as human beings (Liddell and Morris 2010). Thus, there is a high need to work for alternative sources of energy to meet the energy demand and control the pollution in upcoming years. Considering these problems, various researchers, industrialists, and world leaders are working to promote energy sustainability efficiently. The only way is to maximize the use of alternative sources of energy (Mohsin et al. 2021).

A huge increase in biofuel production capacity has been observed worldwide. There has been an increase of 9.7% in biofuel production in the last decade (BPSTATS 2019). The transport sector has been proved the most liquid energy-consuming and emission-producing sector in the world (BP Energy Outlook 2016). Hence, oil has been always a dominating energy source. The current scenario is to focus on the production of biodiesel as an alternative to conventional diesel. Biodiesel is eco-friendly fuel as it is biodegradable, non-toxic, and can be reproduced (Mathew et al. 2021). Biodiesel is a fatty acid alkyl ester produced from vegetable oils and animal fats. There are four general methods to produce biodiesel like micro-emulsification, blending, thermal cracking, and transesterification. In the micro-emulsification method, the feedstock oils are blended with alcohols to decrease the viscosity of the oil to the limit required to be used as a fuel. This method improves the spray and atomization characteristics of the oils. However, in the blending process, the feedstock oils are directly blended with pure diesel fuel. This results in acceptable performance of oils in the diesel engines without any hardware modifications. Thermal cracking is a process of thermal treatment of feedstock seeds or raw biomass to produce the biofuel that contains favorable fuel properties to be used in diesel engines. Transesterification is the method given more attention to producing biodiesel by various researchers as it produces biodiesel with good fuel properties (Garg et al. 2021a, b). In a transesterification reaction, triglycerides react with alcohol in the presence of a catalyst to produce alkyl ester and glycerol as a by-product (Ambaye et al. 2021; Madheshiya and Vedrtnam 2018). The process undergoes three consecutive reversible reactions, where triglyceride is sequentially converted into diglyceride, then monoglyceride, and glycerol. An alkyl ester is produced at each stage (Raheem et al. 2020).

9.2 Fuel Properties of Biodiesel

Fuel properties of any fuel are defined and varied according to the chemical and physical structure of that fuel. Fuel properties of biodiesel mainly depend upon the type of feedstock used and the production process. The fatty acid profile of the feedstocks is the major factor that affects the storage requirements, atomization of fuel, flow properties of the fuel, and reaction mechanism, etc. (Jain and Sharma 2010a, b; Graboski and McCormick 1998). The properties can be altered according to the requirement through blending or pre-heating. The most common fuel properties are cetane index, density, kinematic viscosity, calorific value, and flash point. The biodiesel produced must satisfy the fuel properties within the limits of internationally recognized standards American ASTM D6751 and European EN 14214 (Knothe and Razon 2017). Table 9.1 represents values of various properties in comparison to mineral diesel. Table 9.2 represents fuel properties of biodiesel produced from different feedstocks. This section highlights the reasons why biodiesel is a strong alternative candidate for mineral diesel and its advantages and disadvantages.

Table 9.1 Fuel properties of biodiesel and diesel (Krishnasamy and Bukkarapu 2021; Singh et al. 2019)

Fuel properties	Biodiesel	Diesel
Kinematic viscosity (cSt at 40 °C)	2.0–6.1	1.3–4.2
Density (kg/m^3)	871	840
Boiling point (°C)	330–370	175–360
Flash point (°C)	130	55
Freezing point (°C)	1	−1 to −4
Cloud point (°C)	−3 to 11	−15 to 5
Pour point (°C)	−15 to 15	−34 to −15
Cetane number	51	40–55
Low calorific value (MJ/kg)	38	42.6
High calorific value (MJ/kg)	41	45.8
Carbon content (%)	77.1	85
Hydrogen content (%)	12.1	15
Oxygen content (%)	10.8	0

Table 9.2 Fuel properties of biodiesel from different feedstocks (Verma and Sharma 2016; Yaşar 2020; Garg et al. 2021a, b; Jain and Sharma 2010a, b; Dwivedi and Sharma 2014)

Biodiesel	Density (kg/m^3)	Viscosity (mm^2/s)	Cetane number	Flash point (°C)	Calorific value (MJ/kg)
Sunflower	882	4.30	50	178	–
Soyabean	882	4.37	51	170	–
Karanja	882–935.2	3.99–11.82	32	222	35.27–46
Jatropha	919.5–932	4.52	46–55	211.7	37.01–38.73
Corn	878	4.42	56	172	39.5
Cottonseed	883	4.33	56	175	39.5
Rapeseed	880	4.48	54	172	–
Mahua	920	4.23	57	232	36.85
Neem	929	4.38	41	214	39.84
Algae	881	4.55	59	140	41
Waste cooking oil	937	4.67	52	235	38.27

Advantages of fuel properties of biodiesel over diesel (Agarwal 2007; Agarwal et al. 2017).

- Biodiesel has a high flash point which results in safe storage and transportation of the fuel.
- When biodiesel is blended with mineral diesel, the cetane number of the fuel increases which improves the ignition quality.
- Biodiesel is a sulfur-free fuel and renewable.

Disadvantages of biodiesel over diesel

- Biodiesel possesses high kinematic viscosity which results in poor atomization of the fuel spray.
- Biodiesel has a low calorific value due to which a high amount of fuel is required to generate the same power.
- The presence of any impurities like glycerol or catalyst can lead to wear and corrosion in the engine hardware.

9.3 Using Biodiesel as a Fuel in CI Engines

Numerous studies have been conducted worldwide in the last few decades to evaluate the compatibility of using biodiesel as a fuel in CI engines. These studies have found biodiesel to be a promising fuel in diesel engines without any major hardware modifications. Various combustion characteristics, engine performance, and emissions vary concerning biodiesel feedstock, biodiesel properties, and biodiesel blend concentration as compared to pure diesel. Many researchers have studied various blend concentrations of biodiesel with diesel in the context of long-term usage. According to various studies, blends above 20% generate various wear problems with the hardware and high cost of running biodiesel fuel in engines. Lower blends of biodiesel are successfully used in the existing vehicles (Jiaqiang et al. 2017).

9.3.1 Effect of Biodiesel on Combustion Characteristics

Combustion kinetics and engine parameters have a major impact on the combustion behavior of an engine. An important parameter linked to combustion is heat release rate (HRR). The amount of heat liberated during the ignition of the fuel is the heat release rate. Figure 9.1a–g depicts the variation in heat release rate at different crank angles and brake mean effective pressure. Due to the presence of oxygen concentration in the biodiesel, the combustion process improves which results in proper oxidation of fuel and a high heat release rate (Li et al. 2018). It was observed that the rate of heat release rate was lower initially at the start of ignition and showed an increasing trend with an increase in load conditions. This can be attributed to the increase in the amount of fuel injected into the combustion chamber at higher loads. Biodiesel fuels reported identical trends at all the loads with pure diesel. A negative heat release rate was observed in the beginning due to the accumulation of the vaporized fuel in the combustion chamber during the ignition delay. After the combustion is initiated, an increasing trend in HRR is observed.

High in-cylinder pressure occurs in the combustion chamber on using biodiesel blends as compared to mineral diesel. Due to the high heat release rate, a pressure gradient occurs in the cylinder. It was observed from the literature that there was an increase in-cylinder pressure with an increase in fuel injection pressure and high

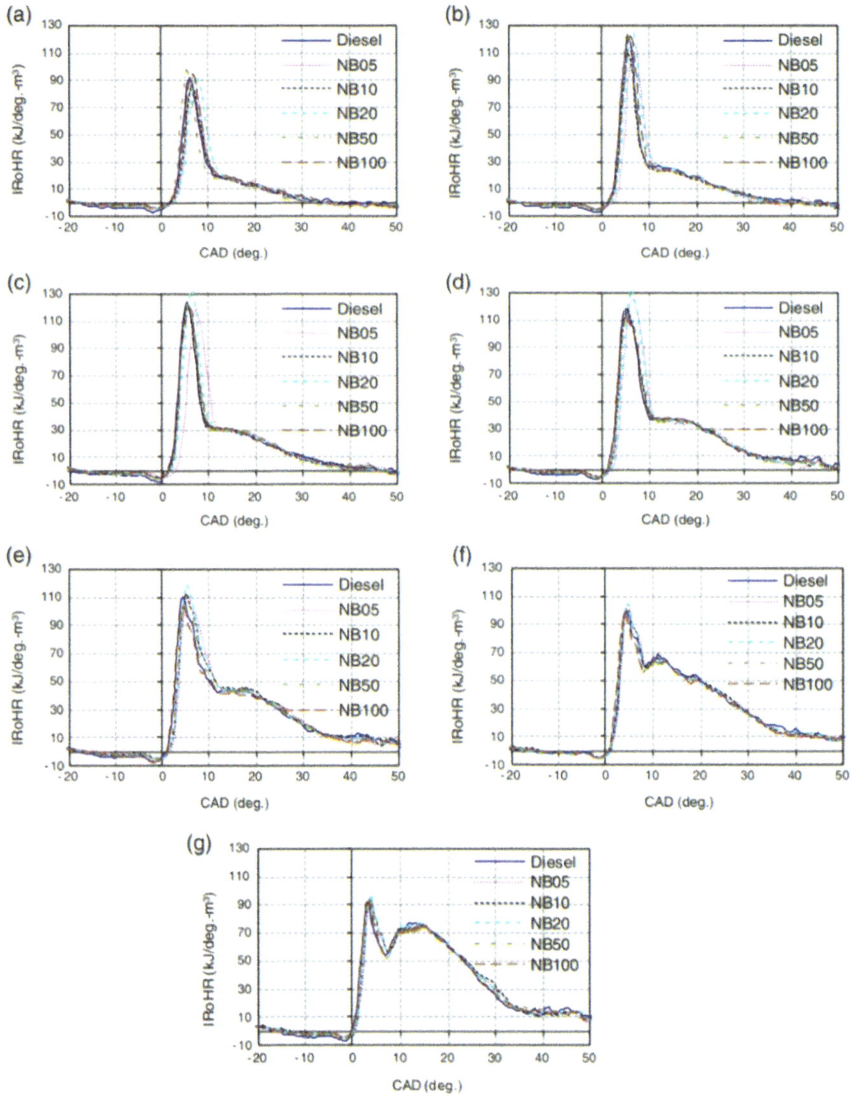

Fig. 9.1 HRR with variation in crank angle at **a** 0, **b** 1.3, **c** 2.2, **d** 3, **e** 4, **f** 5, and **g** 6 bar BMEP (Dhar et al. 2012)

load conditions. Cylinder pressure is controlled by varying the injection pressure of the fuel and high CR (Tamilselvan et al. 2017). Figure 9.2a–g depicts variation in in-cylinder pressure with respect to crank angle at different brake mean effective pressures. It can be observed from the figure that the pressure trends are almost similar for all the test fuels at high engine loads. Due to longer ignition delay, lower biodiesel blends showed a delay in the start of pressure rise whereas, in the case of

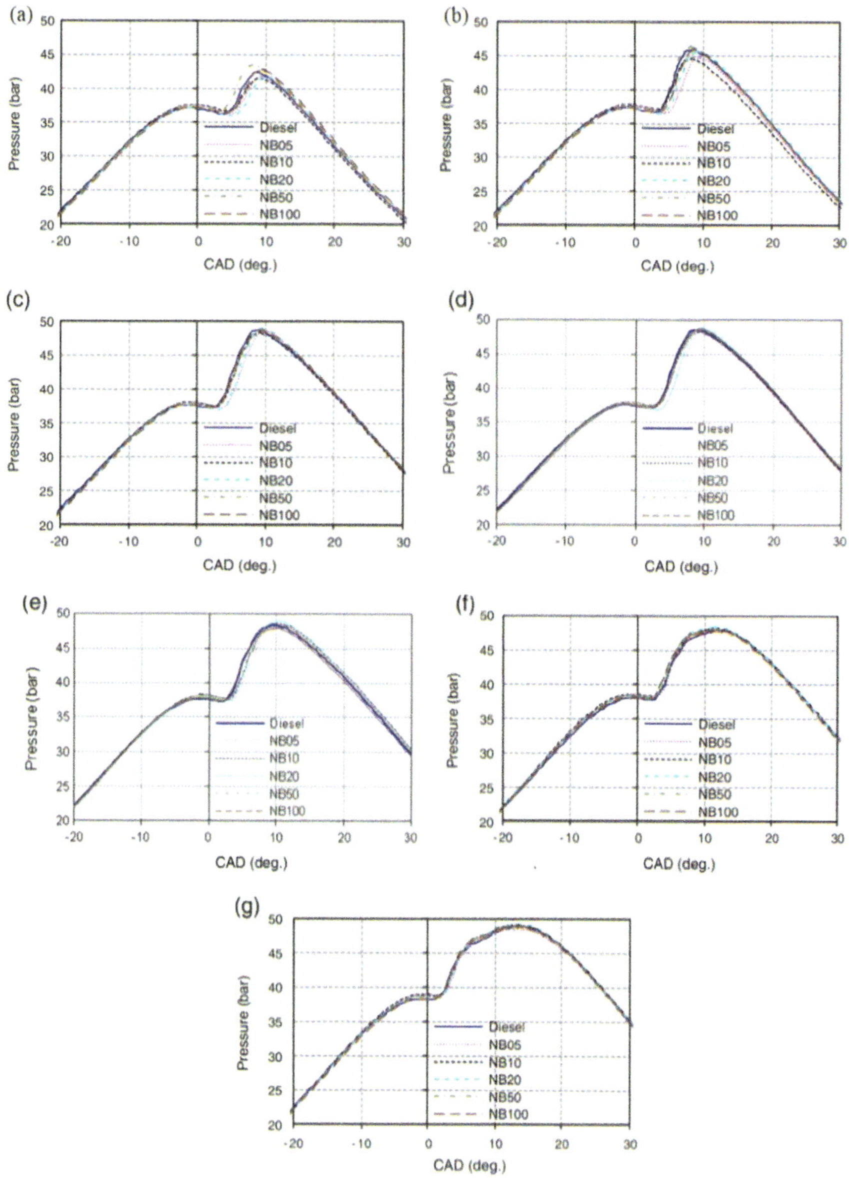

Fig. 9.2 Variation in-cylinder pressure with variation in crank angle at **a** 0, **b** 1.3, **c** 2.2, **d** 3, **e** 4, **f** 5, and **g** 6 bar BMEP (Dhar et al. 2012)

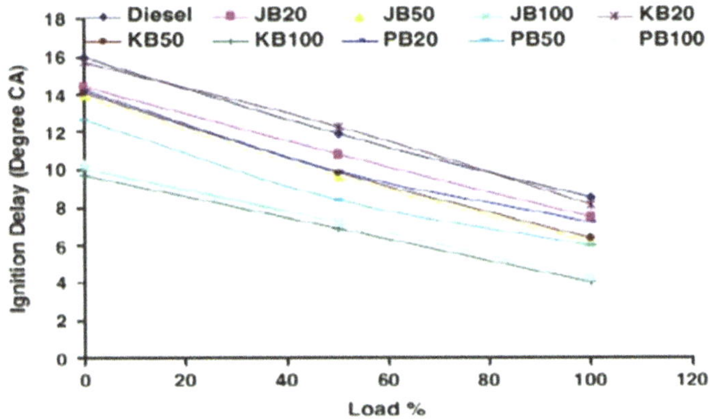

Fig. 9.3 Effect of load on the ignition delay for diesel, biodiesel, and their blends (Shahabuddin et al. 2013)

higher biodiesel blends, the pressure rising trend was comparable to diesel fuel. B20 showed the optimum performance as a blended fuel with a shorter ignition delay. This can be attributed to the presence of oxygen in biodiesel supporting combustion and lower viscosity of mineral diesel, favorable for proper air–fuel mixing.

Ignition delay (ID) is defined as the time interval between the start of injection and the start of combustion. It majorly depends upon the atomization, air–fuel mixing, and pre-combustion characteristics of the fuel. Due to the high viscosity and density of the biodiesel blends, poor atomization of the fuel occurs. On increasing the fuel injection pressure, more blend fuel is injected into the combustion chamber which results in poor atomization, hence significantly affects the ignition delay parameter (Mangus et al. 2015). At higher loads, low ignition delay can be observed in the case of biodiesel blends as compared to mineral diesel. Biodiesel has a higher cetane number as compared to diesel which results in a shorter ID and shorter ID are generally desirable (Knothe and Razon 2017). Figure 9.3 shows on increasing the biodiesel blend%, ignition delay decreases significantly.

9.3.2 Effect of Biodiesel on Engine Performance

Table 9.3 represents the variation in engine performance parameters on using biodiesel blends. From the literature review, it can be concluded that biodiesel has low calorific value than diesel due to the presence of oxygen in the fuel and low carbon% as compared to diesel (Sudalaiyandi et al. 2021). Thermal efficiency is generally better in the case of biodiesel than that of diesel. From the review, it was concluded that the brake thermal efficiency of the engine increases with an increase

Table 9.3 Review of experimental results of biodiesel fuelled CI engines

Investigators	Biodiesel fuel	Engine type	Operating conditions	Performance	Combustion	Emission
Ogunkunle and Ahmed (2020)	Parinari polyandra seeds biodiesel B0, B10, B20, B30	Six-cylinder, 4-stroke, turbocharged diesel engine	Variable torque from 20 to 100 Nm	The highest speed was observed at B10 (↑) BSFC, BTE, EGT, BP with respect to torque	–	(↓) UHC 7.8%, 11.0%, 13.8% (↓) CO 53.8%, 33.5%, 21.7% (↓) SO_2 45.3%, 45.9%, 54.3% (↓) CO 53.4%, 67.8%, 81.7% (↑) NO_x 24.5%, 35.1%, 51.1% for B10, B20, B30 respectively with respect to diesel fuel
Sharma et al. (2020)	Pongamia seed oil biodiesel B0, B10, B20, B30, B40	Single cylinder, 4-stroke diesel engine with CR of 17.5:1 at a constant speed of 1500 rpm	Variable IT, IP, and blend%	(↑) BTE with an increase in blend%, IP, and IT	–	(↑) UHC with an increase in blend%, IP (↓) UHC on increasing IT (↑) NO_x, (↓) *smoke* with an increase in IT, IP, and blend%

(continued)

Table 9.3 (continued)

Investigators	Biodiesel fuel	Engine type	Operating conditions	Performance	Combustion	Emission
Muralidharan and Vasudevan (2011)	Waste cooking oil biodiesel B0, B20, B40, B60, B80	Single cylinder, 4-stroke, water-cooled variable compression ratio (5:1–22:1) diesel engine at a constant speed of 1500 rpm	Variable CR	(↑) BTE, (↓) BP, (↓) IMEP, (↑) mechanical efficiency, (↓) EGT with an increase in CR for all the biodiesel blends B40 showed the optimum performance	(↑) Combustion pressure (↓) HRR with an increase in CR for all the blends	(↑) UHC, (↑) NO_x, (↓) CO, (↓) CO_2 with an increase in CR for all the blends
Sureshkumar et al. (2008)	Pongamia pinnata biodiesel B0, B20, B40, B60, B80, B100	Single cylinder, 4-stroke, water-cooled CR-16.5:1 diesel engine at rated power 6.8 kW and constant speed of 1500 rpm	Variable load	(↓) BSEC for B20 and B40 blends (↓) EGT for all the blends	–	(↓) CO, UHC, NO_x for all the blends (↓) CO_2 for medium blends

(continued)

Table 9.3 (continued)

Investigators	Biodiesel fuel	Engine type	Operating conditions	Performance	Combustion	Emission
Hirner et al. (2019)	Waste cooking oil Karanja biodiesel Jatropha biodiesel	Single cylinder, small bore CRDI diesel engine with CR 16:1	Variable IP of 40, 80, 12 MPa Variable IT-15, -12, -10, -7, -5, -3 CAD aTDC	–	(↓) ID for jatropha and Karanja at 120 MPa (↑) ID for WCO as compared to other biodiesels (↓) ID with an increase in IP (↑) Cylinder pressure with increase in IP (↑) Combustion efficiency with an increase in IT	(↓) NO_x in case of advanced IT and low IP (↑) Smoke for all the blends and showed decreasing trend as the IT was retarded
Patel et al. (2019)	Waste cooking oil (WCOB100) Karanja biodiesel (KB100) Jatropha biodiesel (JB100)	Single cylinder, 4-stroke DI diesel engine at a constant speed of 1500 rpm and rated power 7.4 kW and CR 17.5:1 IT-26° bTDC	Variable load conditions	(↑) BSFC (↓) BTE (↑) Smoke opacity for biodiesel fuels	(↑) In-cylinder pressure, HRR (↓) Combustion duration for biodiesel blends as compared to mineral diesel	(↓) UHC, NO_x (↑) CO for biodiesel fuels

(continued)

Table 9.3 (continued)

Investigators	Biodiesel fuel	Engine type	Operating conditions	Performance	Combustion	Emission
Can et al. (2017)	Canola biodiesel B5, B10, B15 and B20	Single-cylinder, 4-stroke diesel engine of compression ratio 18:1 and a constant engine speed of 3000 rpm	Variable loading condition	(↑) BSFC, BTE on increasing load conditions and blend%	(↓) Cylinder pressure, HRR at full load condition	↓: CO, HC, Smoke ↑: NO_X, CO_2 *With an increase in* load
Gharehghani et al. (2017)	Waste fish oil B25, B50, B75, and B100	Single-cylinder, 4-stroke diesel engine of CR 17:1 and a constant engine speed of 3000 rpm	Variable engine load	(↑) BTE with an increase in blend% and increasing load conditions	(↑) Cylinder pressure (↓) HRR	(↓) CO, UHC (↑) NO_X, CO_2
Khandal et al. (2017)	Honge oil B100	Single-cylinder, 4-stroke diesel engine of CR 17.5:1 and a constant engine speed of 1500 rpm	Variable EGR, fuel IP, and IT	(↑) BTE with an increase in IP	(↑) Peak cylinder pressure, HRR (↓) Ignition delay, combustion duration at high IP	(↓) HC, CO, smoke (↑) NO_X at low EGR conditions and advanced IT
Dhamodaran et al. (2017)	Neem methyl ester B100	Single-cylinder, 4-stroke diesel engine of CR 17.5:1 and a constant engine speed of 1500 rpm	Variable engine load	(↑) EGT (↓) BTE with an increase in load	(↑) HRR (↓) Cylinder pressure with an increase in loads	(↓) CO, UHC (↑) NO_X, smoke

(continued)

Table 9.3 (continued)

Investigators	Biodiesel fuel	Engine type	Operating conditions	Performance	Combustion	Emission
Agarwal et al. (2015)	Karanja oil biodiesel B0, B10, B20, B50	Single-cylinder 4-stroke CRDI diesel engine with rated power 6 kW, CR-17.5:1 at constant speed 1500 rpm	Variable IP of 300, 500, 700, 1000 bars	(↓) BSFC with retarded IT and increasing IP (↑) BTE on increasing IP at constant IT	(↓) Injection duration with an increase in IP (↑) Droplet size (↑) Maximum in-cylinder pressure with an increase in IP (↓) Combustion duration	(↓) CO, UHC (↑) NO_x at constant IT and increased IP
Emiroğlu et al. (2018)	Turkey rendering fat biodiesel B0, B10, B20, B50	Single-cylinder, 4-stroke diesel engine of CR 20.3:1 and a constant engine speed of 3600 rpm	Variable engine load	(↑) BSFC (↓) BTE at high loads with an increase in blend%	(↑) Cylinder peak pressure, HRR for higher blend% with increasing load conditions	(↓) smoke (↑) NO_X With the increase in blend%
Uyumaz (2018)	Mustard oil B0, B10, B20, B30	Single-cylinder, 4-stroke diesel engine of compression ratio 18.01:1 and a constant engine speed of 3000 rpm	Variable load condition	(↑) BSFC (↓) Indicated thermal efficiency with increase in load	(↑) Cylinder pressure at high load conditions for all the blends	(↓) CO, smoke, (↑) NO_X for all the blends as compared to mineral diesel

(continued)

Table 9.3 (continued)

Investigators	Biodiesel fuel	Engine type	Operating conditions	Performance	Combustion	Emission
Tripathi and Subramanian (2017)	Soya soap stock-based acid oil biodiesel B100	Single-cylinder, 4-stroke diesel engine of CR 19.01:1 and a constant engine speed of 3000 rpm	Variable engine speed	↓: Indicated thermal efficiency, BTE on increasing speed	↑: Cylinder peak pressure, cylinder temperature, Injection duration, combustion duration ↓: Heat release rate	↓: UHC, CO, smoke ↑: NO_X at a high speed as compared to diesel fuel
Agarwal and Das (2001)	Linseed biodiesel Diesel, B0, B5, B10, B15, B20, B25, B30, B40, B50, B75, B100	Single-cylinder 4 stroke, water-cooled DI diesel engine at a rated speed of 1500 rpm	Constant speed and variable load	(↑) BSFC at high loads (↑) BTE for all the blends (↑) EGT for all the blends	–	(↓) Smoke opacity (↓) UHC for all the blends at high load conditions
Ruhul et al. (2017)	Millettia pinnata (MP) and Croton megalocarpus (CM) biodiesel MP20, MP15CM5, MP10CM10, MP5CM15, CM20	Single-cylinder, 4 strokes, naturally aspirated, DICI engine with IP 700 bar and IT 15° bTDC	Constant load at 100% variable speed (1000–2400 rpm)	(↑) BP with an increase in speed but lower than diesel in case of all the blends (↑) BSFC for all the blends with an increase in speed (↓) BTE for all the blends on increasing speed	(↓) In-cylinder pressure for all the blends with an increase in speed (↓) HRR with an increase in speed (↑) Ignition delay with an increase in speed (↑) Mass fraction burnt, EGT for all the blends	(↑) NO_x for all the blends (↓) CO, CO_2, UHC for all the blends as compared to mineral diesel

(continued)

Table 9.3 (continued)

Investigators	Biodiesel fuel	Engine type	Operating conditions	Performance	Combustion	Emission
Ruhul et al. (2017)	Millettia pinnata (MP) and Croton megalocarpus (CM) biodiesel MP20, MP15CM5, MP10CM10, MP5CM15, CM20	Single-cylinder, 4 strokes, naturally aspirated, DICI engine with IP 700 bar and IT 15°bTDC	Constant speed of 1800 rpm variable load (50%, 75% and 100%)	(↓) BP 0.53%—3.7% for all blends of biodiesel as compared to diesel (↑) BSFC for all the blends	(↑) In-cylinder pressure for all the blends with an increase in load (↑) HRR with an increase in load (↓) Ignition delay with an increase in load (↑) Mass fraction burnt, EGT for all the blends	(↑) NO_x for all the blends (↓) CO, CO_2, UHC for all the blends as compared to mineral diesel

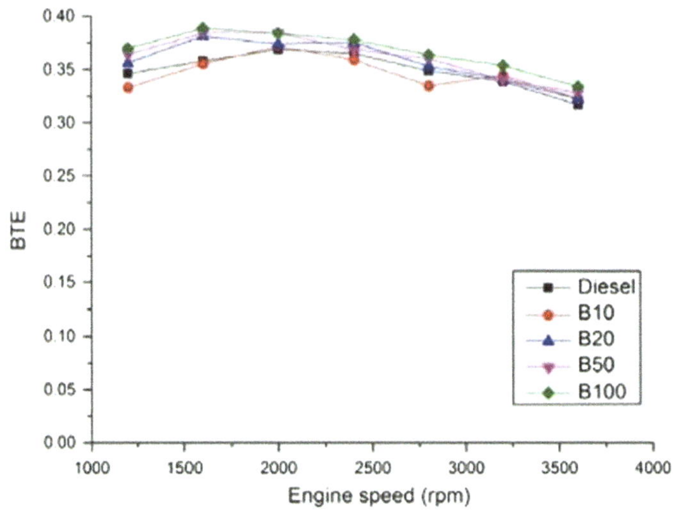

Fig. 9.4 Variation in BTE with respect to engine speed at 100% load (An et al. 2012)

in blend percentage concerning load conditions. On increasing the fuel injection pressure, and compression ratio, the BTE of the engine increases. Figure 9.4 reports similar results from the study.

Low calorific value has a direct impact on Brake Specific Fuel Consumption (BSFC). To produce the same power as diesel fuel, more fuel is required in the case of biodiesel. Due to an increase in injection pressure, poor atomization of biodiesel occurs which results in high fuel consumption (Deep et al. 2017). At high load conditions and high blend% BSFC increases. Figure 9.5 reports similar results from the study.

Exhaust Gas Temperature (EGT) increases with an increase in load in the case of biodiesel as compared to diesel. Due to proper combustion in the combustion chamber, a high in-cylinder temperature is generated which results in higher EGT. Oxygen percentage is the major factor that is responsible for the combustion quality in the engine. Figure 9.6 depicts the variation of EGT with BMEP. Mineral diesel reported the highest EGT as compared to other biodiesel blends. Higher EGT was observed in the case of higher biodiesel blends as compared to lower blends.

9.3.3 Effect of Biodiesel on Engine Emissions

Table 9.3 represents the trends in emissions on using biodiesel as a fuel. Cleaner combustion and reduced exhaust emissions occur in the case of biodiesel due to the presence of oxygen in its chemical structure. Particulate matter (PM) and oxides of nitrogen (NO_x) are the major emissions from diesel engines (McCormick et al. 2001).

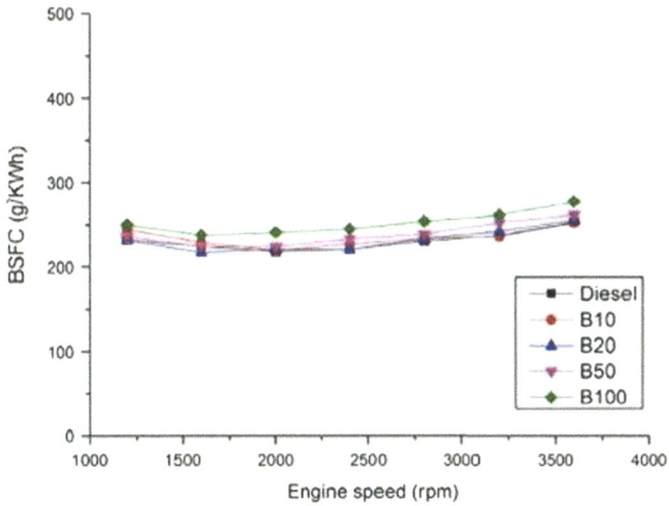

Fig. 9.5 Variation in BSFC with respect to engine speed at 100% load (An et al. 2012)

Fig. 9.6 Exhaust gas temperature with respect to BMEP (Dhar et al. 2012)

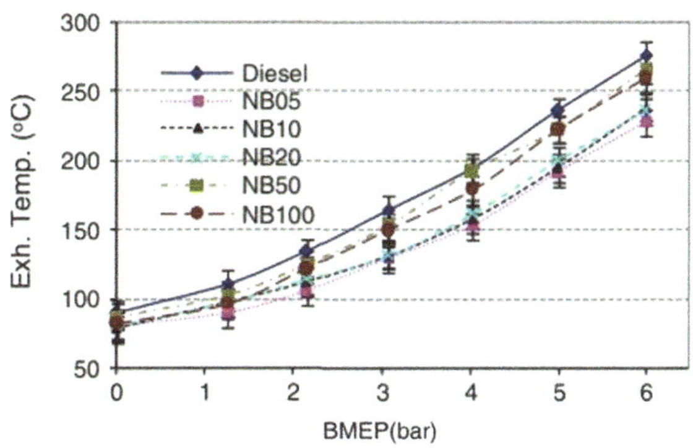

The latest emission regulations of EURO 6 require accounting for mass and number-based PM emissions. In the coming future, diesel engine technologies must be able to meet the limits of unregulated emissions. Numerous studies have been performed to monitor and reduce the emissions in the case of biodiesel. From the literature review, it was concluded that due to clean combustion, there is a significant decrease in carbon monoxide (CO) and unburnt hydrocarbon emissions (UHC) (Agarwal et al. 2018). CO and UHC decrease with an increase in biodiesel blend and load conditions. On increasing the fuel injection pressure more fuel is injected into the combustion chamber, thus unburnt fuel is emitted from the engine which results in high UHC emissions. On advancing the IT, less UHC, CO and CO_2 were observed (Gnanasekaran et al. 2016). On increasing the CR, there was a significant decrease in CO, UHC. But a slight increase in NO_x emissions was observed on using biodiesel

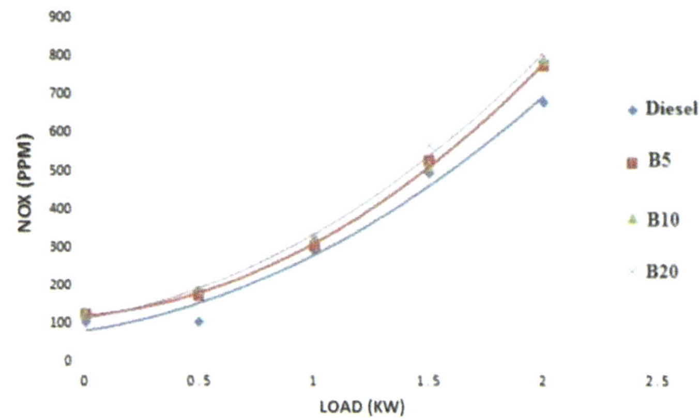

Fig. 9.7 Variation in NO_X with variable load (Kumar et al. 2020)

as compared to diesel. Various technologies like a catalytic converter, exhaust gas recirculation (EGR) system, and increasing CR have been found as potential methods to decrease NO_X emissions from diesel engines (Mirhashemi and Sadrnia 2020). Figures 9.7 and 9.8 show similar results from the study.

9.4 Various Modifications to Improve Engine Characteristics

This section comprises various modifications studied by researchers to improve engine parameters on using biodiesel as a fuel. Scientists from all over the world are studying co-optimization to improve fuel properties and engine hardware to obtain high brake thermal efficiency and reduce emissions at the output. This section gives a brief review of fuel additives/blends and engine modification that have been studied to achieve co-optimization.

9.4.1 Various Fuel Additives and Their Impact on Engine

To improve the fuel properties of biodiesel various modifications like preheating, blending with other fuels like alcohols, blending with other feedstock biodiesels, and addition of nanoparticles, etc., have been studied (Zaharin et al. 2017; Saxena et al. 2017). Agarwal and Agarwal (2007) studied the impact of preheated jatropha oil on engine performance and emissions and reported the optimum injection pressure to be 200 bar. Heating reduced the viscosity of fuel which increased the thermal efficiency and reduced the smoke opacity. Besides, an increase in CO_2 and a small decrease in CO and HC as compared to diesel. To further improve the effects of using

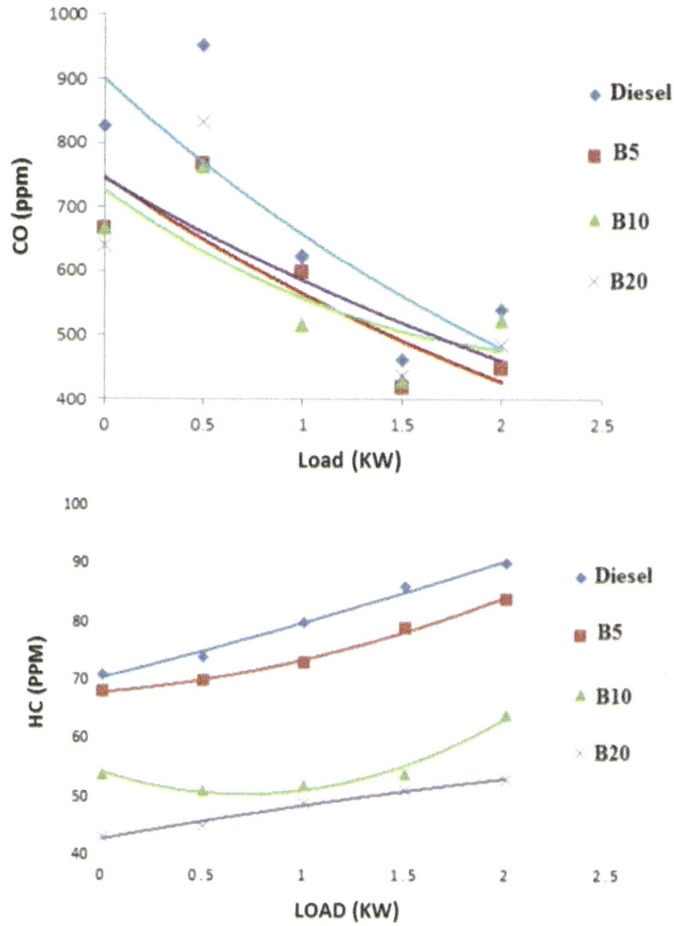

Fig. 9.8 Variation in CO and UHC emissions (Kumar et al. 2020)

direct oil, another study (Agarwal and Dhar 2009) reported blending of jatropha oil with diesel to further reduce the viscosity of fuel and improve the atomization. The study reported a significant increase in BSFC and a decrease in BTE with an increase in oil blend%. Various emissions like CO_2, CO, HC, and smoke opacity were reported to increase with an increase in oil blend% as compared to diesel. To reduce the effects of direct oil use on engine characteristics, numerous studies have reported blending biodiesel with alcohols. Alcohols have been found promising in reducing emissions on burning in diesel engines (Chauhan et al. 2021; Garg et al. 2021a, b). Ma et al. (2021) studied the impact of blending biodiesel with ethanol and pentanol and reported a prolonged ID, increase in-cylinder pressure, and HRR with an increase in speed for the blends for the ethanol-biodiesel blend. Low NO_x, THC emissions were observed with an increase in speed for all the blends whereas

CO emission increased with an increase in speed but lower than diesel. Imdadul et al. (2016) studied the blending of Alexandrian laurel biodiesel with n-butanol and pentanol. Pentanol showed higher performance with a decrease in BSFC by 4.55% and an increase in BP by 4.56%. Emissions like HC reduced by 50–67.7% and CO by 20% as compared to biodiesel/diesel blend. These studies reported that alcohol can be used as a potential additive to biodiesels to improve the overall performance of biodiesel on engine parameters. Blending with other biodiesels was also studied to modify the properties of a single biodiesel feedstock. Kumar et al. (2021) studied the biodiesel prepared from blending jatropha and algal oil. The results reported a decrease in BSFC with an increase in blend% but higher than diesel whereas BTE was lower for all the blends as compared to diesel. The emissions like O_2, HC and CO were lower for all the blends with an increase in load, however, there was a significant increase in NO_x emissions for all the blends.

To improve the combustion characteristics, the addition of nanoparticles in the fuel was studied by various researchers. Nagaraja et al. (2020) studied the effect of adding 300 ppm graphene oxide nanoparticles in B5 and B15 of rice bran biodiesel at IT23°. The results reported a reduction in BSFC by 13.6% and an increase in BTE by 7.62% in the case of the B15 blend as compared to mineral diesel. The HRR and in-cylinder pressure and the emissions like CO, CO_2, NO_x are reduced significantly on using additives. Ağbulut et al. (2020) evaluated the performance of three different nanoparticles i.e., aluminum oxide (Al_2O_3), titanium oxide (TiO_2), and silicon oxide (SiO_2). Nanoparticles were added in the mass fraction of 100 ppm in a B10 blend of waste cooking oil biodiesel. The presence of high oxygen in the nanoparticle fuels improved the combustion process and high thermal efficiency was obtained as compared to biodiesel. CO, HC, NO_x emissions were also reduced by adding nanoparticle blends. Thus, nanoparticles can play a major role in increasing performance and reducing emissions. The major drawback observed in nanoparticle technology is the high production cost and occurrence of agglomeration of nanoparticles in fuel with a long storage time (Soudagar et al. 2018).

9.4.2 Various Engine Modifications

Modifications in combustion chamber geometry, injection parameters, injection equipment, and after-treatment technologies can be used to improve the performance and reduce emissions of diesel engines using biodiesel as a fuel. Combustion chamber geometry is the major factor that influences the performance of the diesel engine. Combustion chambers are classified into two groups:

1. Direct injection combustion chamber
2. Indirect injection combustion chamber.

9.4.2.1 Direct Injection Combustion Chamber

The combustion chamber is located in the main cylinder and the fuel is directly injected into the combustion chamber volume. The combustion chamber has two parameters of design, high and low swirl design. In high swirl conditions, fewer holes are present in the injector and the piston design is has a deep bowl and vice versa in case of low swirl conditions. The fuel is directly injected into the combustion chamber volume. Some common shapes of DICC are square, hemispherical, shallow depth, cylindrical and toroidal combustion chamber. Karthickeyan (2019) compared three CC geometries i.e., toroidal, trapezoidal, and hemispherical shapes to evaluate engine parameters. Pumpkin seed oil and moringa oleifera oil biodiesel were used in the study. Toroidal combustion chamber was found to give optimum results. Higher exergy efficiency, BTE, in-cylinder pressure, and HRR were observed on using toroidal CC as compared to other shapes. Emissions like CO, HC, and NO_x were lower in the case of the toroidal combustion chamber. Similar trends were observed for trapezoidal and hemispherical shapes when compared to diesel fuel.

9.4.2.2 Indirect Injection Combustion Chamber

This category includes a swirl combustion chamber and a pre-combustion chamber. In this technique, the combustion chamber is located in the cylinder head and fuel is injected into that part. In the swirl combustion chamber, the chamber is spherical where the first half of the combustion takes place and the other half occurs in the piston crown. However, in the pre-combustion chamber, the anti-chamber and main chamber are connected with holes. Fuel is injected with pressure below 450 bar (Bari et al. 2020). Hossain et al. (2013) studied the 20% and 30% blends of de-inking pyrolysis oil with biodiesel on an indirect injection combustion engine. The study concluded that 20% blend can be the optimum blend%. On blending with pyrolysis oil, BSFC increased by 4–8% and BTE decreased by 3–6%. CO_2 and NO_x emissions increased whereas CO emissions were significantly low and peak-cylinder pressure was higher than biodiesel.

Inference from the literature was found that modification in the combustion chamber and the injection strategy can significantly improve the performance of biodiesel on CI engines and reduce emissions.

9.4.2.3 After-Treatment Technologies

The most common issue recognized in diesel engines is the NO_x formation, which further increases on adding biodiesel in the fuel. Exhaust gas recirculation system (EGR), and selective catalytic reduction (SCR) technologies have been used to treat the emissions. In this addition, the EGR rate affects the quality of performance and emissions from engines. Solaimuthu et al. (2015) used blends of mahua indica biodiesel to study the effect of EGR and SCR on engine performance and emissions. On using EGR, high HRR, and low BSFC and UHC were reported which were favorable for biodiesel usage. The main drawback of using EGR was reported higher emissions of soot particles whereas SCR resulted in the lowest NO_x and smoke opacity emissions. This study concludes that EGR and SCR can be vital techniques that can be used without any major modifications in the engines.

9.5 Challenges of Using Biodiesel in Diesel Engines

9.5.1 Economic Analysis of Biodiesel Production

Economic evaluation is considered a vital parameter for the commercialization of any newly developed technology. Besides developing technology for energy generation, the economics of the used methods are also essential to check the feasibility of the production process. In the case of biodiesel, the production process, availability of feedstock, geographical locations, and utilization technology are the major factor that influences the economics of fuel. These studies play a major role for policymakers and industrial leaders to select and enhance the usage of biodiesel at low production costs. Farid et al. (2020) estimated the cost of homogeneous base-catalyzed transesterification of waste cooking oil in Malaysia. The net energy ratio was obtained as 1.15 for the plant-based production and the estimated cost per kg was obtained as US$ 0.47. Rajendiran and Gurunathan (Naveenkumar and Baskar 2020) estimated the cost of US$ 0.68 per kg of producing biodiesel from Calophyllum inophyllum oil using Zn doped CaO heterogeneous catalyst. The best possible selling price was considered to be US$ 0.70 per kg in India. Another study reported the comparison of five different feedstocks i.e., palm, jatropha, tallow, microalgae, and waste cooking oil. Raw material cost for the base-catalyzed process was higher than the acid-catalyzed process whereas utility costs are higher in the case of an acid-catalyzed process than for a base-catalyzed process. The processing cost of jatropha biodiesel was US$ 0.15/L and for waste cooking oil was US$ 0.23/L (Rincón et al. 2014).

The availability of feedstock varies the cost of biodiesel production in plants. Biodiesel cost can compete with mineral diesel on increasing the crude oil prices of decreasing the cost of feedstocks, raw materials, and utilities. The development of new technologies like bioreactors, advancement in processes, and cost-effective catalysts can play a vital role in decreasing the cost of biodiesel production (Mohiddin et al. 2021). Production of biodiesel from vegetable oils negatively affects food security, thus new plantation and cultivation technologies must be developed to enhance the availability of non-edible feedstocks decreasing the cost of feedstock. Microalgae have recently gathered attention as a potential feedstock, as it does not require land for cultivation and has high oil content which results in high biodiesel yield (Garg and Jain 2020).

9.5.2 Engine Compatibility with Biodiesel

Besides the potential of biodiesel in reducing emissions, it is necessary to account for the effects of biodiesel on engine material, injection systems, and other components. As biodiesel has a different chemical structure from diesel, its tribological effects on the engine are needed to be addressed. Temizer (2020) carried out an endurance test for 150 h using diesel, canola oil, and sunflower oil biodiesel. The author concluded that more area of wear was observed at the upper region of piston rings also deeper abrasive wear occurred in the area parallel to the sliding direction of piston rings. Reddy et al. (2016) carried out an engine endurance test using straight vegetable oil blends of jatropha and Karanja and biodiesel with diesel. Fuel injection was performed at 750 injections per minute. Fuel injection equipment showed low wear in the case of biodiesel and oil blends as compared to mineral diesel. Maximum dimension and weight loss were observed on using diesel fuel. Chourasia et al. (2018) used mineral diesel and biodiesel-diesel-diethyl ether blend to conduct the endurance test of VCR engine with variation in load and running time. The study concluded that in the case of diesel fuel, more carbon deposition was observed as compared to biodiesel blend whereas more oxides deposition was observed at cylinder head and piston on using biodiesel blend as compared to diesel. Lubrication oil analysis was also conducted in the study and diesel reported higher debris concentration.

Thus, it can be concluded that biodiesel has both positive as well as negative effects on engine materials. The chemical composition of saturated and unsaturated fatty acids in biodiesel varies the behavior of biodiesel in the engine. Thus, it is need of the hour to research feasible feedstocks to produce biodiesel that can be used on the engine with less or no tribological effects.

9.6 Current Status of Biodiesel

Many countries have introduced commercial production of biodiesel and its usage in meeting the energy demand. These efforts were made to develop sustainable sources of renewable energy and for the mitigation of greenhouse gases emissions. Table 9.4 reports the amount of biodiesel produced globally in the last decade, according to BP statistical review of the world energy 2021 report. From the table, it can be observed that there has been significant growth in biodiesel production worldwide. Europe and Asia Pacific regions are the most biodiesel-producing regions all over the world.

The current biodiesel production capacity of India is 520 million liters. The major feedstocks involved in biodiesel production are used cooking oil, imported palm stearin, a small amount of non-edible oil and domestically produced animal fats. According to Indian Oil Corporation Limited (IOCL), India has the potential to produce 2.2 billion liters of biodiesel by using used cooking oil. However, the lack of an appropriate supply chain network of used cooking oil is the major point of

Table 9.4 Biodiesel production globally (Thousand barrels/ day) (BPSTATS 2021)

	Canada and Mexico	US	Brazil	South and Central America	Europe	Asia Pacific
2010	2	19	35	37	168	37
2011	2	54	39	48	160	71
2012	1	55	40	49	171	91
2013	2	76	43	43	183	111
2014	5	71	50	53	205	138
2015	4	70	58	40	203	102
2016	6	87	56	54	200	129
2017	6	89	63	57	230	129
2018	6	104	79	52	248	181
2019	5	96	87	48	255	230
2020	6	101	95	30	245	240

concern. In most countries, the large-scale biodiesel productions units are controlled by large oil companies whereas in India most of the biodiesel market is dominated by entrepreneurs. The current producers in India are small and medium-scale companies and generally depend upon the local supply chain of the feedstocks (Biofuels Annual 2021).

The fuel pricing of biodiesel produced is mainly dependent upon the cost of raw feedstock. The government is giving more emphasis on the plantation of non-edible trees that can play a major role in the transformation of the fossil energy demand of India to renewable fuels. The major feedstocks under consideration are jatropha and used cooking oil (National Policy on Biofuels 2011). Plantation of non-edible sources would lead to preventing the lack of feedstock availability and a proper supply chain network can be a potential method for supply of used cooking oil. The farmers and cultivators are also encouraged to grow non-edible crops that can be a vital source of income. The government of India is also planning to encourage industry corporates to enable contract farming with farmers. The goal of the government is to use the B5 blend by 2030 in on-road vehicles (Biofuels Annual 2021).

9.7 Conclusion

This chapter is an attempt to provide a review of various potential aspects and challenges of biodiesel utilization on diesel engines. The combustion characteristics, performance and emissions of past studies were reviewed and the results reported high thermal efficiency and lower emissions except NO_X. Various modification techniques like blending fuel with other fuels and nanoparticles, and modifications in engine hardware represented potential application. The economic feasibility of biodiesel fuel, and compatibility of engine material with biodiesel were the major challenges acknowledged in commercialization of biodiesel usage. There is a high need to work on biodiesel economics to promote its commercialization, as the cost of production is much higher than diesel. More research is required to engine tribological effects, exhaust gas after-treatment technologies, and modification techniques. Biodiesel production has increased significantly all over the world, but the government and policymakers need to reframe policies and their proper execution. In India, there is a high need for educational and industrial collaborations to promote technology transfer from laboratories to industries.

References

Agarwal AK (2007) Biofuels (alcohols and biodiesel) applications as fuels for internal combustion engines. Prog Energy Combust Sci 33(3):233–271

Agarwal D, Agarwal AK (2007) Performance and emissions characteristics of jatropha oil (preheated and blends) in a direct injection compression ignition engine. Appl Therm Eng 27(13):2314–2323

Agarwal AK, Das LM (2001) Biodiesel development and characterization for use as a fuel in compression ignition engines. J Eng Gas Turbines Power 123(2):440–447

Agarwal AK, Dhar A (2009) Performance, emission and combustion characteristics of jatropha oil blends in a direct injection CI engine. SAE technical papers

Agarwal AK, Dhar A, Gupta JG, Kim WI, Choi K, Lee CS, Park S (2015) Effect of fuel injection pressure and injection timing of Karanja biodiesel blends on fuel spray, engine performance, emissions and combustion characteristics. Energy Convers Manage 91:302–314. https://doi.org/10.1016/j.enconman.2014.12.004

Agarwal AK, Gupta JG, Dhar A (2017) Potential and challenges for large-scale application of biodiesel in automotive sector. Prog Energy Combust Sci 61:113–149. https://doi.org/10.1016/j.pecs.2017.03.002

Agarwal AK, Park S, Dhar A, Lee CS, Park S, Gupta T, Gupta NK (2018) Review of experimental and computational studies on spray, combustion, performance, and emission characteristics of biodiesel fueled engines. J Energy Resour Technol Trans ASME 140(12)

Ağbulut Ü, Karagöz M, Sarıdemir S, Öztürk A (2020) Impact of various metal-oxide based nanoparticles and biodiesel blends on the combustion, performance, emission, vibration and noise characteristics of a CI engine. Fuel 270

Ambaye TG, Vaccari M, Bonilla-Petriciolet A, Prasad S, van Hullebusch ED, Rtimi S (2021) Emerging technologies for biofuel production: a critical review on recent progress, challenges and perspectives. J Environ Manag 290

An H, Yang WM, Chou SK, Chua KJ (2012) Combustion and emissions characteristics of diesel engine fueled by biodiesel at partial load conditions. Appl Energy 99:363–371

Bari S, Hossain SN, Saad I (2020) A review on improving airflow characteristics inside the combustion chamber of CI engines to improve the performance with higher viscous biofuels. Fuel 264(November 2018):116769. http://doi.org/10.1016/j.fuel.2019.116769

Biofuels Annual (2021) Biofuel annual-USDA foreign agricultural services 2021, pp 1–21. https://apps.fas.usda.gov/newgainapi/api/Report/DownloadReportByFileName?fileName=Biofuels%20Annual_New%20Delhi_India_06-07-2021.pdf

BP energy outlook 2016 edn. (2016), p 2035

BPSTATS (2019) BP statistical review of world energy statistical review of world, 68th edn. The Editor BP Statistical Review of World Energy, pp 1–69. https://www.bp.com/content/dam/bp/business-sites/en/global/corporate/pdfs/energy-economics/statistical-review/bp-stats-review-2019-full-report.pdf

BPSTATS (2021) BP statistical review of world energy statistical review of world, 70th edn. The Editor BP Statistical Review of World Energy, pp 1–72. https://www.bp.com/content/dam/bp/business-sites/en/global/corporate/pdfs/energy-economics/statistical-review/bp-stats-review-2021-full-report.pdf

Can Ö, Öztürk E, Serdar Yücesu H (2017) Combustion and exhaust emissions of canola biodiesel blends in a single cylinder DI diesel engine. Renew Energy 109:73–82

Chauhan BVS, Shukla MK, Dhar A (2021) Effect of N-butanol and gasoline blends on SI engine performance and emissions

Chourasia S, Patel PD, Lakdawala A, Patel RN (2018) Study on tribological behavior of biodiesel-diethyl ether (B20A4) blend for long run test on compression ignition engine. Fuel 230(June 2017):64–77. http://doi.org/10.1016/j.fuel.2018.05.055

Deep A, Sandhu SS, Chander S (2017) Experimental investigations on the influence of fuel injection timing and pressure on single cylinder C.I. engine fueled with 20% blend of castor biodiesel in diesel. Fuel 210(July 2016):15–22. http://doi.org/10.1016/j.fuel.2017.08.023

Dhamodaran G, Krishnan R, Pochareddy YK, Pyarelal HM, Sivasubramanian H, Ganeshram AK (2017) A comparative study of combustion, emission, and performance characteristics of rice-bran-, neem-, and cottonseed-oil biodiesels with varying degree of unsaturation. Fuel 187:296–305

Dhar A, Kevin R, Agarwal AK (2012) Production of biodiesel from high-FFA neem oil and its performance, emission and combustion characterization in a single cylinder DICI engine. Fuel Process Technol 97:118–129

Dwivedi G, Sharma MP (2014) Prospects of biodiesel from Pongamia in India. Renew Sustain Energy Rev 32:114–122

Emiroğlu AO, Keskin A, Şen M (2018) Experimental investigation of the effects of turkey rendering fat biodiesel on combustion, performance and exhaust emissions of a diesel engine. Fuel 216:266–273

Farid MAA, Roslan AM, Hassan MA, Hasan MY, Othman MR, Shirai Y (2020) Net energy and techno-economic assessment of biodiesel production from waste cooking oil using a semi-industrial plant: a Malaysia perspective. Sustain Energy Technol Assessments 39:100700. https://doi.org/10.1016/j.seta.2020.100700

Garg A, Jain S (2020) Process parameter optimization of biodiesel production from algal oil by response surface methodology and artificial neural networks. Fuel 277:118254. https://doi.org/10.1016/j.fuel.2020.118254

Garg A, Dwivedi G, Baredar P, Jain S (2021a) Current Status, synthesis, and characterization of biodiesel. In: Biodiesel technology and applications. Wiley, Hoboken

Garg A, Dwivedi G, Jain S, Behura AK (2021b) Impact of methanol on engine performance and emissions. In: Methanol. Springer, Singapore, pp 247–269

Gharehghani A, Mirsalim M, Hosseini R (2017) Effects of waste fish oil biodiesel on diesel engine combustion characteristics and emission. Renew Energy 101:930–936

Gnanasekaran S, Saravanan N, Ilangkumaran M (2016) Influence of injection timing on performance, emission and combustion characteristics of a DI diesel engine running on fish oil biodiesel. Energy 116:1218–1229

Graboski MS, McCormick RL (1998) Combustion of fat and vegetable oil derived fuels in diesel engines. Science 24(97):125–164

Hirner FS, Hwang J, Bae C, Patel C, Gupta T, Agarwal AK (2019) Performance and emission evaluation of a small-bore biodiesel compression-ignition engine. Energy 183:971–982. https://doi.org/10.1016/j.energy.2019.07.015

Hossain AK, Ouadi M, Siddiqui SU, Yang Y, Brammer J, Hornung A, Kay M, Davies PA (2013) Experimental investigation of performance, emission and combustion characteristics of an indirect injection multi-cylinder CI engine fuelled by blends of de-inking sludge pyrolysis oil with biodiesel. Fuel 105:135–142. https://doi.org/10.1016/j.fuel.2012.05.007

Imdadul HK, Masjuki HH, Kalam MA, Zulkifli NWM, Alabdulkarem A, Kamruzzaman M, Rashed MM (2016) A comparative study of C4 and C5 alcohol treated diesel-biodiesel blends in terms of diesel engine performance and exhaust emission. Fuel 179:281–288. https://doi.org/10.1016/j.fuel.2016.04.003

Jain S, Sharma MP (2010a) Stability of biodiesel and its blends: a review. Renew Sustain Energy Rev 14:667–678

Jain S, Sharma MP (2010b) Prospects of biodiesel from Jatropha in India: a review. Renew Sustain Energy Rev 14(2):763–771

Jiaqiang E, Pham M, Zhao D, Deng Y, Le D, Zuo W, Zhu H, Liu T, Peng Q, Zhang Z (2017) Effect of different technologies on combustion and emissions of the diesel engine fueled with biodiesel: a review. Renew Sustain Energy Rev 80:620–647

Karthickeyan V (2019) Effect of combustion chamber bowl geometry modification on engine performance, combustion and emission characteristics of biodiesel fuelled diesel engine with its energy and exergy analysis. Energy 176:830–852. https://doi.org/10.1016/j.energy.2019.04.012

Khandal SV, Banapurmath NR, Gaitonde VN (2017) Effect of exhaust gas recirculation, fuel injection pressure and injection timing on the performance of common rail direct injection engine powered with honge biodiesel (BHO). Energy 139:828–841

Knothe G, Razon LF (2017) Biodiesel fuels. Prog Energy Combust Sci 58:36–59. https://doi.org/10.1016/j.pecs.2016.08.001

Krishnasamy A, Bukkarapu KR (2021) A comprehensive review of biodiesel property prediction models for combustion modeling studies. Fuel 302

Kumar S, Jain S, Kumar H (2020) Experimental study on biodiesel production parameter optimization of jatropha-algae oil mixtures and performance and emission analysis of a diesel engine coupled with a generator fueled with diesel/biodiesel blends. ACS Omega 5(28):17033–17041

Kumar S, Jain S, Kumar H (2021) Application of adaptive neuro-fuzzy inference system and response surface methodology in biodiesel synthesis from jatropha–algae oil and its performance and emission analysis on diesel engine coupled with generator. Energy 226:120428. https://doi.org/10.1016/j.energy.2021.120428

Li H, Yang W, Zhou D, Yu W (2018) Numerical study of the effects of biodiesel unsaturation on combustion and emission characteristics in diesel engine. Appl Thermal Eng 137:310–318. https://doi.org/10.1016/j.applthermaleng.2018.03.066

Liddell C, Morris C (2010) Fuel poverty and human health: a review of recent evidence. Energy Policy 38(6)

Ma Q, Zhang Q, Liang J, Yang C (2021) The performance and emissions characteristics of diesel/biodiesel/alcohol blends in a diesel engine. Energy Rep 7:1016–1024. https://doi.org/10.1016/j.egyr.2021.02.027

Madheshiya AK, Vedrtnam A (2018) Energy-exergy analysis of biodiesel fuels produced from waste cooking oil and mustard oil. Fuel 214:386–408. http://doi.org/10.1016/j.fuel.2017.11.060

Majeed A, Wang L, Zhang X, Kirikkaleli D (2021) Modeling the dynamic links among natural resources, economic globalization, disaggregated energy consumption, and environmental quality: fresh evidence from GCC economies. Resour Policy 73

Mangus M, Kiani F, Mattson J, Tabakh D, Petka J, Depcik C, Peltier E, Stagg-Williams S (2015) Investigating the compression ignition combustion of multiple biodiesel/ULSD (ultra-low sulfur diesel) blends via common-rail injection. Energy 89:932–945. https://doi.org/10.1016/j.energy.2015.06.040

Mathew GM, Raina D, Narisetty V, Kumar V, Saran S, Pugazhendi A, Sindhu R, Pandey A, Binod P (2021) Recent advances in biodiesel production: challenges and solutions. Sci Total Environ 794

McCormick RL, Graboski MS, Alleman TL, Herring AM, Tyson KS (2001) Impact of biodiesel source material and chemical structure on emissions of criteria pollutants from a heavy-duty engine. Environ Sci Technol 35(9):1742–1747

Mirhashemi FS, Sadrnia H (2020) NO_x emissions of compression ignition engines fueled with various biodiesel blends: a review. J Energy Inst 93(1):129–151. https://doi.org/10.1016/j.joei.2019.04.003

Mohiddin MNB, Tan YH, Seow YX, Kansedo J, Mubarak NM, Abdullah MO, San Chan Y, Khalid M (2021) Evaluation on feedstock, technologies, catalyst and reactor for sustainable biodiesel production: a review. J Ind Eng Chem 98

Mohsin M, Kamran HW, Nawaz MA, Hussain MS, Dahri AS (2021) Assessing the impact of transition from nonrenewable to renewable energy consumption on economic growth-environmental nexus from developing asian economies. J Environ Manag 284

Muralidharan K, Vasudevan D (2011) Performance, emission and combustion characteristics of a variable compression ratio engine using methyl esters of waste cooking oil and diesel blends. Appl Energy 88(11):3959–3968

Nagaraja S, Dsilva Winfred Rufuss D, Hossain AK (2020) Microscopic characteristics of biodiesel—graphene oxide nanoparticle blends and their utilisation in a compression ignition engine. Renew Energy 160:830–841. https://doi.org/10.1016/j.renene.2020.07.032

National policy on biofuels (2011) Explor Renew Altern Energy Use India (14):205–214

Naveenkumar R, Baskar G (2020) Optimization and techno-economic analysis of biodiesel production from calophyllum inophyllum oil using heterogeneous nanocatalyst. Bioresour Technol 315

Ogunkunle O, Ahmed NA (2020) Exhaust emissions and engine performance analysis of a marine diesel engine fueled with Parinari polyandra biodiesel-diesel blends. Energy Rep 6:2999–3007. https://doi.org/10.1016/j.egyr.2020.10.070

Patel C, Chandra K, Hwang J, Agarwal RA, Gupta N, Bae C, Gupta T, Agarwal AK (2019) Comparative compression ignition engine performance, combustion, and emission characteristics, and trace metals in particulates from waste cooking oil, jatropha and karanja oil derived biodiesels. Fuel 236(May 2018):1366–1376. http://doi.org/10.1016/j.fuel.2018.08.137

Raheem I, Mohiddin MNB, Tan YH, Kansedo J, Mubarak NM, Abdullah MO, Ibrahim ML (2020) A review on influence of reactor technologies and kinetic studies for biodiesel application. J Ind Eng Chem 91

Reddy MS, Sharma N, Agarwal AK (2016) Effect of straight vegetable oil blends and biodiesel blends on wear of mechanical fuel injection equipment of a constant speed diesel engine. Renew Energy 99:1008–1018. https://doi.org/10.1016/j.renene.2016.07.072

Rincón LE, Jaramillo JJ, Cardona CA (2014) Comparison of feedstocks and technologies for biodiesel production: an environmental and techno-economic evaluation. Renew Energy 69:479–487. https://doi.org/10.1016/j.renene.2014.03.058

Ruhul AM, Kalam MA, Masjuki HH, Shahir SA, Alabdulkarem A, Teoh YH, How HG, Reham SS (2017) Evaluating combustion, performance and emission characteristics of millettia pinnata and croton megalocarpus biodiesel blends in a diesel engine. Energy 141:2362–2376

Saxena V, Kumar N, Saxena VK (2017) A comprehensive review on combustion and stability aspects of metal nanoparticles and its additive effect on diesel and biodiesel fuelled C.I. engine. Renew Sustain Energy Rev 70(November 2016):563–588. http://doi.org/10.1016/j.rser.2016.11.067

Shahabuddin M, Liaquat AM, Masjuki HH, Kalam MA, Mofijur M (2013) Ignition delay, combustion and emission characteristics of diesel engine fueled with biodiesel. Renew Sustain Energy Rev 21:623–632

Sharma A, Maurya NK, Singh Y, Singh NK, Gupta SK (2020) Effect of design parameters on performance and emissions of DI diesel engine running on biodiesel-diesel blends: Taguchi and utility theory. Fuel 281:118765. https://doi.org/10.1016/j.fuel.2020.118765

Singh D, Sharma D, Soni SL, Sharma S, Kumari D (2019) Chemical compositions, properties, and standards for different generation biodiesels: a review. Fuel 253:60–71. https://doi.org/10.1016/j.fuel.2019.04.174

Solaimuthu C, Ganesan V, Senthilkumar D, Ramasamy KK (2015) Emission reductions studies of a biodiesel engine using EGR and SCR for agriculture operations in developing countries. Appl Energy 138:91–98. https://doi.org/10.1016/j.apenergy.2014.04.023

Soudagar MEM, Nik-Ghazali NN, Kalam MA, Badruddin IA, Banapurmath NR, Akram N (2018) The effect of nano-additives in diesel-biodiesel fuel blends: a comprehensive review on stability, engine performance and emission characteristics. Energy Convers Manage 178:146–177. https://doi.org/10.1016/j.enconman.2018.10.019

Sudalaiyandi K, Alagar K, Manoj Praveen VJ, Madhu P (2021) Performance and emission characteristics of diesel engine fueled with ternary blends of linseed and rubber seed oil biodiesel. Fuel 285(September 2020):119255. http://doi.org/10.1016/j.fuel.2020.119255

Sureshkumar K, Velraj R, Ganesan R (2008) Performance and exhaust emission characteristics of a CI engine fueled with Pongamia Pinnata Methyl Ester (PPME) and its blends with diesel. Renew Energy 33(10):2294–2302

Tamilselvan P, Nallusamy N, Rajkumar S (2017) A comprehensive review on performance, combustion and emission characteristics of biodiesel fuelled diesel engines. Renew Sustain Energy Rev 79:1134–1159. https://doi.org/10.1016/j.rser.2017.05.176

Temizer İ (2020) The combustion analysis and wear effect of biodiesel fuel used in a diesel engine. Fuel 270

Tripathi S, Subramanian KA (2017) Experimental investigation of utilization of soya soap stock based acid oil biodiesel in an automotive compression ignition engine. Appl Energy 198:332–346

Uyumaz A (2018) Combustion, performance and emission characteristics of a DI diesel engine fueled with mustard oil biodiesel fuel blends at different engine loads. Fuel 212:256–267

Verma P, Sharma MP (2016) Review of process parameters for biodiesel production from different feedstocks. Renew Sustain Energy Rev 62:1063–1071

Yaşar F (2020) Comparision of fuel properties of biodiesel fuels produced from different oils to determine the most suitable feedstock type. Fuel 264:116817

Zaharin MSM, Abdullah NR, Najafi G, Sharudin H, Yusaf T (2017) Effects of physicochemical properties of biodiesel fuel blends with alcohol on diesel engine performance and exhaust emissions: a review. Renew Sustain Energy Rev 79:475–493

Chapter 10
Biodiesel and Renewable Diesel as a Drop-in Fuel for Decarbonized Maritime Transportation

Cagatayhan Sevim and Burak Zincir

Abstract Some of the biggest problems facing our world on a global scale are global warming, climate change, and air pollution. The transport and logistics sector causes these problems to become even more intractable with the internal combustion engines used, and the role of maritime transport in this regard is substantial. The International Maritime Organization (IMO), which is the United Nations' (UN) specialized agency for regulating maritime transportation, began its studies on the topic of greenhouse gas within the scope of the UN's sustainable goals for the future and announced the initial greenhouse gas (GHG) strategy plan in April 2018. In these ongoing studies, the importance of alternative marine fuels becomes more and more critical day by day. There are many types of alternative marine fuels with different properties, such as liquefied petroleum gas, liquefied natural gas, methanol, ethanol, and biofuels. Each alternative fuel has its own characteristics, which leads to some advantages and disadvantages in terms of use. During the transition to alternative fuels from conventional fuels, drop-in fuels such as biodiesel and renewable diesel can be used to meet the reduction target of the IMO Initial GHG Strategy. The advantage of these biofuels is that it can be used with the existing fuel system and a diesel engine without any retrofit requirement or can be used with minor modifications. The purpose of this study is to state the steps that IMO has taken for decarbonization and then to explain the properties, raw materials, and production methods of biodiesel and renewable diesel, an alternative marine fuel, and to determine its usability and suitability by comparing it with other promising alternative fuels and revealing their pros and cons for decarbonizing the maritime industry.

Keywords Biodiesel · Renewable diesel · Alternative fuel · Decarbonization · Lifecycle assessment · Maritime transportation

---

C. Sevim
Naval Architecture and Maritime Faculty, Yildiz Technical University, Istanbul, Turkey

B. Zincir (✉)
Maritime Faculty, Istanbul Technical University, Istanbul, Turkey
e-mail: bzincir@itu.edu.tr

A. V. Agarwal and H. Valera (eds.), *Potential and Challenges of Low Carbon Fuels for Sustainable Transport*, Energy, Environment, and Sustainability,
https://doi.org/10.1007/978-981-16-8414-2_10

10.1 Introduction

Global warming, climate change, and air pollution, which are directly related to each other, have reached a critical position for the whole world and humanity, especially in the last decades. Over the years, increasing fossil fuel consumption for transportation's energy demand is directly responsible for changes in atmospheric composition and air pollution (Kampa and Castanas 2008). Emissions resulting from the combustion of fossil fuels have negative effects on the human as well as the environment. Among the emissions of internal combustion engines, the most harmful to human health and the ecosystem are carbon monoxide (CO), carbon dioxide (CO_2), sulfur oxides (SO_X), nitrogen oxides (NO_X), and particulate matter (PM) (Sadeghinezhad et al. 2013). Excessive exposure to these gases causes acute or chronic diseases such as lung cancer, bronchitis, pneumonia, asthma, etc. in humans (Kampa and Castanas 2008). Emissions of these gases are mainly caused by industrial facilities, agriculture and animal husbandry, accommodation and heating needs of people, and transportation sectors. Transportation is one of the most important parts of the global economy and international trade. Maritime transportation, on the other hand, constitutes nearly 90% of the world's trade transportation (Deniz and Zincir 2016). Approximately 90% of the European Union's (EU) foreign trade and 40% of its domestic trade were carried out by maritime transportation (Van Fan et al. 2018).

10.1.1 Status of Maritime Transportation

Maritime transportation is carried out with diesel engines, as diesel engines are more efficient than other engines. For this reason, it can be deduced that the exhaust gases emitted by marine diesel engines are one of the primary reasons of marine air pollution. As indicated by the 2020 information of the United Nations Conference on Trade and Development (UNCTAD), 98,140 ships are running which is equivalent to 2.062 billion dead-weight tons (dwt). Between 1 January 2019 and 1 January 2020, the worldwide maritime transportation fleet has expanded by 4.1%, which was estimated averagely by 3.4% from 2019 through 2024, enrolling the most noteworthy development since 2014 (Han et al. 2020; Zincir and Deniz 2021).

As for the fuel consumption amounts in maritime transport, according to the fourth greenhouse gas study in 2020 (GHG) of the International Maritime Organization (IMO), the total fuel consumption for 2018 on an annual basis is 339 million tons including fishing and domestic shipping, which is approximately 309 million tons for 2012 (The Fourth IMO Greenhouse Gas Study 2020). According to the data, there is a small decrease in heavy fuel oil (HFO) consumption, while the increase is observed in marine diesel oil (MDO), liquified Natural Gas (LNG), and methanol consumption. The reason for the increase in the consumption of these fuel types was the new emission rules and regulations that were put into practice by IMO.

10.1.2 Emissions from Global Maritime Transportation

The CO_2 emission of all these fuels consumed in the maritime sector increased by 9.3% compared to 2012 and reached 1056 million tons in 2018. 1056 million tons of CO_2 emissions correspond to 2.89% of global CO_2 emissions. Carbon dioxide equivalent (CO_2e) greenhouse gas emissions including CO_2, NO_X, and methane (CH_4) were recorded as 1076 million tons with a 9.6% increase compared to 2012 (The Fourth IMO Greenhouse Gas Study 2020). According to a study conducted by the European Environment Agency (EEA) (European Environment Agency (EEA) 2021), shipping contributes 11.05% of global SO_X emissions, 19.35% of global NO_X emissions, 4.41% and 8.39% of global PM10 and PM2.5 emissions sequentially, and finally, 1.89% of global CO emissions. All these values are visually expressed in Fig. 10.1.

General information about the state of maritime transportation is mentioned in this section with official data. The number of ships in maritime transport is increasing day by day to meet the global logistics demand. As can be seen from the data above, this growth in the number of ships directly increases the total fuel consumption and the amount of carbon emitted to nature. The regulations and rules put in place for the decarbonization of shipping are becoming more stringent over time. Within the scope of IMO's targets, it is planned to operate new ships with fossil-free or zero-carbon fuels such as hydrogen or ammonia in the future. However, using such fuels for existing ships, major and costly modifications are required. It is an important issue to use the existing diesel engine-powered ships efficiently without making major and costly modifications in the transition process to alternative fuels. Especially the ships built in the last 10 years have a critical importance in this context. Because the commercial life of a ship is 30 years on average. This means that even if it was built 10 years ago, it is an operable ship to 2040. According to UNCTAD's 2020 maritime

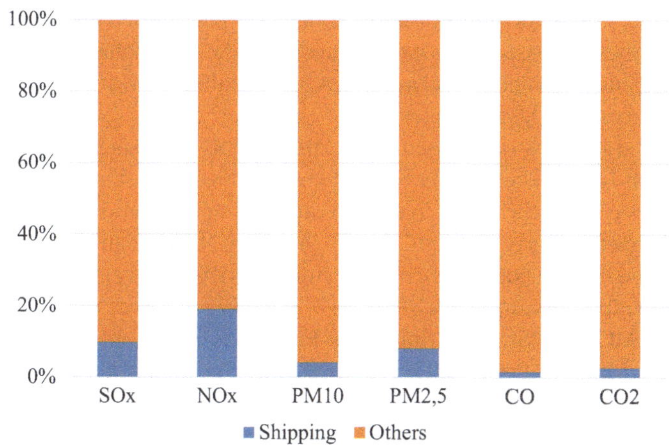

Fig. 10.1 Global shipping emissions (European Environment Agency (EEA) 2021)

Table 10.1 Age distribution percentage of ships worldwide (Han et al. 2020)

Ship type		Age of ship				
		0–4	5–9	10–14	15–19	More than 20
Bulk carriers	Percentage	20.22	42.17	18.70	8.99	9.93
Container ships	Percentage	15.60	20.39	32.79	14.67	16.55
General cargo ships	Percentage	4.64	12.34	15.67	7.99	59.36
Oil tankers	Percentage	14.45	18.95	20.19	11.11	35.32
Others	Percentage	11.21	18.05	15.53	8.28	46.93
All ships	Percentage	11.64	20.11	17.42	8.98	41.85

transportation report, around half of all ships globally are 14 years old or younger. In Table 10.1, the ratio of the ages of the ships is given in detail according to the ship types (Han et al. 2020).

As can be seen from Table 10.1, 11.64% of all ships are 4 years old or below, and 21.11% are 5–9 years old. Especially these ships should be run efficiently for many years so that the spent capital is not wasted. If these existing ships can be operated without going through long shipyard processes, this issue will be overcome without any disruption in maritime transportation. In the next section, the steps taken to decarbonize maritime transport and the determined strategies and targets will be introduced.

10.2 International Maritime Organization Actions

Decarbonization in maritime transportation has gained more importance with the impact of the Kyoto Protocol and the Paris Agreement in the last decades. Today, almost all fuels used for transportation contain carbon atoms, and as a result of combustion, these carbon atoms are emitted into the atmosphere in the form of carbon monoxide (CO) and hydrocarbon (HC) molecules, the vast majority of which are carbon dioxide (CO_2) molecules.

IMO started its work on CO_2 emission after MARPOL Annex-VI was adopted in 1997 which includes resolution 8 on CO_2 emissions from ships. This resolution further requested that the IMO conduct a study in collaboration with the United Nations Framework Convention on Climate Change (UNFCCC) in order to determine the volume and relative percentage of CO_2 emissions from ships as part of the global inventory of CO_2 emissions.

The First IMO GHG Study on Ship GHG Emissions was released to the forty-fifth session of the MEPC in 2000, and it projected that ships linked in international trade are responsible for roughly 1.8% of global CO_2 emissions in 1996 (International Maritime Organization (IMO) 2021b).With this study, studies on greenhouse gases emitted by ships gain momentum.

After the first GHG study, the IMO published its second GHG study in 2009. In the second GHG study, it was concluded that carbon dioxide is the most important greenhouse gas emitted by ships. According to IMO's second greenhouse gas study data, total shipping emitted 1050 million tons of CO_2 in 2007, which corresponds to 3.3% of global CO_2 emissions. Emissions scenarios show that shipboard emissions could increase by 150–250% (compared to emissions in 2007) by 2050 as a result of enlargement in shipping if no action is taken. By modifying ship design and operation, a variety of alternatives for enhancing energy efficiency and lowering emissions have been described in the study such as novel design hull and superstructure, novel power and propulsion system, optimizing speed and capability, voyage optimization, and using low carbon fuels, etc. On the other hand, a series of policy options have been discussed in order to ensure that these alternative methods, which are defined to reduce emissions, are put into practice in real terms. Some of these policy options are, Energy Efficiency Operational Indicator (EEOI), Energy Efficiency Design Index (EEDI), and Ship Energy Efficiency Management Plan (SEEMP). EEDI and SEEMP are mandatory from these policies and those responsible will be punished if they are not complied with. However, EEOI is optional and its implementation makes a positive contribution to energy efficiency (The Second IMO Greenhouse Gas Study 2009). Among these policies, EEDI was adopted in 2011 for all new ships of 400 gross tons and above and existing ships which have undergone a major transformation at the 62nd session of the Maritime Environment Protection Committee (MEPC) of IMO. Additionally, each ship must have its own SEEMP on board (The Marine Environment Protection Committee 2011).

EEDI is an index that aims to equip new ships or ships undergoing a major transformation with energy-efficient machinery, auxiliary engines, and equipment. More efficient machinery and equipment means less fuel consumption, which directly reduces the CO_2 emissions released into the atmosphere. EEDI is a mechanism that completely frees the selection and application of the technology, machinery, and equipment to be used, as long as it meets the specified energy efficiency level to that ship type from the design process of shipbuilding. For this reason, EEDI is an incentive technical mechanism for the shipbuilding and ship machinery industry to produce more efficient products. EEDI, which was put into practice on January 1, 2013, consists of four phases. Phase 0 which covers the first 2 years necessitates the new ship which will be built to meet the reference level determined for that type of ship. Other EEDI phases were determined based on the first determined reference value so that the limit decreases continuously and is given in Table 10.2 (Zincir and Deniz 2021; Psaraftis 2019).

Table 10.2 EEDI phases

Phase	Year	Amount of reduction (%)
1	2015–2020	10
2	2020–2025	20
3	2025–	30

The SEEMP was created to reduce CO_2 emissions from current ships that are used for international shipping in a cost-effective manner. It is a compulsory strategy for all ships, intending to encourage and increase energy-efficient ship operations. The SEEMP also provides the international shipping sector a way to track and manage vessel and fleet efficiency index, using optional tools like the EEOI as a monitoring and evaluation tool. The EEOI paves the way for operators to assess a vessel's current fuel efficiency and the effect of any operational modifications, such as more frequent propeller cleaning or enhanced voyage planning, or the implementation of technological solutions such as a new propeller design or waste heat recovery systems (Zincir and Deniz 2021).

IMO then published its third GHG study in 2014.According to IMO third GHG study, the maritime industry emitted an average of 1015 million tons of CO_2 annually between 2007 and 2012, which corresponds to 3.1% of global CO_2 emissions (The Third IMO Greenhouse Gas Study 2014). This study also showed that CO_2 emissions from the shipping industry would continue to increase due to increased logistics and energy demand depending on global economic growth if no measure is taken.

The European Union believes that the IMO regulations are insufficient to mitigate emissions from international shipping to acceptable levels. As a result, in 2009 the EU attempted to implement a novel strategy based on CO_2 emissions monitoring, reporting, and verification, known as the MRV regulation which entered into force on July 2015. The European Union attempted to pave the way for the IMO through MRV, a new regulatory framework aimed at correctly estimating ship fuel consumption and CO_2 emissions (Boviatsis and Tselentis 2019).

On the other side, IMO agreed in October 2014 to build a data collection system (DCS) for ship fuel consumption and it has three steps; data collection, data analysis, and decision-making. By resolution MEPC.278(70) approved in October 2016, the IMO made compulsory MARPOL Annex VI requirements for ships to record and report their fuel oil use, which goes into effect on March 1, 2018 (Boviatsis and Tselentis 2019). The goal of this system is to record the annual fuel consumption of ships of 5000 gross tonnage and above and to lay the groundwork for further measures to be taken. The difference between these systems is that EU MRV covers only EU and European Free Trade Association (EFTA) ports while IMO DCS covers all ports worldwide (Zincir and Deniz 2021).

After these studies, IMO adopted a strategy plan to mitigate GHG emissions from ships in April 2018 (IMO initial GHG Strategy Plan) and the MEPC approved this strategy plan in its 72nd session and put it into effect. The goal of IMO's initial GHG strategy plan is that a minimum 40% reduction in CO_2 emissions per unit transport work by 2030, a 70% reduction by 2050 compared to 2008, and sustained efforts at this point. In addition, it aims to reduce total annual global CO_2 emissions by 50% by 2050 compared to 2008 (Psaraftis 2019; Rutherford and Comer 2018). For the implementation of the strategy plan, some guiding principles and candidate measures in the short, mid, and long term have been studied and shared as a roadmap. Short-term plans cover the period 2018–2023, mid-term plans cover the years between 2023 and 2030, lastly long-term plans cover 2030 and beyond. In Table 10.3, candidate measures have been demonstrated in detail (IMO 2018).

Table 10.3 The initial GHG strategy plan candidate measures of IMO (IMO 2018)

Term	Measure	Target
Short	Improving the existing energy efficiency framework by developing EEDI and SEEMP	New and existing vessels
Short	Developing technical and operational energy efficiency measures, e.g. Annual Efficiency Ratio (AER), Energy Efficiency Per Service Hour (EESH), Individual Ship Performance Indicator (ISPI), Fuel Reduction Strategy (FORS)	New and existing vessels
Short	The creation of an existing Fleet Development Program	Existing vessels
Short	Evaluate and analyze speed optimization and speed reduction	Existing vessels
Short	Measures for methane and volatile organic compound emissions	Engines
Short	Development and updating of national action plans to address greenhouse gas emissions from shipping	Governments
Short	Proceed and upgrade technical collaboration and capacity-building actions	Whole sector
Short	Measures for port developments globally by utilizing renewable sources and alternative low carbon fuels including infrastructure	Ports
Short	Initiate research addressing marine propulsion, alternative low-carbon and zero-carbon fuels. Initiate innovative technologies to further increase the energy efficiency of ships	Shipyards and fuels
Short	Impetuses for first movers to create and take up new advances	Governments
Short	For all types of fuels, develop robust lifecycle GHG/carbon intensity guidelines. (especially for low-carbon and zero-carbon fuels)	Fuels
Mid	Implementation program for the effective uptake of alternative low-carbon and zero-carbon fuels, including update of national actions plans	Fuels
Mid	Operational energy efficiency measures that can be used to demonstrate and improve the energy efficiency performance of vessels	New and existing vessels
Mid	Further continue and enhance technical cooperation and capacity-building activities	Whole sector
Mid	Advancement of an feedback mechanism to empower illustrations learned on execution of measures to be examined and shared through information exchange	Whole sector
Long	Pursue research and development of zero-carbon or fossil-free fuels to render possible decarbonized shipping	Fuels
Long	Encourage and facilitate the general adoption of other possible new/innovative emission reduction mechanisms	Whole sector

There are major international projects that support IMO's Greenhouse Gas strategy goals. These projects; Global Industry Alliance to Support Low Carbon Shipping (GIA), Global Maritime Technology Cooperation Centers network (GMN), GreenVoyage2050, and Multi-donor trust fund on GHG (GHG TC-Trust Fund). Each project makes a significant contribution to achieving this goal (International Maritime Organization (IMO) 2021a). These major projects make significant contributions to the IMO's initial strategy plan in the fields of technological cooperation, capacity building, technology development, data collection, technological solution generation, and financial support.

As can be seen from the IMO first GHG strategy plan candidate measures, it is emphasized to focus on research studies in the short term for the transition to alternative low-carbon and zero-carbon fuels, establish and start the use of alternative fuels application programs in the mid and long term. It is seen that alternative fuels come to the fore in the whole process. Alternative fuels for maritime transportation are liquefied petroleum gas (LPG), liquefied natural gas (LNG), ethanol, methanol, dimethyl ether (DME), Fischer–Tropsch (FT) diesel, pyrolysis oil, hydrogen and fuel cells, ammonia, biodiesel, renewable diesel (HVO), and other biofuels. Each alternative fuel has different advantages and disadvantages due to its components and properties. Although there are many different alternative fuels, the ones that are widely used today are LNG, methanol and biodiesel, respectively, due to their supply chain and production capacity (Moirangthem 2016). Among these alternative fuels, biodiesel and renewable diesel stand out with their unique features such as being able to be used in diesel engines without major modifications and being mixed with petroleum-derived marine fuels, and that means they are drop-in fuel for maritime transportation. These features are extremely important for utilizing existing vessels for a prolonged time by meeting new emission regulations. In the next section, detailed information about selected biofuels is given.

10.3 Biofuel

Biofuel is defined as the fuel obtained by various methods from the organic biomass of plants basically. Biofuels can be produced from many different raw materials by different processes, which causes the end product to have different properties. In the next sections, detailed information about biofuels is given.

10.3.1 Generations of Biofuels

There are three biofuel generations which are first, second and third. The basic raw material of first generation biofuel is edible food crops such as soybean, palm oil, rapeseed, sunflower sugar cane etc. It has scaled up to commercial manufacturing and is now being used in the transportation sector including maritime transportation. Biodiesel (FAME) is in the first generation but renewable diesel (HVO) can be assessed in the first and second generation biofuels depending on the raw materials. Second-generation biofuels are mainly generated from lingo-cellulosic materials and non-edible vegetable oils such as forest residues, jatropha, castor and pongamia as well as the same raw materials for first generation. Examples of second generation biofuels are FT-diesel (Fischer–Tropsch), Pyrolysis Oil and HTL biocrude (Hydrothermal Liquefaction). Improved processing technologies are used in second-generation biofuels because of the difficulty to break down the lignocellulose of the plants. These procedures provide a more stable fuel, but they necessitate specific technologies that vary depending on the type of biomass used as a result the production cost increase (Kesharwani et al. 2019; Yuanrong Zhou 2020). First-generation biofuels are not likely to meet all industry demands and be sustainable due to the large-scale use of farmland and competition with food. Nonetheless, second generation biofuel sources is not commercially viable at the present time because the manufacturing process necessitates pricey and cutting-edge technology. For these reasons, researchers focused on third generation biofuel studies, and started to work on producing biodiesel from microalgae. Microalgae biomass is more likely to be a sustainable biofuel due to its advantages such as not requiring quality farmland for production, obtaining more biomass than conventional crops in the same area, high growth rate and short harvest cycle. The third generation biofuel studies are still ongoing (Chowdhury and Loganathan 2019).

10.3.2 Biodiesel and Renewable Diesel

Biofuels are basically combustible fuels that created from biomass or bio-waste. There are many different types of biofuels, depending on the production process and raw materials. Biodiesel, chemically known as Fatty Acid Methyl Ester (FAME), is the most mature of these biofuels, both its production process and its supply chain. In the biodiesel producing process, fats and oils are converted into chemical materials called FAME. Esterification is the name given to this process. The esterification process is demonstrated in basic form in Fig. 10.2 (Alleman and McCormick 2016).

Another prominent biofuel is renewable diesel which named Hydrotreated Vegetable Oil (HVO) due to its similarity with conventional diesel fuel (Kesharwani et al. 2019). Hydrotreated vegetable oil is called renewable diesel and it is a product of fats, vegetable oil and used cooking oil (UCO) refined by a hydrotreating process known as fatty acids to hydrocarbon hydrotreatment. In the hydrotreatment process hydrogen is utilized to extract oxygen from triglyceride vegetable oil molecules and divide the triglyceride into three chains. This results in hydrocarbons that are identical to current diesel fuel components. This enable to you to combine in any ratio you want or use pure form without worrying about quality in the diesel engines

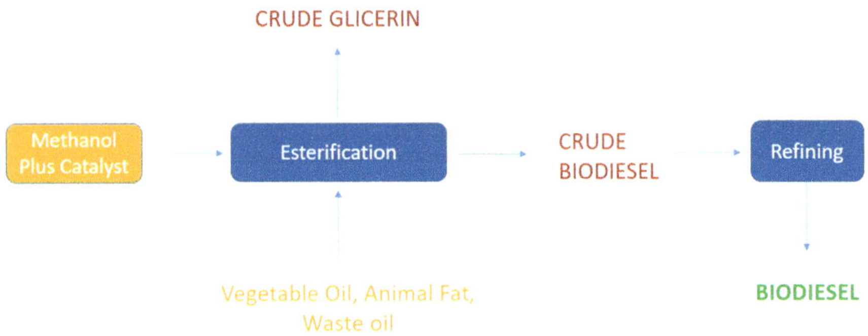

Fig. 10.2 Basic esterification process (figure reproduced and adapted) (Alleman and McCormick 2016)

(Engman et al. 2014). The basic hydrotreatment process is given in Fig. 10.3 (Moreno Fernández-Villamil and Hurtado De Mendoza Paniagua 2018).

HVO is highly compatible with new and existing diesel engines and fuel systems. HVO also meets the diesel fuel standards EN 590 excluding density which is lower and meets ASTM D975 fully (Engman et al. 2014).

Biodiesel and HVO are renewable, clean-burning diesel replacement fuels that can be manufactured from a growing number of different raw materials in the group of plant oil, waste cooking oil, animal fats. Biodiesel and petroleum-derived diesel blends are shown as BXX. The "XX" part in the BXX expression represents the percentage of biodiesel. According to this, it is expressed as B100 of pure biodiesel (Mohd Noor et al. 2017). There are several standards for biodiesel fuel and the most widely accepted of these standards are EN14214 and ASTM D6751. ASTM 6751 is set by the United States Energy Department and is the most internationally

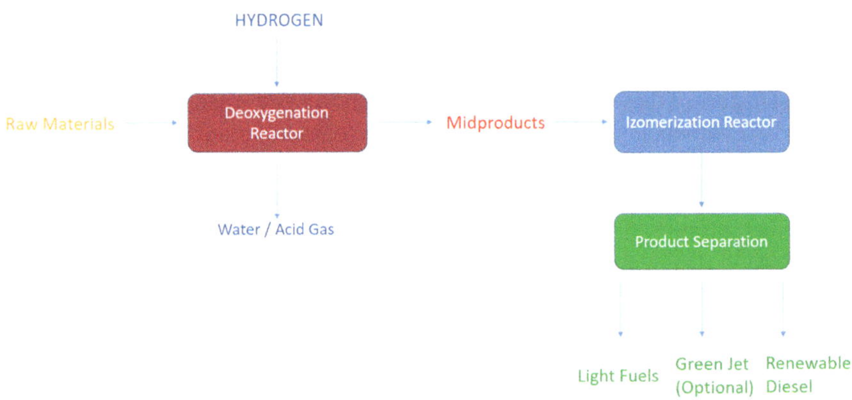

Fig. 10.3 Basic hydrotreatment process (figure reproduced and adapted) (Moreno Fernández-Villamil and Hurtado De Mendoza Paniagua 2018)

accepted. EN14214 is the standard created by the European Standards Committee (CEN) (European Standards 2019; ASTM International 2020).

10.3.3 Raw Materials

Biofuel raw materials are basically divided into four main groups. These four main groups are edible vegetable oils, non-edible vegetable oil, waste oil or used cooking oil (UCO) and animal fats. Edible vegetable oils are obtained from peanut, soybean, sunflower, palm, corn, rapeseed, coconut, canola, sugar cane, wheat, barley, rice and barley, etc. Non-edible vegetable oils are extracted from algae, jatropha, castor, camelia, karanja, cumaru, cottonseed, pongamia, mahua, moringa, forest residue etc. (Mohd Noor et al. 2017). In Table 10.4, major raw materials are given by region and country (ABS 2021; Mohd Noor et al. 2017).

Table 10.4 Major raw materials

Region	Country	Major raw materials
European	Turkey	Rapeseed and waste oil
	Germany	Rapeseed
	United Kingdom	Sunflower and rapeseed
	Spain	Sunflower and linseed
	Italy	Sunflower and rapeseed
	Sweden	Rapeseed
	Greece	Cottonseed
	France	Sunflower and rapeseed
North America	United States of America	Waste oil and soybeans
	Canada	Canola
	Mexico	Waste oil and animal fat
South America	Brazil	Soybeans and palm oil
	Argentina	Soybeans
Asia	Indonesia	Palm oil
	Malaysia	Palm oil
	Japan	Waste oil
	China	Rapeseed and waste oil
	Philippines	Coconut oil
	Thailand	Palm oil and coconut oil
	India	Jatropha and pongamia
Africa/Australia	South Africa	Jatropha
	Australia	Jatropha
	New Zealand	Waste oil and animal fat

10.3.4 Properties of Biodiesel

The properties of biodiesel vary according to the feedstock used, the production method and mixing ratios. For this reason, the properties of biodiesel fuel are shown in comparison with diesel fuel by both American Society for Testing and Materials (ASTM) and European (EN) standards in Table 10.5 (Atabani et al. 2012).

Table 10.5 Diesel and biodiesel fuel standards (Atabani et al. 2012)

Properties of fuels	Diesel	Biodiesel	
	ASTM D975	ASTM D6751	EN 14214
Density 15 °C (kg/m^3)	850	880	860–900
Viscosity at 40 °C (cst)	2.6	1.9–6.0	3.5–5.0
Cetane number	40–55	Min 47	Min 51
Iodine number	38.3	–	Max120
Calorific value (MJ/kg)	42–46	–	35
Acid value (mg KOH/g)	0.062	Max 0.50	Max 0.50
Pour point (°C)	−35	−15 to −16	–
Flash point (°C)	60–80	Min 100–170	>120
Cloud point (°C)	−20	−3 to −12	–
Cold filter plugging point (°C)	−25	19	Max 5
Copper strip corrosion (3 h at 50 °C)	1	Max 3	Min 1
Carbon (wt%)	84–87	77	–
Hydrogen (wt%)	12–16	12	–
Oxygen (wt%)	0–0.31	11	–
Methanol content % (m/m)	–	–	Max 0.20
Water and sediment content (vol%)	0.05	Max 0.05	Max 0.05
Ash content % (w/w)	0.01	0.02	0.02
Sulfur % (m/m)	0.05	Max 0.05	Max 0.001
Sulfated ash % (m/m)	–	Max 0.02	Max 0.02
Phosphorus content	–	Max 0.001	Max 0.001
Free glycerin % (m/m)	–	Max 0.02	Max 0.02
Total glycerin % (m/m)	–	Max 0.24	Max 0.24
Monoglyceride % (m/m)	–	0.52	0.8
Diglyceride % (m/m)	–	–	0.2
Triglyceride % (m/m)	–	–	0.2
CCR 100% (mass%)	0.17	Max 0.05	Max 0.03
Distillation temperature (°C)	–	Max 360	–
Oxidation stability (h, 110 °C)	–	3 min	6 min
Lubricity (HFFR; μm)	685	314	–

Density

Fuel density is an important value for injection quantity, injection timing and perfect spraying for a diesel engine. Depending on the type of feedstock, the fuel density of biodiesel ranges between 830 and 960 kg/m^3. In generally, the density of biodiesel is higher than petroleum-derived diesel (Mohd Noor et al. 2017; Atabani et al. 2012).

Viscosity

Viscosity also is a significant factor that directly affects the injection property and combustion quality just like density. Too low viscosity leads to leaks in the pipelines and seals, while too high viscosity causes the deterioration of combustion quality and difficulty in pumping for transfers. Due to chemical structure, biodiesel' viscosity is generally higher than petroleum-derived diesel (Mohd Noor et al. 2017; Atabani et al. 2012). The high viscosity of biodiesel compared to diesel results in reduced atomization quality and poor spray cone angle. This problem can be overcome when the engine is run with the optimum operating temperature of biodiesel for atomization (Aldhaidhawi et al. 2017).

Calorific Value

The calorific value is also known as the heating value. It is a crucial element for fuel because it directly determines the amount of energy released by combustion. The calorific value of biodiesel varies according to the raw material used but is averagely 12% less than diesel derived from petroleum (Oliveira and Silva 2013).

Cetane Number

Another critical property for fuel is its cetane number. The cetane number determines the ignition delay and the ignition delay are directly related to the efficiency of the combustion. Ignition delay can be defined as the time elapsed from the start of the fuel being injected into the cylinder until the combustion starts in the cylinder. Ignition delay consists of two phases, physical and chemical. The structural properties of the fuel are effective in both these phases. Longer ignition delays cause excessive pressure increases in the cylinder and knocking operation (Aldhaidhawi et al. 2017). A high cetane number of biodiesel causes a short ignition delay, thereby increasing combustion efficiency especially in cold weather. The cetane number of biodiesel is generally higher than fossil diesel fuels (Lapuerta et al. 2008).

Flash Point

Flash point is the lowest temperature that fuel will burn when exposed to open flame. It is a substantial value for the safe handling of fuel. Biodiesel flash point is blatantly higher than that of fossil diesel which makes it safer. Biodiesel typically has a flash point of more than 150 °C, whereas conventional diesel fuel has a flash point of 55–66 °C (Mohd Noor et al. 2017; Atabani et al. 2012).

Oxidation Stability

Oxidation stability can be defined that it is disposition of a fuel to chemical react with oxygen, which is not a good property for storage. The existence of unsaturated fatty acid chains and a double bond in the parent molecule cause oxidation because they react with oxygen as soon as it is exposed to air. Because of its composition, biodiesel is more prone to stability degradation than fossil diesel (Atabani et al. 2012; Silitonga et al. 2011).

Acid Value

Amount of free fatty acid content in the oil can be defined as an acid value of a fuel. The high acid value causes corrosion in the pipelines, fuel supply system and fuel injectors, so it is an undesirable feature. Biodiesel acid value is much more than the petroleum-derived diesel (Mohd Noor et al. 2017; Atabani et al. 2012).

Cloud and Pour Point

The temperature at which the fuel begins to solidify is the cloud point. At this stage, wax crystals start to develop in the fuel, which is undesirable. The lowest temperature at which fuel starts to flow is called the pour point. The low temperature characteristic of fuel is an important criterion. It causes the fuel blockage of the transfer lines and filters, and also causing initial starting difficulties of the engine (Atabani et al. 2012; Canakci and Sanli 2008). Biodiesel often has a greater cloud point than conventional diesel fuel (Alleman and McCormick 2016).

Lubrication Property

The lubricating feature of biodiesel is considerably higher than that of conventional diesel, and this feature extends the lifespan of the engine (Atadashi et al. 2010). The high lubricating property of biodiesel reduces friction losses between engine parts and increases the mechanical efficiency of the engine (Xue et al. 2011).

10.3.5 Safety and Storage

On a vessel, marine fuels must have a flashpoint greater than 60 °C, according to the IMO (1974). Diesel fuel flash point is generally between 55 and 96 °C and is considered flammable (ABS 2021). Biodiesel has a higher flash point than diesel. As can be seen from ASTM and EN standards, it has a minimum value of 100 and 120 °C. The flash point of pure biodiesel, which has varying values according to the raw material used, rises from 93 to 200 °C. When mixed with diesel, it increases the flash point of the fuel obtained depending on the mixing ratio compared to pure diesel (Alviso et al. 2020). Therefore, it is less dangerous than diesel fuel.

Biodiesel is an eco-friendly and non-toxic fuel because it is a less polluting fuel. Compared to petroleum-derived diesel, reductions in sulfur, carbon monoxide, carbon dioxide and particulate matters were observed in the use of biodiesel at varying rates depending on the engine condition, type and quality of biodiesel (Živković and Veljković 2018). In addition, biodiesel is a biodegradable fuel in nature due to its organic structure. Biodiesel biodegrades significantly faster than petrolium-derived diesel (Alleman and McCormick 2016). This means that in case of any spill, the decomposition process starts as soon as it comes into contact with sea water. On the other hand, biodiesels spilled in a maritime environment behave similarly to petroleum fuels until entirely degraded, despite the fact that they are non-toxic on land (ABS 2021).

On the other hand, biodiesel is more prone to microbial contamination than diesel due to its organic structure. Bacterial and fungal growth may occur, which is a undesirable, when water is present in biodiesel or in storage tanks. The most important point in dealing with the problem of microbial contamination is to remove the water contained in the fuel. During the operation, it is necessary to regularly drain the tanks and test the samples taken and keep them under constant observation. In addition, the problem can be eliminated by using fuel additives such as biocide or anti-microbial additives to the biodiesel or it's blend to avoid microbial build-up (ABS 2021; Alleman and McCormick 2016).

The burning characteristics of biodiesel blends are similar to those of petroleum-derived drop-in fuels. For this reason while transporting, handling, and burning biodiesel blends, standard industrial fire safety standards and implementations should be followed (ABS 2021). In case of fire, dry chemical, foam, halon, CO_2, or water O_2 spray extinguishers can be used to put out a pure (B100) biodiesel fire (ABS 2021; Alleman and McCormick 2016).

It is crucial to ensure proper temperature values when storing and using fuels because temperature affects viscosity. Viscosity, on the other hand, affects many situations such as the flow characteristics of the fuel, the clogging of the filters, injection into the cylinder and perfect atomization. Basically, in order to avoid operational problems, the storage temperature of biodiesel should be kept 10–15 °C above the cloud point. The storage temperature of biodiesel varies according to the type of biodiesel, the raw material used and the mixing ratio (ABS 2021). Especially low temperatures can be a problem when using biodiesel. Biodiesels tend to gel and cloud at low temperatures. This leads to clogging of the filters. The infrastructure of existing ships that can use HFO is capable of keeping the fuel temperature above the cold filter plugging point. Most of the vessels' fuel storage tanks are already fitted with heating systems. If the temperature cannot be kept above the cold filter plugging point, anti-gel and cold flow additives must be used (ABS 2021). Compared to diesel, biodiesel is more prone to oxidation and degradation when stored for a long time because of its poor oxidation stability. There are different reasons for biodiesel oxidation, but the most important reason is the unsaturated fatty acid methyl esters in it (Knothe and Steidley 2018). It is not compatible with all materials due to its poor oxidation stability and high probability of interaction with them. Table 10.6 lists

Table 10.6 Suitable/unsuitable materials with biodiesel (ABS 2021; Alleman and McCormick 2016; Mohd Noor et al. 2017)

Suitable materials with biodiesel Recommended	Unsuitable materials with biodiesel Not recommended
Steel	Bronze
Aluminium	Zinc
Fiberglass	Tin
Teflon	Brass
Fluorinated polyethylene	Copper
Fluorinated polypropylene	Lead

some recommended materials and incompatible with biodiesel (ABS 2021; Alleman and McCormick 2016; Mohd Noor et al. 2017).

As for HVO, HVO is so similar to conventional diesel fuel. Therefore, no extra precautions are required to store renewable diesel. Since it is compatible with all materials that can be used with diesel fuel, there is no need to change materials on any pipelines, tank, system or engine (Engman et al. 2014).

10.3.6 Engine Performance and Engine Manufacturer Warranty

Many researchers agree that there is a decrease in engine power when using pure biodiesel. The main reason for this decrease in engine power is the low calorific value of biodiesel and this ratio varies according to the raw material used (Xue et al. 2011). Generally biodiesel contains about 10% less energy than petroleum-derived diesel fuels (Nayyar 2010). Biodiesel blends obtained from waste cooking oil were tested in an experimental study with a single-cylinder 4-stroke diesel engine. The tested blends are B10, B20, and B30. According to the result of the study, the specific fuel consumption increased compared to diesel oil at all mixing ratios. Likewise, thermal efficiency was observed slightly lower than diesel oil in all mixing ratios (Abed et al. 2019). In an experimental study to analyze the performance and emission values of pure HVO, similar but better results were obtained than biodiesel. Compared to diesel oil, specific fuel consumption increased up to 15% in biodiesel. However, it was detected 6% in HVO (Ushakov and Lefebvre 2019).

Most of the global engine manufacturers have not been unresponsive to the widespread use of biodiesel and have given engine warranty regarding the use of biodiesel and its blends in their engines. The warranty limits given by the engine manufacturers at varying rates are given in Table 10.7 (Mohd Noor et al. 2017; Nayyar 2010; Cummins 2021).

Table 10.7 Warranty limits of major engine manufacturers (Mohd Noor et al. 2017; Nayyar 2010; Cummins 2021)

Manufacturer	Biodiesel standards	Allowable blend ratio
MAN diesel	EN 14214	Up to 100%
Wartsila	EN 14214	Up to 100%
Fairbanks Morse	ASTM D6751	Up to 100%
Caterpillar	ASTM D6751/EN 14214	Up to 30%
Cummins	ASTM D6751/EN 14214	Up to 20%
Detroit diesel	ASTM D6751	Up to 5%
Volvo Penta	ASTM D6751	Up to 30%
Yanmar	ASTM D6751/EN 14214	Up to 20%

As for renewable diesel (HVO), since HVO meets ASTM D975 standards, it can be used as a drop-in fuel in all engines that can use diesel fuel without any modification, and this will not cause any problems on the engine warranty.

10.3.7 Emissions

Most researchers agree that carbon monoxide, carbon dioxide, hydrocarbon, and particulate matter emissions are reduced when pure biodiesel or biodiesel blends are used instead of diesel oil. In the literature, it has been stated that depending on the mixing ratio of biodiesel, CO and PM emissions are reduced by approximately 50% and HC emissions by up to 70% (Nayyar 2010). In another study, the emissions of B10 and B20 biodiesel obtained from jatropha oil, palm oil and algae were observed at different engine loads and a clear reduction was found in emissions except NO_x emissions (Abed et al. 2019). In another study, waste oil-based biodiesels (B20, B50, B100) and diesel oil emissions were measured by comparing. Different results have been obtained depending on the engine load and mixing ratio. According to the results of the study, CO emissions decreased in all biodiesel blends, and also HC emissions decreased in general. Nonetheless, while CO_2 emissions decreased in the B20 blend, they increased slightly in the B100 (Kaya and Kökkülünk 2020). Besides, FAME biodiesel is a fuel that has very low sulfur. It ensures mitigation in SO_x emissions up to 90% compared to fossil diesel fuel (Yuanrong Zhou 2020). Another large-scale study includes the results of the experimental studies in the literature on the engine performance and emissions of pure biodiesel compared to the petroleum-derived diesel. These statistical information are given in Table 10.8 (Xue et al. 2011).

The emission studies mentioned above refer to the emissions resulting from the combustion of the fuel in the engine. The current regulatory framework, such as IMO DCS and EU MRV, is more concerned with ship emissions from the combustion of a fuel than with the overall life cycle emissions of a fuel. Although there is no guideline on life cycle emissions put into effect by the IMO, another point that

Table 10.8 Statistical results of experimental studies (Xue et al. 2011)

	Number of references	Increment number	Increment %	Similar number	Similar %	Decreasement number	Decreasement %
CO_2 emissions	13	6	46.2	2	15.4	5	38.5
CO emissions	66	7	46.2	3	3.0	57	84.4
HC emissions	57	3	5.3	2	5.3	51	89.5
PM emissions	73	7	9.6	2	2.7	64	87.7
NO_x emissions	69	45	65.2	4	5.8	20	29.0
Economy performance	62	54	87.1	2	3.2	6	9.7
Power performance	27	2	7.4	6	22.2	19	70.4

needs to be taken into account in nature's carbon cycle is the life cycle emission of a fuel. For the most complete expression of a fuel's environmental effects, the entire carbon footprint of the fuel needs to be considered. A life cycle assessment (LCA) approach should be adopted to calculate the entire carbon footprint of the fuel. The life cycle assessment approach aims to detect carbon emissions during the production, transportation and combustion phases. Basically, it can be examined in two parts, which are well-to-tank emissions and tank-to-wake emissions. Well-to-tank emissions include emissions from production processes and transportation. Tank-to-wake emissions involve emissions from the combustion of fuel or its use as an energy source. By adding these two emission values, the net emission effect of the fuel is measured, which is called well-to-wake (ABS 2021).

For biofuels, the production phase of life cycle emissions should include feedstock extraction, which covers farming and collection of plants. Biofuels have a significant advantage over petroleum-derived fuels in life cycle emission assessment. Plants to be used in the production of raw materials for biofuels grow by photosynthesis until harvest time. In this process, they reduce the amount of CO_2 in the atmosphere by adding the carbon dioxide they absorb from the air to their structure, and this creates a counterbalance.

Another point to consider when measuring the carbon life cycle of biofuels is indirect land use change (ILUC) factor. ILUC can be defined as follows. Cultivation of the plant to be used as a raw material in a biofuel production in the existing agricultural land prompts the farmers to search for new agricultural land, since the food and feed crops needs of the living things remain the same. The demand for new farmland can sometimes even lead to the conversion of forests to farmland. The use of machinery on new farmland and the conversion of forests to farmland makes a biofuel an indirect negative contribution to the carbon life cycle (European Commission 2012).

A recent study conducted by the International Council on Clean Transportation (ICCT) revealed the life cycle carbon footprints of some alternative marine fuels by comparing them with marine gas oil (MGO). In this study of the ICCT, ILUC factors were also taken into account in the carbon life cycle assessment of fuel. According to the results of the study, especially the biofuels that classified as 2nd generation obtained from waste oil and lignocellulosic biomass provide well-to-wake carbon emission gains from 70% to almost 100% compared to MGO due to their very low ILUC factors as well as their clean-burning characteristic. Conventional biodiesel (FAME) and renewable diesel (HVO) from soy and palm oil have high carbon emissions comparable to MGO in terms of the carbon life cycle when ILUC factors are included. Particularly, using oilseeds used in the food industry in the production of biofuels leads to new farmland conversion to ensure food supply and stability (Yuanrong Zhou 2020). In another study, well-to-wake GHG emissions of FAME and HVO fuels derived from soybean oil were estimated to be approximately 50% and 60% lower, respectively, than marine diesel fuel. ILUC factors are not included in this study (ABS 2021).

10.3.8 Biofuel Marine Applications

Applications of biodiesel for the marine industry began in 1998 in the North American Great Lakes region (Mohd Noor et al. 2017). The National Oceanic and Atmospheric Administration (NOAA) conducted the test study by operating Great Lakes Environmental Research Laboratory's (GLERL) 10 small research vessels on board with B100 biodiesel obtained from soybean oil. After switching to B100 biodiesel, GLERL reported that its fleet significantly reduced emissions compared to petrolium-derived diesel. The rate of decrease was reported as 48% in CO, 59% in PM, 7% in NO_x, 74% in SO_x and 77% in HC. In addition, it has been reported that there is a 20–40% reduction in maintenance costs due to the cleaning properties of biodiesel (ABS 2021).

Mediterranean Shipping Company (MSC) announced in 2019 that it will benefit from biofuels and has tested B10 biodiesel. The biofuel used was biodiesel obtained from used cooking oil and a total of 100,000 tons of fuel was supplied. MSC estimates it saved 30,000 tonnes of CO_2. After the successful test process, MSC aims to reduce CO_2 emissions by 15% by 20% by using 30% biodiesel blends (ABS 2021; MSC 2019).

The Alexander von Humboldt is another biofuel test vessel. Jan De Nul Group announced in 2020 that they operated 2000 h on this ship with second generation 100% biofuel. It has been reported that a significant CO_2 emission of 85% is achieved. In addition, Jan De Nul Group also stated that the second generation biofuel is ready to be used as a drop-in fuel that can meet the emissions targets of the maritime industry (Deruyck et al. 2020).

Another biofuel application was made by Stena Bulk shipping company for 10 days in April 2020 in cooperation with Good Fuels. The 49,646 deadweight ton tanker vessel has operated with 100% biofuel its normal operation and it has been reported that 83% gain in greenhouse gas emissions has been achieved. It also showed that low-carbon shipping could be achieved in a short time (Stena Bulk 2020).

A successful trial was also made by Ocean Network Express (ONE) company on the M/V MOL vessel, again with supplier Good Fuels, and its report was published in April 2021. In the application, 100% biofuel obtained from waste oil and residue was used. The application performance demonstrates the future possibility of sustainable biofuels. ONE thus underlined the importance of biofuels in meeting IMO's 2030 and 2050 targets (Ocean Network Express Pte. Ltd. 2021).

10.3.9 Comparison of FAME and HVO with Other Alternative Marine Fuels

Even though there are many alternative fuels for maritime transportation today, LNG and methanol are the most accepted alternative fuels by the marine industry and their commercial activities have started (Zincir and Deniz 2021). In its fourth GHG study, IMO highlighted LNG and methanol and made future predictions about these alternative fuels (The Fourth IMO Greenhouse Gas Study 2020). Apart from LNG and methanol, another prominent alternative marine fuel that has a mature production technology and supply chain and has started its commercial activities is biofuels. Biofuels vary among themselves according to the production process and raw material. Among the biofuels, FAME biodiesel and HVO renewable diesel are among the most prominent and produced fuels. For a maritime company that carries on commercial activities, the cost of fuel consumed by its fleet is one of the most important expenses. The prices of new alternative fuels will play a decisive role in the transition period to alternative fuels. The current market prices of alternative fuels, which we have discussed in our study, are given in Table 10.9 together with their types.

At this stage, FAME biodiesel and HVO renewable diesel were compared with LNG and methanol with reference to marine diesel oil. The evaluative criteria used in the comparison are safety, compatibility with existing ships, cost, bunkering availability, voyage distance and GHG emissions. Cost was considered according to the price of a metric ton, and voyage distance was considered according to volumetric efficiency. GHG emissions were evaluated based on the life cycle assessment model in the form of well-to-tank, tank-to-wake and well-to-wake. Evaluation criteria were compared with reference fuel MDO on four scales; very good (++), good (+), poor (−), and very poor (− −). Reference fuel MDO was considered averagely between good and poor. The results of our comparison are summarized in Table 10.10.

This table can be used to infer a number of conclusions. Starting with FAME biodiesel, biodiesel is a safe fuel with mature production technology. The cost is higher than the MDO and the cruising distance is slightly less due to the lower calorific value compared to the same volume of MDO. However, having better emission results than MDO is an important criterion for decarbonized maritime transport. It can be

Table 10.9 Current market prices

Fuel	Price ($/MT)	Price date
MDO	673[a] (Ship & Bunker 2021)	September 2021
FAME biodiesel	1630 (Neste 2021)	September 2021
SME biodiesel	1480 (Neste 2021)	September 2021
UCOME biodiesel	1719 (Greenea 2021)	August 2021
TME biodiesel	1625 (Greenea 2021)	August 2021
HVO renewable diesel class I	1819 (Argus 2021)	June 2021
HVO renewable diesel class II	2204 (Argus 2021)	June 2021
HVO renewable diesel class III	2082 (Argus 2021)	June 2021
Methanol	511[a] (Methanex 2021)	September 2021
LNG	1053 (Mabux Marine Bunker Exchange 2021)	September 2021

[a] Global average bunker price
FAME biodiesel; it is blend of various biodiesel, it can comprise different amounts of neat vegetable oil and/or tallow oil methyl esters
SME biodiesel; soybean methyl ester
UCOME biodiesel; Used cooking oil methyl ester
TME biodiesel; tallow methyl ester
HVO class I; produced from food and feed crops
HVO class II; produced from used cooking oil (UCO)
HVO class III; produced from tallow

used as a drop-in with existing ships, but when it is desired to be used in high ratio blends or pure, minor changes such as sealing elements and fuel systems are required in the engine.

HVO renewable diesel is also a safe fuel, it is highly compatible with existing ships and pure HVO can be used as a drop-in to existing ships. The cost of HVO, which

Table 10.10 Evaluation of promising alternative fuels with the selected criteria

Reference fuel	Fuel	Evaluation criteria							
		Safety	Compatibility with existing ships	Cost	Bunkering availability	Voyage distance	Well-to-tank emission	Tank-to-wake emission	Well-to-wake emission
MDO	FAME biodiesel	+	+	–	–	–	++	+	+
	HVO renewable diesel	+	++	– –	– –	–	++	++	++
	LNG	– –	– –	–	–	–	–	+	+
	Methanol	–	–	+	– –	– –	–	+	–

also meets IMO's emission rules, is currently higher than other fuels and its biggest disadvantage is cost. It is the fuel with the highest heating value among biofuels (Neste Corporation 2020). At the same volume, the voyage distance is slightly shorter because the density of HVO is lower than that of MDO. However, this difference is not much since HVO has a slightly higher lower heating value (LHV) than MDO.

LNG vapor is a highly flammable gas so it requires extra storage and operational safety measures compared to MDO. It is not compatible with existing ships and requires major modifications. The production technology is very mature, but the cost is high compared to MDO. It is an alternative fuel that comes to the fore today as it meets IMO's short and medium term emission targets.

As for methanol, it performs worse than MDO in terms of safety due to its lower flash point than diesel. It cannot be used as a drop-in on existing ships, requires major modifications. Although it makes a positive contribution to tank-to-wake emissions, the result is negative when well-to-tank emissions are included. However, if renewable energy sources such as wind and solar are used in the methanol production process, the net emission effect, which is well-to-wake, will be positive.

10.4 Summary

This chapter includes the status of international maritime transport, emissions from global maritime transport, the steps taken by IMO in the target of decarbonization and the initial IMO GHG plan, biofuels and generations, characteristics of FAME and HVO fuels, raw material and production processes, life cycle emission assessment model, biofuel marine applications. In addition, it covers comparisons of FAME, HVO, LNG and methanol with criteria selected with reference to MDO and their current market prices.

Maritime transportation is crucial for international trade. 98,140 ships on the world's seas contribute to these trade activities, which constitutes approximately 90% of world trade. Since ships are equipped with large engines, they consume large amounts of fuel and as a result they emit high amounts of polluting exhaust gases into the atmosphere. The most important of these are the gases that create the greenhouse effect. In the fight against global warming and climate change, these gases emitted to the atmosphere should be reduced gradually. IMO has started its work in this framework and has published four GHG studies and its initial GHG strategy plan. The strategy plan includes three-term measures as short, mid and long-term. The main objective of the strategy plan is the decarbonization of maritime transport. In this process, which brought many new rules and changes to decarbonize maritime transportation, alternative fuels became prominent in all periods, especially in the mid and long terms. New ships to be built can be equipped with engines and types of equipment that are fully compatible with the selected specific alternative fuel. However, the transition process of ships with existing diesel engines to alternative fuels will not be so easy due to the necessary modification problems. In the process of decarbonization of maritime transportation, it is of great importance for the world economy to make efficient use of existing diesel-powered ships.

Today, there are many alternative fuels for maritime transportation. Each of these alternative fuels has its own characteristics and these properties can act as an advantage or a disadvantage. In the transition period to alternative fuels, biofuels come to the fore as they can be used without making any changes in the engine or with minor changes and meet the emission rules of IMO. Among the biofuels, FAME biodiesel, which has the most mature production technology and the largest supply chain. FAME biodiesel is an ideal fuel to meet IMO's short-term targets. It can be used in blends up to 30% without making any changes in the engine. However, in order to use pure biodiesel, it is necessary to modify with proper material of the sealing elements and fuel system pipelines in the engine. Aside from that, due to its poor oxidation stability and microbial contamination problem, it still needs extra fuel additives for long-term storage.

HVO renewable diesel, which meets the marine diesel fuel standards and can be used in pure form as drop-in directly in the engine, were examined. Existing ships need to adapt quickly to ensure significant short-term progress in decarbonizing shipping. HVO has a great advantage in this context, as it can be used as a drop-in. However, in order for it to be a sustainable fuel, its costs must decrease and the availability of raw materials must be sufficient. The biggest disadvantage of HVO is its price. In addition, there is also the problem of insufficient raw materials when using edible vegetable oil. However, since the 2nd generation biofuels can be produced from inedible plants, wastes or forest residues, they do not compete with the food industry. The 3rd generation biofuel studies are still ongoing.

Emission values obtained in experimental studies with biodiesel blends and HVO reduce GHG emissions compared to MDO. These experimental studies measure only tank-to-wake emissions. It would be a much more logical approach to evaluate the emissions not only as gases released as a result of combustion, but also during the life cycle. For this reason, the life cycle evaluation method of emissions has been explained. The raw materials of biofuels can make a positive contribution to the carbon cycle until the harvest time due to their organic structure and photosynthesis capabilities, making the net emission effects of the final product from 70 to 100% better compared to fossil-derived fuels.

The application of biofuels on ships began in 1998. The positive results obtained in the projects and ships prove that biofuels are ready to be used as drop-in or blend to meet new emission targets especially for short-term targets. Today, it is still using in some ships that continue their commercial activities.

Biofuels are promising fuels for decarbonized shipping. Because biofuels can be used in pure form or mixed with petroleum-derived fuels in ships with diesel engines. FAME and HVO, two of the biofuels whose use is increasing day by day, will be used much more in the future as production costs decrease and the supply chain increases.

References

Abed KA, Gad MS, El Morsi AK, Sayed MM, Elyazeed SA (2019) Effect of biodiesel fuels on diesel engine emissions. Egypt J Pet 28(2):183–188. https://doi.org/10.1016/j.ejpe.2019.03.001

ABS (2021) Sustainability whitepaper—biofuels as marine fuel, p 28

Aldhaidhawi M, Chiriac R, Badescu V (2017) Ignition delay, combustion and emission characteristics of diesel engine fueled with rapeseed biodiesel—a literature review. Renew Sustain Energy Rev 73:178–186. https://doi.org/10.1016/j.rser.2017.01.129

Alleman TL, McCormick RL (2016) Biodiesel handling and use guide

Alviso D, Saab E, Clevenot P, Romano SD (2020) Flash point, kinematic viscosity and refractive index: variations and correlations of biodiesel–diesel blends. J Brazilian Soc Mech Sci Eng 42(6):1–15. https://doi.org/10.1007/s40430-020-02428-w

Argus (2021). https://www.argusmedia.com/en/news/2220328-nwe-hvo-saf-at-records-on-feedstocks-prices-demand

ASTM International (2020). https://www.astm.org/Standards/D6751.htm

Atabani AE, Silitonga AS, Badruddin IA, Mahlia TMI, Masjuki HH, Mekhilef S (2012) A comprehensive review on biodiesel as an alternative energy resource and its characteristics. Renew Sustain Energy Rev 16(4):2070–2093. https://doi.org/10.1016/j.rser.2012.01.003

Atadashi IM, Aroua MK, Aziz AA (2010) High quality biodiesel and its diesel engine application: a review. Renew Sustain Energy Rev 14(7):1999–2008. https://doi.org/10.1016/j.rser.2010.03.020

Boviatsis M, Tselentis B (2019) A comparative analysis between EU MRV and IMO DCS—the need to adopt a harmonised regulatory system. In: 16th international conference on environmental science and technology, pp 2018–2019

Canakci M, Sanli H (2008) Biodiesel production from various feedstocks and their effects on the fuel properties. J Ind Microbiol Biotechnol 35(5):431–441. https://doi.org/10.1007/s10295-008-0337-6

Chowdhury H, Loganathan B (2019) Third-generation biofuels from microalgae: a review. Curr Opin Green Sustain Chem 20:39–44. https://doi.org/10.1016/j.cogsc.2019.09.003

Cummins (2021). https://www.cummins.com/engines/biodiesel-faqs

Deniz C, Zincir B (2016) Environmental and economical assessment of alternative marine fuels. J Clean Prod 113(X):438–449. http://doi.org/10.1016/j.jclepro.2015.11.089

Deruyck M, Mauger P, De Nul J (2020) Hopper Dredger first to sail 2,000 hours on 100% sustainable marine biofuel, pp 2–4

Engman A, Hartikka T, Honkanen M, Kiiski U, Kuronen M, Mikkonen S, Saikkonen P (2014) Hydrotreated vegetable oil (HVO) premium renewable biofuel for diesel engines. Neste oil technical paper

European Commission (2012) Indirect land use change (ILUC) what are biofuels?

European Environment Agency (EEA) (2021) Emissions of air pollutants from transport. https://www.eea.europa.eu/data-and-maps/indicators/transport-emissions-of-air-pollutants-8/transport-emissions-of-air-pollutants-8

European Standards (2019). https://www.en-standard.eu/din-en-14214-liquid-petroleum-products-fatty-acid-methyl-esters-fame-for-use-in-diesel-engines-and-heating-applications-requirements-and-test-methods-includes-amendment-2019/

Greenea (2021). http://www.greenea.com/en/market-analysis/2021

Han ES, Goleman D, Boyatzis RE, Mckee A (2020) Review of maritime transport 2020, vol 53, no 9

IMO (2018) Adoption of the initial IMO strategy on reduction of GHG emissions from ships and existing IMO activity related to reducing GHG emissions in the shipping sector. Mepc 72, p 27 [Online]. Available: https://unfccc.int/sites/default/files/resource/250_IMO%20submission_Talanoa%20Dialogue_April%202018.pdf

International Maritime Organization (IMO) (1974) International convention for the safety of life at sea (SOLAS)

International Maritime Organization (IMO) (2021a) Initial IMO GHG strategy. https://www.imo.org/en/MediaCentre/HotTopics/Pages/Reducing-greenhouse-gas-emissions-from-ships.aspx

International Maritime Organization (IMO) (2021b) Historic background. https://www.imo.org/en/OurWork/Environment/Pages/Historic%20Background%20GHG.aspx
Kampa M, Castanas E (2008) Human health effects of air pollution. Environ Pollut 151(2):362–367. https://doi.org/10.1016/j.envpol.2007.06.012
Kaya C, Kökkülünk G (2020) Biodiesel as alternative additive fuel for diesel engines: an experimental and theoretical investigation on emissions and performance characteristics. Energy Sources Part A Recover Util Environ Eff 1–23. https://doi.org/10.1080/15567036.2020.1774685
Kesharwani R, Sun Z, Dagli C, Xiong H (2019) Moving second generation biofuel manufacturing forward: investigating economic viability and environmental sustainability considering two strategies for supply chain restructuring. Appl Energy 242:1467–1496. https://doi.org/10.1016/j.apenergy.2019.03.098
Knothe G, Steidley KR (2018) The effect of metals and metal oxides on biodiesel oxidative stability from promotion to inhibition. Fuel Process Technol 177:75–80. https://doi.org/10.1016/j.fuproc.2018.04.009
Lapuerta M, Armas O, Rodríguez-Fernández J (2008) Effect of biodiesel fuels on diesel engine emissions. Prog Energy Combust Sci 34(2):198–223. https://doi.org/10.1016/j.pecs.2007.07.001
Mabux Marine Bunker Exchange. https://www.mabux.com/
Methanex (2021). https://www.methanex.com/
Mohd Noor CW, Noor MM, Mamat R (2018) Biodiesel as alternative fuel for marine diesel engine applications: a review. Renew Sustain Energy Rev 94:127–142. http://doi.org/10.1016/j.rser.2018.05.031
Moirangthem K (2016) Alternative fuels for marine and inland waterways
Moreno Fernández-Villamil J, Hurtado De Mendoza Paniagua A (2018) Preliminary design of the green diesel production process by hydrotreatment of vegetable oils
MSC (2019). https://www.msc.com/brb/news/2019-december/msc-becomes-first-major-shipping-line-to-use-30-bi
Nayyar M (2010) The use of biodiesel fuels in the US marine industry. Bulletin, pp 1–88 [Online]. Available: http://origin-www.marad.dot.gov/documents/The_Use_of_Biodiesel_Fuels_in_the_US_Marine_Industry.pdf
Neste Corporation (2020) Neste renewable diesel handbook. Neste, pp 1–33
Neste (2021). https://www.neste.com/investors/market-data/biodiesel-prices-sme-fame
Ocean Network Express Pte. Ltd. (2021) ONE conducts successful second trial use of sustainable biofuel for decarbonization
Oliveira LE, Da Silva MLCP (2013) Comparative study of calorific value of rapeseed, soybean, jatropha curcas and crambe biodiesel. Renew Energy Power Qual J 1(11):679–682. https://doi.org/10.24084/repqj11.411
Psaraftis HN (2019) Sustainable shipping: a cross-disciplinary view
Rutherford D, Comer B (2018) International Maritime Organization's initial greenhouse gas strategy. International Council on Clean Transportation, p 8 [Online]. Available: https://theicct.org/publications/IMO-initial-GHG-strategy
Sadeghinezhad E, Kazi SN, Badarudin A, Oon CS, Zubir MNM, Mehrali M (2013) A comprehensive review of bio-diesel as alternative fuel for compression ignition engines. Renew Sustain Energy Rev 28:410–424. https://doi.org/10.1016/j.rser.2013.08.003
Ship & Bunker (2021). World bunker prices. https://shipandbunker.com/prices/#MGO2021
Silitonga AS, Atabani AE, Mahlia TMI, Masjuki HH, Badruddin IA, Mekhilef S (2011) A review on prospect of Jatropha curcas for biodiesel in Indonesia. Renew Sustain Energy Rev 15(8):3733–3756. https://doi.org/10.1016/j.rser.2011.07.011
Stena Bulk (2020) Stena Bulk and good fuels successfully complete trial of sustainable marine biofuel, p 2 [Online]. Available: https://www.stenabulk.com/sites/default/files/press_releases/stena_bulk_and_goodfuels_successfully_complete_trial_of_sustainable_marine_biofuel.pdf
The fourth IMO greenhouse gas study (2020). International Maritime Organization

The Marine Environment Protection Committee (2011) Annex 19: resolution mepc.203(62), vol 203, no July, pp 1–17
The second IMO greenhouse gas study (2009). International Maritime Organization, p 220
The third IMO greenhouse gas study (2014). International Maritime Organization, p 327. http://doi.org/10.1007/s10584-013-0912-3
Ushakov S, Lefebvre N (2019) Assessment of hydrotreated vegetable oil (HVO) applicability as an alternative marine fuel based on its performance and emissions characteristics. SAE Int J Fuels Lubr 12(2):109–120. https://doi.org/10.4271/04-12-02-0007
Van Fan Y, Perry S, Klemeš JJ, Lee CT (2018) A review on air emissions assessment: transportation. J Clean Prod 194:673–684. https://doi.org/10.1016/j.jclepro.2018.05.151
Xue J, Grift TE, Hansen AC (2011) Effect of biodiesel on engine performances and emissions. Renew Sustain Energy Rev 15(2):1098–1116. https://doi.org/10.1016/j.rser.2010.11.016
Yuanrong Zhou NP (2020) The potential of liquid biofuels in reducing ship emissions | International Council on Clean Transportation [Online]. Available: https://theicct.org/publications/marine-biofuels-sept2020
Zincir B, Deniz C (2021) Methanol as a fuel for marine diesel engines. In: Shukla PC, Belgiorno G, Di Blasio G, Agarwal AK (eds) Alcohol as an alternative fuel for internal combustion engines. Springer, Singapore, pp 45–85
Živković S, Veljković M (2018) Environmental impacts the of production and use of biodiesel. Environ Sci Pollut Res 25(1):191–199. https://doi.org/10.1007/s11356-017-0649-z

The manufacturer's authorised representative in the EU is Springer Nature Customer Service Centre GmbH, Europaplatz 3, 69115 Heidelberg, Germany. If you have any concerns regarding our products, please contact ProductSafety@springernature.com

Printed and bound by CPI Group (UK) Ltd, Croydon, CR0 4YY
30/09/2026
02225291-0012